Biotechnology and Culture

Biotechnology and Culture: Bodies, Anxieties, Ethics
is Volume 25 in the series
THEORIES OF CONTEMPORARY CULTURE
CENTER FOR 21st CENTURY STUDIES
UNIVERSITY OF WISCONSIN-MILWAUKEE
Kathleen Woodward, General Editor

Nothing in Itself: Complexions of Fashion Herbert Blau

Figuring Age: Women, Bodies, Generations Edited by Kathleen Woodward

Electronic Culture: Fictions of the Present Tense in Television, Media Art, and Virtual Worlds Margaret Morse

Ethics after Idealism: Theory—Culture—Ethnicity—Reading Rey Chow

Too Soon Too Late: History in Popular Culture, 1972–1996 Meaghan Morris

The Mirror and the Killer-Queen: Otherness in Literary Language Gabriele Schwab

Re-viewing Reception: Television, Gender, and Postmodern Culture Lynne Joyrich

Pedagogy: The Question of Impersonation Edited by Jane Gallop

Fugitive Images: From Photography to Video Edited by Patrice Petro

Displacements: Cultural Identities in Question Edited by Angelika Bammer

Libidinal Economy Jean-François Lyotard

Aging and Its Discontents: Freud and Other Fictions Kathleen Woodward

Indiscretions: Avant-garde Film, Video, and Feminism Patricia Mellencamp

Logics of Television: Essays in Cultural Criticism Edited by Patricia Mellencamp

The Remasculinization of America: Gender and the Vietnam War Susan Jeffords

The Eye of Prey: Subversions of the Postmodern Herbert Blau

Feminist Studies/Critical Studies Edited by Teresa de Lauretis

Studies in Entertainment: Critical Approaches to Mass Culture Edited by Tania Modleski

Memory and Desire: Aging—Literature—Psychoanalysis
Edited by Kathleen Woodward and Murray M. Schwartz

Displacement: Derrida and After Edited by Mark Krupnick

Innovation/Renovation: New Perspectives on the Humanities
Edited by Ihab Hassan and Sally Hassan

The Technological Imagination: Theories and Fictions
Edited by Teresa de Lauretis, Andreas Huyssen, and Kathleen Woodward

The Myths of Information: Technology and Postindustrial Culture Edited by Kathleen Woodward

Performance and Postmodern Culture Edited by Michel Benamou and Charles Caramello

Biotechnology and Culture

Bodies, Anxieties, Ethics

EDITED BY **PAUL E. BRODWIN**

Indiana University Press / Bloomington and Indianapolis

This book is a publication of

Indiana University Press

601 North Morton Street

Bloomington, IN 47404-3797 USA

http://www.indiana.edu/~iupress

Telephone orders 800-842-6796
Fax orders 812-855-7931
Orders by e-mail iuporder@indiana.edu

© 2000 by The Regents of the University of Wisconsin System

The paper used in this publication meets the minimum
requirements of American National Standard for Information
Sciences—Permanence of Paper for Printed Library
Materials, ANSI Z39.48-1984.

Manufactured in the United States of America

Library of Congress Cataloging-in-Publication Data

Biotechnology and culture : bodies, anxieties, ethics / edited by Paul E. Brodwin.
     p.   cm.—(Theories of contemporary culture ; v. 25)
Includes bibliographical references and index.
ISBN 0-253-33831-X (cl : alk. paper)—ISBN 0-253-21428-9 (pa : alk. paper)
     1. Biotechnology—Social aspects.   I. Brodwin, Paul.   II. Series.
TP248.2.B55117   2000
303.48'3—dc21   00-040970

1   2   3   4   5   05   04   03   02   01   00

CONTENTS

Part IV: Biotechnology and Globalization

ACKNOWLEDGMENTS

This book is truly a collaborative effort. Most of the contributors presented earlier versions of their chapters at the conference Biotechnology, Culture, and the Body held in April 1997 at the Center for 21st Century Studies, a postdoctoral research institute at the University of Wisconsin-Milwaukee. The center director, Professor Kathleen Woodward, provided the leadership and resources to make the conference a success. I am also indebted to the editorial care and wisdom of Dr. Carol Tennessen, associate director, and Dr. Nigel Rothfels, associate editor for center publications, who oversaw the passage from conference to book.

In addition to the essays collected here, the conference featured exciting work by several other scholars: Richard Doyle, Kathryn Montgomery, Mary Mahowald, Patricia Marshall, Panivong Norindr, Angela Wall, and José van Dijck. During the conference, the artwork of physician Eric Avery kept us mindful of our visual and visceral engagements with biotechnologies. Finally, the insights from discussants Michael Fischer and Helen Longino and an anonymous reviewer have found their way into the introduction as well as several of the chapters. I would like to thank all the participants in this process for their hard work and creative dedication.

The chapter by Gillian M. Goslinga-Roy was originally published in *Feminist Studies* 26.1 (2000), and it is reprinted here by permission of the publisher, *Feminist Studies*, Inc.

Biotechnology and Culture

INTRODUCTION

PAUL E. BRODWIN

Since 1970, a host of new medical technologies has transformed the experience of birth, illness, and death in Euroamerican society.[1] The technologies have created new images of the body—perhaps even undercut "the body" as a cultural category—and they have changed the ways we think about human identity, connectedness, and the limits of the life span. This book takes up the ramifying cultural effects of recent biotechnologies. It explores the personal and political stakes of several clinical procedures: surrogacy, organ transplantation, genetic screening, artificial respiration, ultrasound and digitized images of the body, as well as the precursor field of tissue-culture research. These technologies have emerged from years of specialized laboratory and clinical research. They come with the aura of objective science and the prestige of a highly trained and credentialed class of experts. The meanings of these technologies, however, quickly escape professional control and infiltrate the diverse domains of everyday life. This process begins again with every new media frenzy over genetic testing or human cloning, and it limns the passage from science to popular culture and from professional medicine to the intimate realms of bodily experience. This is the background for *Biotechnology and Culture: Bodies, Anxieties, Ethics*.

The book examines how people debate, criticize, and re-imagine these contemporary interventions into the human body. It clarifies the fears and longings that surround biotechnologies: for instance, the fantasies of immortality connected with organ transplantation, or the desire to know the likely cause of one's death through genetic diagnosis. Of course, these cultural implications are not only a matter of personal reflection. Biotechnologies also acquire compelling political meanings, and the book explores how these both subvert and reinforce the dominant, legitimizing categories of contemporary life. The objects and procedures studied here are recasting the central debates in our society about the authority of scientific medicine, the personal and political control of the body, and the desirable relations among technology, nature, and human agency (compare Downey, Dumit, and Williams). Such far-reaching effects demand an inter-disciplinary response, represented here by anthropology, literature, philos-ophy, history, and cultural studies.

The word "biotechnology" denotes much more than material devices, designed for specific medical functions. It also includes the techniques for using them: the background practices and treatment rituals in which a given device acquires its meanings. The value and meaning of biotechnolo-gies, like those of all manufactured objects, are not inherent properties but rather judgments made by people who use them (Appadurai). These judgments arise first among the laboratory researchers who fashion the technology and their colleagues and financial supporters (for example, Rabinow). The judgments continue in the clinic as people gradually master a new instrument or machine, routinize it as a standard therapy, and embed it in the complex negotiations among health workers, patients, their families, insurers, and others (for example, Koenig). Finally certain biotechnologies become powerful public symbols even for those who never directly encounter them. In the contemporary United States, for instance, surrogacy and organ transplantation are highly charged terms in various cultural debates. They provide a convenient, tangible focus for arguments about the fundamental qualities of human nature as well as specific contra-dictions of gender, class, and professional authority (compare Woolgar). Biotechnologies thus become collective representations, encoding diverse anxieties and motivating political action.

The broad effects of biotechnology emerge in amazingly diverse sites and registers. People struggle over their implications in laboratories, court-rooms, fertility centers, intensive care units, and intimate domestic spaces. Debates and speculations are carried out in the mass media, treatment protocols, textbooks of bioethics, legal cases, and science fiction; and they transform the language we use to talk about personhood and social ties. We now wonder about our "genetic futures"; a woman becomes a "gesta-tional" or "genetic" mother; a recently deceased man is a "neo-mort" awaiting the "harvest" of his organs. Precisely how do such cultural effects come about? Out of what struggles and compromises? These are the cen-tral questions of *Biotechnology and Culture*. In some cases, people use bio-technologies to re-inscribe old assumptions in new contexts, as when surrogacy arrangements tie women more directly to the reproductive func-tion (chapters 3, 4, and 5). In other cases, people enlist biotechnology to subvert conventional wisdom and re-imagine so-called natural limits to life. In the face of tissue-culture research (chapters 1 and 2) and medical im-aging (chapters 6, 7, 9, and 11), for instance, people revise their notions of individual identity, privacy, and mortality. The unforeseen and contra-dictory reception to biotechnology—the "cultural imaginary" of specific devices and procedures—is the thread that runs through the chapters of this book.

Biotechnologies thus set off a cascade of ideological repercussions (see Lock). They alter the course of individual lives as much as the machinery of state regulation, and no single volume can encompass all their effects. The approach taken here features complex social dramas of the refusal, celebration, or ambivalent acceptance of new medical technologies. These "representative anecdotes" (Laqueur's words, chapter 3) illustrate both the persuasive power of biotechnologies and the way that power collides with the contingent strategies of everyday life. By privileging the pragmatic and embodied responses to biotechnology, the contributors to *Biotechnology and Culture* complement much of the theorizing on this topic in feminism, cultural studies of science, and medical anthropology. This book moves in the same direction as Sarah Franklin and Helena Ragoné, who argue for a critical anthropological empiricism to ground "overly speculative, abstract, and decontextualized accounts of the 'impact' of new technology" (5). Biotechnology does not act as an agent of social change in its own right. Its ramifying effects flow from the infinite decisions about how to use it and the specific personal and political stakes these decisions raise. This approach leads us to reconsider the ideological and counter-ideological uses of biotechnology in contemporary society.

The ideological power of biotechnology flows from its role in stabilizing and legitimating dominant social institutions, while simultaneously hiding this supporting function. The technological mediation of birth, illness, and death has become so pervasive that, in many cases, we no longer regard it as contingent and open to political discussion. The technology is falsely understood as necessary, and "what serves particular interests is seen, without reflection, as of universal interest" (Pippin 46; compare Geuss; Schroyer). Taking its cue from critical social theory, this argument begins by singling out science as the ideological institution par excellence in advanced industrial societies. Science claims sole authority to represent the objective world of nature, even though its tacit assumptions about reality are shot through with social interests (Comaroff). What we popularly accept as scientific facts—immutable aspects of the natural worlds—are fabricated in particular sites, through a handful of rhetorical devices, by people following institutional and professional agendas, and this whole network fits into specific political and economic regimes. Although scientific facts are fabricated, they are simultaneously certified as not fabricated, and therein lies the key to their ideological power (Treichler, Cartwright, and Penley, *Visible* 9).

Moreover, the scientists who invent and (initially) control biotechnologies subscribe to the worldview of biomedicine, with its commitments to naturalism and individualism. This worldview classifies the human body, the object of technical intervention, as a natural thing-in-itself, indifferent to human purpose and relationships (Gordon 25). It considers disease an individual problem that is essentially unrelated to structural inequalities and historical forces. This orientation easily certifies biotechnologies as mere techniques and disguises the massive social interests which produce them and which they also advance. (The particular interests singled out by this critique include professional biomedicine, health care corporations, national and global economic elites, and patriarchy, among others). Biotechnologies thus become effective instruments of medicalization, since they create more occasions for expert intervention into the human body. As exhaustively documented in the sociological literature, medicalization is a peculiarly effective instrument of social control (see Conrad and Schneider; Kutchins and Kirk). By silencing competing definitions of bodily experience, it individualizes problems which have wider social causes and which could easily indict particular contradictions and inequities (see Taussig). Biotechnologies advance the project of medicalization in different ways. Because they so comfortably become a routinized part of clinical work and have immediate and dramatic effects, they justify expert authority and further interventions (for example, ultrasound, see Taylor, "Image"; this volume). In some cases, biotechnologies dovetail perfectly with the politically conservative preference for individualist explanations of behavior and destiny (for example, genetic screening and diagnosis, see Cranor).

The feminist engagement with new reproductive technologies (NRTs) has set the agenda for the critical study of biotechnology, and this vast body of work explores the ideological cross-currents of various instruments and procedures. Under the guise of therapeutic benevolence, NRTs have the potential to medicalize female existence and strengthen the control of female bodies within particular networks of power and value. At the most general level, this process advances the masculine desire to control and create life (Rowland; Corea et al.; Lublin). More recent and contextualized research analyzes NRTs through theories of biopower and surveillance. Laparoscopy and fetal photography, for instance, furnish ever more invasive and naturalized depictions of the fetus, which perform the crucial ideological work (in the context of American New Right politics) of visually separating mother and fetus, asserting fetal autonomy, and reducing women to passive reproductive machines (Stabile). Similar arguments about the ideological effects of biotechnologies and the struggles over authority and control which they engender run through books edited and authored by Ginsburg and Rapp; Franklin and Ragoné; Lock and Kaufert; Treichler, Cartwright, and Penley; Davis-Floyd and Dumit; and Kahn.

The contributors to the present volume build on this literature, and many share the project of denaturalizing the truth claims accompanying medical procedures. Nonetheless, as all the authors cited above make clear, contemporary biotechnology does not exert a single ideological effect. It

is instead a site of struggle and redefinition, a crossroads where different
social actors and institutions follow contradictory agendas toward an uncertain outcome. In ferreting out the ambivalent influence and reception of
biotechnologies, we thus must avoid replicating the conventional opposition between technophilia and technophobia (a danger also noted by
Treichler, Cartwright, and Penley, *Visible* 11; and Lock, "Displacing"). This
polarized discourse is a staple of popular media accounts as well as scholarly analysis, at least in the modernist vein (Pfaffenberger). The celebratory
rhetoric about biotechnology fits the "philia" slot: the confident hope that
technologies will ameliorate the human condition and decrease pain and
suffering. Unfortunately, the critique of ideology can easily slip into the
phobia, that is, the dystopian fear that technology uniformly strengthens
certain forms of domination and destroys the subject's autonomy. This opposition structures Andrew Kimbrell's polemic against bioengineering, to
choose one of many examples. Kimbrell compares the sale of blood, organs, and semen and the "renting" of the surrogate mother's womb to the
exploitation of labor in Blake's satanic mills of nineteenth-century England. In this scheme, the ideological effects of biotechnology are only
the most recent triumph of the capitalist logics of commodification and
alienation. In a historical echo of the Industrial Revolution, biotechnologies substitute efficiency for empathy and destroy older totalities—before,
the local community, and, now, the bounded and natural body (Kimbrell).

Kimbrell's argument has the virtue of connecting the analysis of contemporary medicine to the classic themes in technology criticism from Jacques Ellul, Lewis Mumford, and others. But it largely reproduces one side
of the technophilia/technophobia divide, and its overly general approach
misses how particular biotechnologies get caught up in dense webs of social
influence. Compare his broad pessimism about organ transplantation to
the recent ethnographic and critical writings of Margaret Lock and Veena
Das. The very notion of brain death (a prerequisite for organ retrieval and
transplantation) became thinkable only in the 1960s, and, as authoritative
knowledge, it is manufactured in particular national settings. It thus differs
remarkably in Japan compared to the United States; in Japan, a national
dispute centers precisely on the "Western" ideologies attached to brain
death criteria and the untrustworthiness of medical professionals (Lock,
"Contesting"; Ohnuki-Tierney). Supporting organ transplantation in this
setting enlists one in certain ideologies and opposing it, in others. Some
individuals consider the procedure valuable and spend great sums to procure it, others call it profane, and the state has criminalized it. The conjunctures which make up the political significance of organ transplantation
become even more complex at the local level. For instance, medical workers in the transplant units of Indian hospitals deploy scientific criteria
within the local worlds of poverty and civic duty. For Indian organ donors,
the technology offers ways to transform marginalization and structural vio-

lence into opportunity, yet of an ambivalent and even tragic sort (Das). In these rich accounts, there is no single "ideology" of biotechnology. The analytic focus shifts instead to the "signatures" of nation, citizenship, and local economies on the transplant procedure (compare Hogle).

Biotechnologies participate in several competing ideological projects at once, and this theme structures contemporary studies of reproductive technologies in the United States. For instance, many Americans regard screening for maternal serum alpha fetoprotein as a vehicle toward greater autonomy and reproductive choice (this test screens for fetal neural tube defects and, secondarily, Down's syndrome). At the same time, this procedure unproblematically supports the values of self-reliance and physical independence which stigmatize the disabled (Press et al.; compare Dumit and Davis-Floyd). In the motivation of women who refuse amniocentesis, Rayna Rapp finds a complex admixture of male prerogative over family decision making, African-American distrust of professional medicine, and religious-based opposition to abortion (*Testing* 165–91). The technological transformation of pregnancy, like the world of organ transplantation, is a crossroads for many discrete groups' interests: an arena for people to pursue diverse strategies of acceptance, refusal, and resistance, among many other shades of political action.

Moving beyond a deterministic critique of ideology also has important effects on claims to scholarly authority as well as overall approaches to the production and consumption of biotechnologies. Who possesses the privileged perspective on ideology, since even the critic is caught up in the operation of power? Those who assess the political effects of particular biotechnologies necessarily do so from a particular position which takes for granted its agendas and capacities. Rapp captures this impasse for middle-class feminist critics of reproductive technologies. As a group, they enjoy both privileged access to high-tech medicine and the economic autonomy to reject it. This class position is seamlessly incorporated into the analysis, and it thereby marginalizes interpretations from other class, ethnic, and religious locations (Rapp, "Real-Time" 33). The view from the margins can differ considerably; it relies on a different calculus of costs and benefits and is calibrated to different experiences of autonomy and constraint (for example, Lopez).

Do medical technologies actually emerge out of a unified, scientific worldview and a confident set of propositions about the body? The notion of ideology sometimes assumes such a monolithic worldview, but this runs against the grain of multidisciplinary research in science, technology, and society (STS) (see Hess). Scientific facts have a social history (Fleck), and they emerge out of local, contingent decisions made by scientific workers (Latour and Woolgar; Latour), despite the reified and inevitable quality these facts later acquire. The day-to-day practice of technoscience is crisscrossed by subtle rhetorics and tactics, and closure is not necessarily reached by the time biotechnologies leave the lab and enter social life in the clinic or become objects of media fascination. Their cultural imaginary is heterogeneous right from the start, as exemplified in recent studies of fetal surgery (Casper) and screening tests for cervical cancer (Singleton). At a level below the everyday clinical routine, we find instability of both concept and method. Different actors (technicians, researchers, clinicians, patients) compete for control over particular procedures, and, consequently, the meanings and uses of technical objects can change. These studies demonstrate that medicine—supposedly the unified ideological touchstone for biotechnologies—is not a coherent whole, but instead "an amalgam of thoughts, a mixture of habits, an assemblage of techniques" (Mol and Berg 3).

In this light, the producers as well as the users of any given biotechnology each respond from a unique position, including the resources they control or try to obtain, the identities they choose or have thrust upon them, and their investments in particular personal or professional futures. To represent such contingencies, we must privilege the pragmatic and situated quality of people's engagements with specific biotechnologies. To engage with biotechnologies makes immediate demands and alters long-range life possibilities. It pushes people into new territory without maps, reliable guides, or stable expectations (see Franklin; Lock and Kaufert). The way people inhabit and negotiate this terrain is a prime concern of *Biotechnology and Culture*. While Haraway ("Cyborg"; "Promises") and others pursue these engagements in the voice of high theory, this book provides detailed social dramas in diverse local dialects.

However, the uncertainties produced by engaging with biotechnologies concern not only personal commitments and the way individual bodies are managed but also how we regard the body as a category of thought and debate. As a collective symbol, the human body provides the most basic metaphors for order and disorder. It furnishes a natural alibi for cultural conventions and political arrangements; people continually appeal to the biological body to ratify these as objective, foreordained aspects of the world.[2] The figure of the natural body, much like the concept of nature, is a moral arbiter of acceptable behavior and legitimate political relations (see Lock, "Contesting"). The instruments and procedures examined in this volume potentially subvert this general strategy. As we currently use them, biotechnologies tend to denaturalize the body. They put under human control those processes that once symbolized the very limits of such control. For instance, they make conception, birth, and death—the traditional "facts of life"—into matters of expert judgment and partisan debate. Manipulated images of the body's surface and interior run through the mass media and our collective imagination, and we have learned to live with a permeable boundary between bodies and machines (see Davis-Floyd and Dumit; Gray; Balsamo). Biotechnologies undercut the "effect of the natural" which accompanies many long-standing discourses and practices (Doane). The figure of the organic, integrated body no longer serves as the gold standard for conventional arrangements of gender, families, and generations.

This loss creates both anxiety and exhilaration in the public sphere, and it also motivates several longer-term strategic responses. Some groups of people attempt to shore up the natural body as a blueprint for the social order, others celebrate its disruption along with the verities it once supported, and most move uncertainly and ambivalently between these two poles. *Biotechnology and Culture* explores such strategies and their possible outcomes. In the social dramas described here, the use of biotechnologies does not necessarily act as an ideological cover for conventional social and political norms. It also exposes contradictions that are hidden when we accept the integrity and naturalness of the body, and, in some cases, it even creates new contradictions and instabilities. In the long run, the use of biotechnologies can shift the ground of social life by transforming dominant cultural presuppositions from doxa to orthodoxy.

Pierre Bourdieu defines doxa as that which is beyond question and orthodoxy as straightened opinion supported by explicit arguments or canons (169ff). These terms allow us to compare the different modes of authority which support particular social arrangements. But they also preserve the insight, introduced by classic writings on ideology, that a given social order becomes compelling because its principles infiltrate the most immediate and banal aspects of daily life. Doxic authority may endure for long stretches of time: out of sheer repetition, particular constructions of the natural and universal start to appear uniquely true, and the contingent social order seems rooted in the objective structure of the world. Bourdieu once called this the "illusion of spontaneous understanding" (Bourdieu and Wacquant 73). Our perceptions of the objective world—including our decision to label it as "natural"—are a product of repeated social learning; so, of course, these perceptions tend to mirror and silently justify the immediate social arrangements, divisions, and hierarchies. Moreover, enforcing and normalizing certain uses of the body—so that they become "second nature"—are arguably the most effective ways of projecting a particular social order into the quotidian and unquestioned precincts of consciousness.

Nonetheless, countervailing forces always arise through the action of marginal groups, the influence of alternative world maps, as well as material changes in the objects once deemed outside human control (see Comaroff and Comaroff 28ff). As these forces rise to a certain pitch, doxa is no longer ratified by people's immediate perceptions of the world. Losing the gold standard of the natural body is one such countervailing force. It does not, of course, automatically subvert all the cultural knowledge and social relations predicated upon this iconic body. It simply renders them less tacit and more in need of explicit justification. This is what biotechnologies potentially accomplish. These instruments and procedures intervene in the material body that mediates our experience of the world (see Csordas; this volume). By altering this body, its limits, capacities, and boundaries, biotechnologies make the social world seem that much less self-evident and indisputable. For instance, in the presence of surrogacy technologies, it makes sense to ask why children need two parents, and not one or three, and this line of question extends even deeper into cultural models of the family and sexuality. In the presence of organ transplantation, we begin to inquire whether personal identity lingers after death in one's body parts, and the inquiry eventually implicates capitalist logics of exchange and rhetorics of personal autonomy.

The answers are a matter of struggle between various orthodoxies and heterodoxies, but that is just the point. Biotechnologies do not exert a determinate effect, ideological or counter-ideological, conservative or progressive. Nonetheless, using them does demand that we pursue certain questions, some of these for the first time in history, and the outcomes directly implicate the legitimating categories of social life. For those caught up in the biotechnological embrace, tacit distinctions separating humans from machines, kin from non-kin, female from male, and even nature from culture are exposed and radically questioned and defended. The ways people enter these issues are a recurring theme in *Biotechnology and Culture: Bodies, Anxieties, Ethics,* and the subtitle suggests one way the questioning proceeds.

As the gold standard of the body recedes into the historical past, profound anxieties emerge about the conventional social hierarchies which are rooted in the body and which the body's truth silently certifies. I deliberately use the word "anxiety": a state of vague apprehension, a coloration of experience in which much seems at stake but which lacks clear paths of action. Entering the worlds of surrogacy, organ transplantation, genetic testing, and the like, most people cycle through this emotional state to a greater or lesser extent. The representations of biotechnology in mass media, policy debates, fiction, et cetera often display a similar unease. It is a diffuse but immediate and unsettling experience, accompanied by the suspicion or fear that further unknown changes are still to come. Obviously, the key terms used in this introduction—ideology, doxa, and the natural body—come from a rational theoretical discourse that is foreign to people's immediate feelings of apprehension, but this only supports the main methodological point. The call to study the situated responses, pragmatic knowledge, and local idioms of biotechnology (see Lock and Kaufert) demands that we take seriously the particular anxiety it produces.

How to gain access to this baseline anxiety, and how to represent it? How to connect it with the realignment of power relations, or at least their newly visible operation, discussed above? The anxiety and its political stakes, I suggest, get articulated through ethical discourses: the talk about moral quandaries which are inseparable from our engagement with specific biotechnologies. These quandaries range from the most personal (should I accept that amniocentesis?) to the political (should the state pay for it?). They concern the duties, obligations, and ideals of justice between individuals and families (for example, in the case of surrogacy) and nation-states (for example, in the global trade in donated organs; see Scheper-Hughes). These are the sorts of sanctioned debates, and the major public arena, in which people voice their current apprehensions and attempt to predict and control the technological future.

The binding of anxiety is a psychoanalytic notion, but here it refers to both a personal and collective process. In American society and elsewhere, anxieties over the reformulation of birth, illness, and death motivate rich debates about social duties and prohibitions. The debates exhibit an intensity and popularity far out of proportion to the number of people who actually make surrogacy contracts, receive donated organs, or are affected by the other currently controversial procedures. Moreover, the language which people use in such debates varies enormously. Most well known is the elaborate discourse about principles and their applications which dominates American bioethics (see Jonsen). However, this rational and experience-far framework is as foreign to people's fundamental vague apprehension as is the jargon of social theory. Much more common, but less noticed by scholars, are local ethical dialects which are calibrated to specific technologies and their immediate social stakes. These emerge from the moral imagination of a given time and place as well as the conjunctures of interests, values, and resources particular producers and users of biotechnology face. This is the ethical language which the contributors to this volume attempt to document and understand.

A range of controversies reappear throughout this book, and readers can group the chapters in several ways. In keeping with the spirit of the 1997 conference Biotechnology, Culture, and the Body at the Center for Twentieth Century Studies, University of Wisconsin-Milwaukee, I have arrayed the different disciplines around core issues. The book begins with discussions of the specific technologies of tissue-culture research (chapters 1 and 2) and surrogacy and ultrasound (chapters 3–6). The third section (chapters 7–10) explores how biotechnology both disrupts and drives our moral sensibility; and the fourth section (chapters 11 and 12) interrogates the circulation of biotechnology across divisions of class and nation. In each section, authors make their case using their discipline's conceptual tools and rules of evidence, and they complement each other in surprising ways. Anthropologists, historians, sociologists, ethnographers, and literary critics each pick up what the other has missed. However, these chapters also exemplify the blurred genres and breaching of disciplinary conventions that mark cutting edge scholarship in this field. Faced with the multiple effects of biotechnologies, anthropologists enter cyberspace, historians puzzle over California sperm banks, and literary critics analyze surrogacy contracts, to cite just a few examples. The use of biotechnology disrupts the current academic division of labor among all the other unarticulated cultural verities in its path.

I. GENEALOGIES

How should we write the history of contemporary biotechnologies? This section draws on insider accounts of specific laboratories and clinical sites, but it rejects triumphal narratives of technical mastery or medical benefits. These studies of tissue-culture research instead favor a genealogical approach that allows for discontinuities and avoids the search for origins. These chapters examine the personal motives and preoccupations of bioscientists and the cultural resonance of their work. From this perspective, the history of a given biotechnology lies in the shifting cultural meanings, political effects, and legal contests which grow up around it in different places and times.

The chapters follow particular techniques and interventions as they move from the guarded sites of professional expertise into more fluid, open public domains. They show that scientists' official statements—cast in a language of universalism and objectivity—co-exist with other sorts of texts, filled with dense allusions and metaphors and aimed at the dominant anxieties of their time. In these documents, tissue cultures have agency, cells have racial traits and are aggressive or malicious, and scientists become immortal through their work. These studies thus complicate scientists' self-image as cautious truth seekers, but they also subvert the determinist cliché about biotechnologies. Biotechnologies are not responsible for challenging cherished notions such as "natural" family ties or the integrity of the body. The challenge comes from the meanings we attach to particular technologies, and the precursor technologies have raised such challenges many times over the past hundred years.

First accomplished in 1907, tissue culture involves isolating a small fragment of tissue from the body and keeping it alive in the laboratory. This technique paved the way for current practices of in vitro fertilization (IVF) and organ transplantation. However, tissue-culture research was not only a technical precursor. As Susan M. Squier shows, it also became one of the first occasions to renegotiate the boundaries of the human species and life span in contemporary bioscience. In lectures, scholarly and popular articles, and poems and fiction, people weighed the potential of tissue culture for the creation of new life and the fantasy of immortality. Squier's chapter explores this through the optic of the Strangeways Research Laboratory of Cambridge, the first tissue-culture laboratory in Great Britain. She surveys both formal scientific texts and the poems written by researchers through which "the tissue-culture point of view" gradually emerged. This is a distinct set of assumptions which call into question the definition of the individual, the boundaries of the body, and the relations among species. These texts are an alternate form of scientific truth-telling, in which researchers freely wonder about the personal and metaphysical implications of their work.

Hannah Landecker also looks at scientists' personal investment in their research, but she writes in a more suspicious voice. In 1951, a piece of cancerous cervical tissue was cut from a terminally ill woman named Henrietta Lacks. Live cells grown from this biopsy continue to divide in laboratories around the world, and Landecker traces the history of this cell line. The stories of its life and origins reflect successive cultural anxieties in American society. Lacks was first depicted as heroic and immortal, especially when her cells (known as HeLa) were used to develop the Salk polio vaccine in the 1950s. Researchers also utilized the seemingly ageless cell lines to claim professional immortality of their own. These benevolent images dramatically shifted in 1966 with the discovery that HeLa cells had contaminated many other kinds of cell cultures. At nearly the same time, the narrative took on a racial grammar of miscegenation and heredity pollution when Henrietta Lacks's identity as African American became widely known. Now her cells were represented as aggressive, surreptitious, and a luxuriant "monster in the Pyrex." Landecker deftly explores the changing cultural response to HeLa cells, from sentimental personification to fear of racialized threats to scientific order. She shows how the cell line has become, in popular imagination, a microcosm of the human body—an immortal but uncanny double which catalyzes dominant American anxieties over individual identity and race.

II. MATERNITY IN QUESTION

Biotechnologies have profoundly altered reproduction—not only physiologically but also as a way to build connections between people and to strengthen, or contest, larger social arrangements. No longer a purely organic process bounded by the mother's body, reproduction now involves repeated technological interventions and is monitored by legions of laboratory workers. This is especially true for surrogacy: women bearing children who are not genetically related to them or who are contracted to another couple. In surrogacy, reproduction is diffused among many people: those who contribute genetic material, gestate the fetus, nurture the child, and are awarded parental rights. This stunning shift in the social management and political stakes of reproduction has inspired both utopian and dystopian responses. We hear technocratic claims about finally overcoming infertility and emancipatory hopes about recasting rigid models for gender and the family, once falsely inscribed as "natural." We also hear warnings about medicalization and the repressive control of women's bodies. In the dystopian scenario, reproductive technologies threaten to erode women's rights in a nightmarish quest for control over the maternal function (Lublin).

Maternity—as personal experience and public symbol, domestic relation and legal role—is surely changing in response to reproductive technologies. These chapters offer not speculative scenarios but instead detailed social dramas of conception, gestation, childbirth, and the production of new families. The use of surrogacy technologies shifts the way that bodily connections are culturally acknowledged, and this occurs daily in the mundane settings of courtrooms, fertility clinics, and baby showers. The people portrayed here negotiate the conceptual dilemmas about maternity in supremely practical and embodied ways. They debate the historically ancient distinction between plan and labor as the basis for legitimate parenthood. They reinvent maternity as a shared and public rather than private process. Finally, they struggle over the status of the fetus and newborn child as both a person and commodity. In the contemporary United States, debate rages over precisely what reproduction creates: a new biological organism, an item for exchange, a new set of social relations, an opportunity for consumption and display, or all four at the same time?

For Thomas W. Laqueur, the legal and moral dilemmas recently sparked by surrogacy expose the historic ambivalence in the West about the grounds of human connectedness. In Laqueur's analysis of sperm banks, ovum brokerages, and legal contests surrounding lesbian motherhood, the notion of primordial blood ties turns out to be remarkably unstable and contingent. This evidence bears on a larger question: how do we recognize generational relatedness? The naturalness of family relations has long been under attack, and the practice of surrogacy only intensifies fundamental conceptual shifts in the grounds of human connectedness which began in eighteenth-century Europe. Who are accepted as parents: those who plan for the child or those who contribute genetic material? Those whose bodies gestate and give birth or those who raise the child? These are ancient ambiguities in the West, and manifestly novel technologies simply rehearse them in new contexts. The novelty of surrogacy thus lies in new cultural possibilities, not the technology per se. Like other reproductive technologies, surrogacy can precipitate a crisis of nature, but it also demonstrates how vexed the meaning of "flesh of my flesh" has always been.

Deborah Grayson traces similar debates in the well-known legal controversy over gestational surrogacy, *Johnson v. Calvert*. In 1990, Anna Johnson, an African-American woman, gestated and bore a child from the implanted embryo of Mark and Crispina Calvert, a white/Filipina couple. She then sued to terminate the surrogacy contract and become the baby's legal parent. This conflict turned on the ambiguous process by which intimate bodily connections create families. Should the embodied experience of pregnancy influence our judgments about "natural" motherhood? Can a child have two mothers? Grayson analyzes the legal attempts to shore up the privatized nuclear family and assert the primacy of genetic ties. The debates about this case, however, connect the ambiguities of surrogacy to the restrictive meanings of race and the obstacles to black women's autonomy in America. To make Anna Johnson the legal mother of a white child would undermine our metaphysical model of racial separateness. Grayson argues strongly for an alternative notion of communal maternity ("motherwork"), based on practices in African-American communities. This category disrupts naturalizing assumptions about family and maternity, and it is thus uniquely suited to the public and denaturalized drama that reproduction has become.

Gillian M. Goslinga-Roy presents another tale of surrogacy, one where the realities of class as well as race drive the rhetoric of exclusive genetic ties. As an ethnographic filmmaker, Goslinga-Roy accompanied Julie Thayer and Paul and Pamela Martin (all names are pseudonyms), the couple whose embryo Julie bore, through the process from initial contract to childbirth. Although Julie freely joked about being a "cow" and an "incubator," she embraced the idea of surrendering the privacy of her womb. She cast surrogacy as a redemptive and emotionally powerful response to several losses in her life. To say that Julie commodified her body, therefore, ignores her embodied experience and her efforts to control the social meanings of surrogacy. Julie did not locate reproduction narrowly within her body. In a fascinating application of Grayson's proposal, Julie envisioned sharing her body as a collective space and actively involving the Martins in her pregnancy. The Martins, however, insisted on framing Julie's experience through the rhetoric of genetic relatedness. For them, surrogacy represented the triumph of their willpower, rational action, and wealth over the problem of infertility. For the Martins, Julie simply supplied the means to this end, and their attitude subverted Julie's desires for a shared pregnancy. Goslinga-Roy thus traces the competing ways to narrate the surrogacy process and the play of personal interests, imagination, and class authority on its outcome.

Clearly, many women who gestate another couple's embryo resist the categories of "breeder" and "alienated worker." But what accounts for the cultural anxiety over these categories in the first place? Janelle S. Taylor directly interrogates the fear of commodified reproduction in her ethnographic study of obstetrical ultrasound in American society. Two assumptions underlie this fear: (1) reproduction, in a world of surrogacy, fetal imaging, and the like, now dangerously resembles industrial production; and (2) we can and should distinguish persons from commodities. Taylor finds scant evidence for either assumption in the way middle-class Americans regard pregnancy and maternity. In particular, Americans construe ultrasound in terms of consumption rather than production. In the warm pleasure they take in displaying and viewing ultrasound images, people construct the fetus more and more as a commodity at the same time—and through the same means—as they construct it as a person. Commodification and consumption, after all, are central to the way we understand personhood in the contemporary United States. Obstetrical ultrasound and, perhaps, other reproductive technologies do not necessarily disrupt core cultural categories. They enact these categories, along with their previously hidden tensions and contradictions.

III. ETHICS AND THE TECHNOLOGICAL SUBJECT

Contemporary bioethics arose in the 1960s in the wake of innovations in dialysis, transplants, artificial organs, and assisted reproduction. These technologies sparked debates about the allocation of scarce resources and the quality and limits of life. In response, ethicists developed a set of abstract normative principles—autonomy, beneficence, and distributive justice—which largely structure professional debates to this day. These principles also infiltrate everyday life. In signing the organ donation line on our driver's license or requesting prenatal tests, our actions are already shaped by a particular ethical discourse. What assumptions drive this dominant discourse? Whose voices does it include or rule out? How does it compete with the diffuse, non-codified ways Americans discuss personal morality and responsibility? To investigate these questions, the chapters in this section contrast formal ethical principles against the personal ambivalences and interests of people engaged with specific biotechnologies.

The core principles of American bioethics take a generic concept of the person and make it the basis for a universal morality. This is, of course, the autonomous individual of Western liberalism: the sovereign individual who acts freely according to a self-chosen plan. However, the very technologies which sparked early bioethics unsettle this tacit understanding of the person. They have created new ways of exerting one's will and gauging one's present identity and future fate. These transformations, as much as the conflict between abstract principles, motivate our deepest ethical concerns. For instance, predictive genetic testing alters the way people calculate their life prospects—the likely mixture of happiness and suffering they will encounter—and it can erase or intensify certain aspects of their identity. New strategies to assess the subjectivity of technologically altered bodies are demanded by medical imaging, transplantation, and ventilation. The authors in this section thus wrestle with complex ethical questions, but the figure of the autonomous, self-governing subject is less and less available to help answer them.

The dilemma appears in Thomas J. Csordas's study of the Visible Human Project, a massive database of digitized anatomical photographs of a man and a woman. Computer manipulations of these images give the sense of traveling through a virtual body, and this promises to revolutionize medical education. However, the technology also transforms the experience of selfhood and embodiment among its users. The Visible Man, now wholly a resident of cyberspace, was originally Joseph Paul Jernigan, an executed prisoner who donated his body to science. The person is not only dead but also dissolved as a physical being (due to the transverse slicing required to prepare the images). What is the relation between Jernigan and the database he became? The question echoes Landecker's moral anxiety about Henrietta Lacks, but for Csordas the inquiry turns on the changes in our embodiment and self-perception this technology induces. The Visible Human Project evokes visual wonderment as well as physical revulsion at the freezing and slicing of bodies. It creates a fragmented and objectified sense of the body but also a more physically intimate one. It trains medical workers in greater compassion but also omnipotence. The Visible Man—a shade inhabiting cyberspace—affects our embodied subjectivity, and here lie the ethical stakes of this particular biotechnology.

Predictive testing for the DNA markers of disease alters subjectivity in different ways. Alice Ruth Wexler probes the layered meanings of the genetic test for Huntington's disease (HD) as they emerged for herself, her family, as well as others who accept, or refuse, this peculiar knowledge of the future. In general, predictive genetic tests threaten to marginalize those found gene-positive, but they may also produce new kinds of personal and collective identity. Despite the eugenic threat and the absence of effective treatment, many people at risk want to be tested in order "to escape uncertainty." What lies behind this wish? Cultural pressures are mounting to take the test; genetic counselors consider refusal a sort of denial and they construct the need to know as a basic psychological drive. Wexler explores this discourse, as well as the testing narratives from the HD community, not as windows onto raw experience but rather as complexly mediated by diverse medical, religious, and popular idioms. In this community, a positive genetic test gives one an aura, a transparency, as though others could see inside one's body to the gene itself. Moreover, people link the decision to take the test with valued moral traits in American society: courage, truth seeking, and (ironically) the drive for mastery over the future. Identity terms thus increasingly circulate around genetic status, and powerful cultural ideas infiltrate how people calculate the stakes of predictive testing.

The next chapters take up two vexing tendencies in current use of biotechnologies: (1) the replacement of moral discourse by technical expertise, and (2) the insertion of the human body into commodity exchange, governed by free-market logic. In each case, people's ethical concerns revolve around the subjectivity of technologically altered bodies. Robert M. Nelson describes the conflict of interpretation in a pediatric clinic between medical staff and the mother of "Michael" (a pseudonym), a ventilator-dependent child less than a year old. For his mother, Michael's tears signified profound and unending emotional distress, and for this reason she asked to remove the ventilator. She regarded the technology as an invasive foreign object which prolonged his suffering. The medical staff, however, considered Michael as simultaneously body and machine and his tears a simple physiological response. For this reason, they refused the mother's request. Michael's subjective experience, and the authority to read it one way or another, is at the heart of this dilemma. Nelson's case study illustrates how rational, technical expertise can foreclose genuine moral debate. By regarding technology as a mere collection of devices—a morally neutral means to the ultimate good of prolonging life—certain procedures become a standard, unquestionable component of care (see Koenig). This, in turn, justifies technical discourse as the sole guide for treatment decisions. Rational, technical expertise provides certainty for medical workers, but it also rules out other ways to figure the subjectivity of near-cyborgs, such as Michael, and our ethical obligations to them.

In the United States and elsewhere, the cultural status of the cadaver is shot through with moral concerns. Transplant technology, which requires a supply of cadaveric organs, inevitably generates anguished ethical debates. Some regard removing organs as a symbolic act, celebrating the generosity of the deceased; but for others it is a commercial act, reflecting individuals' choice to dispose of their body as they please. Donald Joralemon argues strongly against the commercial interpretation and the schemes of financial incentives for organ retrieval, and this offers a striking contrast to Taylor's essay. However, the debate exposes cultural presuppositions about precisely what happens to the self and body at the moment of death. Does one's identity somehow inhere in the body and hence subsist in body parts after death? Or does the body become a simple storehouse of *materia medica*? Based on cross-cultural evidence, Joralemon argues that the cadaver is usually an important site for personal and collective mourning. In the contemporary West, however, the symbolic link between deceased individuals and their bodies is morally diverse, and this accounts for our ambivalence about treating the body as a reservoir of spare parts. Nonetheless, notions of embodied identity and communal obligations to the dead have not yet entered conventional bioethics, and thus Joralemon turns to comparative studies of death rituals to refute current efforts to monetize the dead body.

IV. BIOTECHNOLOGY AND GLOBALIZATION

To export biotechnologies across the boundaries of nation and class is not an innocent exercise in technology transfer. It instead participates in the general inequalities between the industrialized and developing worlds and between the core and periphery within the West. The regional and global movement of biotechnologies can thus be used to advance projects of social control and to subvert alternative ways of representing and treating the body. Indeed, their cachet of modernity, rationality, and instrumental effectiveness permeates the rhetoric of health planners and potential patients alike. But the circulation of biotechnologies does not reprise the use of medical services as tools of domination in the classic era of European colonialism (Arnold). It takes place in a more complicated terrain, connecting not empires and colonized subjects but welfare agencies and immigrants and computer specialists, transnational corporations, and people marginalized by class, ethnicity, and other forces.

The reception to biotechnology in such settings involves moments of both accommodation and resistance. Out of sheer pragmatism, people in most places freely mix high-tech and low-tech medicine. At times, they are coerced to do so, but, at other times, they appropriate technologically sophisticated health services to pursue diverse political ends. This section examines the political stakes of biotechnology through specific contests over material resources, citizenship, and moral status, and it points to more general questions. What comprises the aura of cosmopolitan modernity surrounding biotechnology? How is it produced? What is its effect and fate in marginal communities, that is, among people relegated to a minor role in standard narratives of the West?

Telemedicine allows physicians to treat patients across great distances through the combined use of the Internet, satellites, video, and digital imaging. To its supporters, telemedicine fulfills the liberal dream of using medical knowledge to benefit underserved populations. Lisa Cartwright, however, shows that it chiefly advances the privatization of health care and justifies reducing services to remote areas. Telemedicine electronically aggregates a dispersed population into centralized facilities and subjects people to routine surveillance and interventions. It produces virtual communities of patients which a combination of administrative strength and biopower controls. However, Cartwright also describes indigenous communities in the United States and Canada which have appropriated the technology to increase local autonomy and agency, although this is so far a rare occurrence. Her chapter explores how these new ways of visualizing the body carry political effects which travel in complicated circuits between centers and peripheries. Telemedicine creates novel, de-localized communities, but they are communities that do not thereby escape the structural inequalities in global flows of technology and capital.

In the postcolonial Caribbean, people routinely criticize biotechnology, but they do so for reasons that have little to do with sickness and health. Paul E. Brodwin examines how residents of the Haitian diaspora living in Guadeloupe, French West Indies, debate the effectiveness of medical technology and thereby counteract their uncertain civil status and devalued reputation. Haitian migrants in Guadeloupe are marginalized in practical and ideological ways by a hostile majority society. This situation motivates their rhetorical contrasts between technologically sophisticated French biomedicine and Haitian religious healing. The latter is more effective, they argue, and this allows them to claim an alternative and higher moral status. This strategy also plays on the ambivalence of Guadeloupeans to the French civilizing process and its promise of modernity, as symbolized by imported medical technology. Haitian migrants are typically slotted into the subordinate position in the Guadeloupean narrative of progress, and criticizing French medicine becomes an important sort of symbolic resistance. Brodwin's ethnography thus illustrates the seductive force of bio-technology in Western narratives of modernity as well as their ironic local uses.

NOTES

1. In 1978, the first human child was born through in vitro fertilization (IVF), and, only a few years later, IVF and surrogacy programs were well established in the United States (Stacey; Ragoné). The development of recombinant DNA technology in the 1970s enabled fetal screening for hereditary disorders (Kan), and these techniques soon became routine in prenatal care and the counseling of prospective parents (Rapp, "Accounting"; Bosk). Knowledge of genetic susceptibility to disease has increased dramatically since the Human Genome Project began in 1990, and this new information has radically altered how people decide about reproduction and calculate their personal vulnerability to disease (Kevles and Hood). The discovery in 1972 of cyclosporine, a powerful immunosuppressive drug, immediately transformed the worldwide scale of organ transplantation (Fox and Swazey). Indeed, the demand for cadaveric organs now outstrips the supply, and this market condition has spurred new attempts to commodify body parts (Joralemon; Kimbrell) and to specify the meaning and markers of death (Ohnuki-Tierney; Lock, "Contesting"). Finally, medical imaging technologies developed in the past twenty years—ultrasound, CAT scans, PET scans—have produced new information about ourselves as bodies and as persons, and these images have migrated out of clinics and research labs and into courtrooms, advertisements, and everyday life (Dumit; Taylor, "Public").

2. Cultural historians and anthropologists have long examined how body representations legitimate social order; see, among others, articles in Feher; Csordas; and Featherstone, Hepworth, and Turner.

WORKS CITED

Appadurai, Arjun. "Introduction: Commodities and the Politics of Value." *The Social Life of Things: Commodities in Cultural Perspective.* Ed. Arjun Appadurai. Cambridge: Cambridge University Press, 1986. 3–63.

Arnold, David, ed. *Imperial Medicine and Indigenous Societies.* Manchester: Manchester University Press, 1988.

Balsamo, Anne. *Technologies of the Gendered Body: Reading Cyborg Women.* Durham: Duke University Press, 1996.

Berg, Marc, and Annemarie Mol, eds. *Differences in Medicine: Unraveling Practices, Techniques, and Bodies.* Durham: Duke University Press, 1998.

Bosk, Charles L. *All God's Mistakes: Genetic Counseling in a Pediatric Hospital.* Chicago: University of Chicago Press, 1992.

Bourdieu, Pierre. *Outline of a Theory of Practice.* Trans. Richard Nice. Cambridge: Cambridge University Press, 1977.

Bourdieu, Pierre, and Loic J. D. Wacquant. *An Invitation to Reflexive Sociology.* Chicago: University of Chicago Press, 1992.

Casper, Monica J. "Working on and around Human Fetuses: The Contested Domain of Fetal Surgery." Berg and Mol 28–52.

Comaroff, Jean. "Medicine—Symbol and Ideology." *The Problem of Medical Knowledge: Examining the Social Construction of Medicine.* Ed. P. Wright and A. Treacher. Edinburgh: Edinburgh University Press, 1982. 49–68.

Comaroff, John L., and Jean Comaroff. *Ethnography and the Historical Imagination.* Boulder: Westview, 1992.

Conrad, Peter, and Joseph W. Schneider, eds. *Deviance and Medicalization: From Badness to Sickness.* Philadelphia: Temple University Press, 1992.

Corea, Gena, et al. *Man-made Women: How New Reproductive Technologies Affect Women.* Bloomington: Indiana University Press, 1987.

Cranor, Carl F., ed. *Are Genes Us? The Social Consequences of the New Genetics.* New Brunswick: Rutgers University Press, 1994.

Csordas, Thomas J. "Embodiment as a Paradigm for Anthropology." *Ethos* 18.1 (1990): 5–47.

———, ed. *Embodiment and Experience: The Existential Ground of Culture and Self.* Cambridge: Cambridge University Press, 1994.

Cussins, Charis. "Producing Reproduction: Techniques of Normalization and Naturalization in Infertility Clinics." Franklin and Ragoné 66–101.

Das, Veena. "Organ Transplants: Gift, Sale, Theft?" *Living and Working with New Biomedical Technologies: Intersections of Inquiry.* Ed. Margaret Lock, Allan Young, and Alberto Cambrosio. Cambridge: Cambridge University Press, 2000.

Davis-Floyd, Robbie, and Joseph Dumit, ed. *Cyborg Babies: From Techno-Sex to Techno-Tots.* New York: Routledge, 1998.

Doane, Mary Ann. "Technophilia: Technology, Representation, and the Feminine." *Body/Politics: Women and the Discourses of Science.* Ed. M. Jacobus, E. F. Keller, and S. Shuttleworth. New York: Routledge, 1990. 163–76.

Downey, Gary Lee, and Joseph Dumit, eds. *Cyborgs and Citadels: Anthropological Interventions in Emerging Sciences and Technologies.* Santa Fe: School of American Research, 1997.

Downey, Gary Lee, Joseph Dumit, and Sarah Williams. "Cyborg Anthropology." Gray 341–46.

Dumit, Joseph. "Twenty-first Century PET: Looking for Mind and Morality through the Eye of Technology." *Technoscientific Imaginaries: Conversations, Profiles, and Memoirs.* Ed. G. E. Marcus. Chicago: University of Chicago Press, 1995. 87–128.

Dumit, Joseph, and Robbie Davis-Floyd. "Cyborg Babies: Children of the Third Millennium." Davis-Floyd and Dumit 1–18.

Featherstone, Mike, Mike Hepworth, and Bryan Turner, eds. *The Body: Social Process and Cultural Theory.* London: Sage, 1991.

Feher, Michel, ed. *Fragments for a History of the Human Body.* New York: Zone, 1989.

Fleck, Ludwig. *Genesis and Development of a Scientific Fact*. 1935. Chicago: University of Chicago Press, 1979.

Fox, Renée C. "Afterthoughts: Continuing Reflections on Organ Transplantation." *Organ Transplantation: Meanings and Realities*. Ed. S. J. Youngner, R. C. Fox, and L. J. O'Connell. Madison: University of Wisconsin Press, 1996. 252–72.

Fox, Renée C., and Judith P. Swazey. *Spare Parts: Organ Replacement in American Society*. New York: Oxford University Press, 1992.

Franklin, Sarah. *Embodied Progress: A Cultural Account of Assisted Reproduction*. London: Routledge, 1997.

Franklin, Sarah, and Helena Ragoné, eds. *Reproducing Reproduction: Kinship, Power, and Technological Innovation*. Philadelphia: University of Pennsylvania Press, 1998.

Geuss, Raymond. *The Idea of a Critical Theory: Habermas and the Frankfurt School*. Cambridge: Cambridge University Press, 1981.

Ginsburg, Faye D., and Rayna Rapp, eds. *Conceiving the New World Order: The Global Politics of Reproduction*. Berkeley: University of California Press, 1995.

Gordon, Deborah R. "Tenacious Assumptions in Western Medicine." Lock and Gordon 19–56.

Gray, Chris Hables, ed. *The Cyborg Handbook*. New York: Routledge, 1995.

Haraway, Donna J. "A Cyborg Manifesto: Science, Technology, and Socialist Feminism in the Late Twentieth Century." *Simians, Cyborgs, and Women: The Reinvention of Nature*. New York: Routledge, 1991. 149–81.

———. "The Promises of Monsters: A Regenerative Politics of Inappropriated Others." *Cultural Studies*. Ed. L. Grossberg, C. Nelson, and P. A. Treichler. New York: Routledge, 1992. 295–337.

Hess, David J. "If You're Thinking of Living in STS: A Guide for the Perplexed." Downey and Dumit 143–64.

Hogle, Linda F. *Recovering the Nation's Body: Cultural Memory, Medicine, and the Politics of Redemption*. New Brunswick: Rutgers University Press, 1999.

Jonsen, Albert. *The Birth of Bioethics*. New York: Oxford University Press, 1998.

Joralemon, Donald. "Organ Wars: The Battle for Body Parts." *Medical Anthropology Quarterly* 9.3 (1995): 335–56.

Kahn, Susan M. *Reproducing Jews: A Cultural Account of Assisted Conception in Israel*. Durham: Duke University Press, 2000.

Kan, Yuet Wai. "New Methods for the Diagnosis of Genetic Diseases." *A Revolution in Biotechnology*. Ed. J. L. Max. Cambridge: Cambridge University Press, 1989. 172–85.

Kevles, Daniel J., and Leroy Hood, eds. *The Code of Codes: Scientific and Social Issues in the Human Genome Project*. Cambridge: Harvard University Press, 1992.

Kimbrell, Andrew. *The Human Body Shop: The Engineering and Marketing of Life*. New York: Harper, 1993.

Koenig, Barbara A. "The Technological Imperative in Medical Practice: The Social Creation of a 'Routine' Treatment." Lock and Gordon 465–96.

Kutchins, Herb, and Stuart A. Kirk. *Making Us Crazy: DSM: The Psychiatric Bible and the Creation of Mental Disorders*. New York: Free, 1997.

Latour, Bruno. *Science in Action: How to Follow Scientists and Engineers through Society*. Cambridge: Harvard University Press, 1987.

Latour, Bruno, and Steven Woolgar. *Laboratory Life: The Construction of Scientific Facts*. Princeton: Princeton University Press, 1986.

Lock, Margaret. "Contesting the Natural in Japan." *Culture, Medicine, and Psychiatry* 19.1 (1995): 1–38.

———. "Displacing Suffering: The Reconstruction of Death in North America and

Japan." *Social Suffering*. Ed. Arthur Kleinman, Veena Das, and Margaret Lock. Berkeley: University of California Press, 1997. 207–44.

Lock, Margaret, and Deborah Gordon, eds. *Biomedicine Examined*. Dordrecht: Kluwer, 1988.

Lock, Margaret, and Patricia A. Kaufert, eds. *Pragmatic Women and Body Politics*. Cambridge: Cambridge University Press, 1998.

Lopez, Iris. "An Ethnography of the Medicalization of Puerto Rican Women's Reproduction." Lock and Kaufert 240–59.

Lublin, Nancy. *Pandora's Box: Feminism Confronts Reproductive Technology*. Lanham: Rowman, 1998.

Martin, Emily. *The Woman in the Body: A Cultural Analysis of Reproduction*. Boston: Beacon, 1987.

Mol, Annemarie, and Marc Berg. "Differences in Medicine: An Introduction." Berg and Mol 1–12.

Ohnuki-Tierney, Emiko. "Brain Death and Organ Transplantation: Cultural Bases of Medical Technology." *Current Anthropology* 35.3 (1994): 233–42.

Pfaffenberger, Bryan. "Social Anthropology of Technology." *Annual Review of Anthropology* 21 (1992): 491–516.

Pippen, Robert B. "On the Notion of Ideology as Technology." *Technology and the Politics of Knowledge*. Ed. Andrew Feenberg and Alastair Hannay. Bloomington: Indiana University Press, 1995. 43–61.

Press, Nancy, et al. "Provisional Normalcy and 'Perfect Babies': Pregnant Women's Attitudes toward Disability in the Context of Prenatal Testing." Franklin and Ragoné 46–65.

Rabinow, Paul. *Making PCR: A Story of Biotechnology*. Chicago: University of Chicago Press, 1996.

Ragoné, Helena. *Surrogate Motherhood: Conception in the Heart*. Boulder: Westview, 1994.

Rapp, Rayna. "Accounting for Amniocentesis." *Knowledge, Power, and Practice: The Anthropology of Medicine and Everyday Life*. Ed. S. Lindenbaum and M. Lock. Berkeley: University of California Press, 1993. 55–76.

———. "Real-Time Fetus: The Role of the Sonogram in the Age of Monitored Reproduction." Downey and Dumit 31–48.

———. *Testing Women, Testing the Fetus: The Social Impact of Amniocentesis in America*. New York: Routledge, 1999.

Rowland, Robyn. *Living Laboratories: Women and Reproductive Technologies*. Bloomington: Indiana University Press, 1992.

Scheper-Hughes, Nancy. "Truth and Rumor on the Organ Trail." *Natural History* 107.1 (1998): 48–57.

Schroyer, Trent. *The Critique of Domination: The Origins and Development of Critical Theory*. New York: Braziller, 1973.

Singleton, Vicky. "Stabilizing Instabilities: The Role of the Laboratory in the United Kingdom Cervical Screening Programme." Berg and Mol 86–104.

Stabile, Carol. "Shooting the Mother: Fetal Photography and the Politics of Disappearance." Treichler, Cartwright, and Penley 171–97.

Stacey, Meg. "Social Dimensions of Assisted Reproduction." *Changing Human Reproduction: Social Science Perspectives*. Ed. M. Stacey. London: Sage, 1992. 9–47.

Taussig, Michael. "Reification and the Consciousness of the Patient." *The Nervous System*. New York: Routledge, 1992. 83–109.

Taylor, Janelle S. "The Public Fetus and the Family Car: From Abortion Politics to a Volvo Advertisement." *Public Culture* 4.2 (1992): 67–80.

———. "Image of Contradiction: Obstetrical Ultrasound in American Culture." Franklin and Ragoné 15–45.

Treichler, Paula A., Lisa Cartwright, and Constance Penley. "Introduction: Paradoxes of Visibility." Treichler, Cartwright, and Penley 1–17.

Treichler, Paula A., Lisa Cartwright, and Constance Penley, eds. *The Visible Woman: Imaging Technologies, Gender, and Science.* New York: New York University Press, 1998.

Woolgar, Steven. "Reconstructing Man and Machine: A Note on Sociological Critiques of Cognitivism." *The Social Construction of Technological Systems: New Directions in the Sociology and History of Technology.* Ed. W. E. Bijker, T. P. Hughes, and T. J. Pinch. Cambridge: MIT Press, 1987. 311–28.

Part I: Genealogies

Chapter 1

Life and Death at Strangeways

The Tissue-Culture Point of View

SUSAN M. SQUIER

I do not love thee, Doctor Fell
The reason why, I cannot tell
But this I know, and know full well
I do not love thee, Doctor Fell.
—Tom Brown

The gray walls, black gowns, masks and hoods; the shining, twisted glass and pulsating colored fluids; the gleaming stainless steel, hidden steam jets, enclosed microscopes and huge witches' caldrons of the "great" laboratories of "tissue culture" have led far too many persons to consider cell culture too abstruse, recondite, and sacrosanct a field to be invaded by mere *hoi polloi*! . . . But every biology student should at some time have the dramatic experience of seeing the rhythmic beat of heart muscle, the sweep of the cilia of pulmonary epithelium, the twitching of skeletal muscle, the peristalsis of chorioallantoic or intestinal vesicles, the migration of fibroblasts, and the spread of nerve fibres. And every student can not only see these things but have the thrill of preparing them himself.
—Philip R. White, *Cultivation*

"Transplant of Pig Hearts to Be Banned: Blow for Patients Awaiting Surgery" was the headline of the *London Times* I picked up at the news agent in Green Park. It was 16 January 1997, and I had come straight in from Heathrow, heading up to the hospice in Saint John's Wood where my friend was dying of end-stage lung cancer. The story took up all the top left of page 1: the British government was banning "pioneering surgery to transplant pig hearts to people after a government inquiry concluded that the procedure was too risky," reported the *Times*'s health correspondent (Laurance).[1] The context for this decision was a "global organ donor crisis—at least 6,000 people are waiting for transplants in Britain and five times that number are on waiting lists in America." Set into the story was a cartoon of an argyll-sweatered man swinging from a light fixture; a twinset-clad woman in the foreground, holding a newspaper with the headline

"PIG HEART TRANSPLANTS UNSAFE," explained to her male companion, "He opted for the monkey heart transplant instead." The government recommended the creation of an "interim authority"—a "regulatory body to control experiments," the correspondent reported, noting that "a similar arrangement was made to control in-vitro fertilisation clinics in the 1980s before the Human Fertilisation and Embryology Act was passed in 1990." And now the in vitro fertilization (IVF) industry is in full swing in England, and perhaps xenotransplantation too is on its way to implementation, I thought.[2]

Over the next several days, I forgot the newspaper story, as I sat with my friend in the hospice and as he died. But the story has come back to me now, with its tangled web of seemingly stable oppositions—birth and death, institutional power and personal choice, profit and gift, free will and bondage, First World and Third World, nature and culture, human and animal—as the route to my topic: the analysis of a biomedical technique, developed in the first three decades of this century, that has occupied a pivotal position in the construction of our contemporary understanding of life and death. This technique is tissue culture.[3] As defined in *Animal Tissue into Humans,* the report of the Advisory Group on the Ethics of Xenotransplanation issued the day I arrived in London, tissue is "an organised aggregate of similar cells that perform a particular function . . . used to refer also to organs," and tissue culture is the technique of "isolating a small fragment of tissue from the body and cultivating it in a glass vessel containing suitable nutritive medium" (Great Britain 163). First successfully accomplished by Ross G. Harrison, who in 1907 published his discovery that amphibian nerve fiber could be made to live and grow in nutrient outside the body, the technique of tissue culture was given additional impetus during World War I.[4] In the postwar period, Nobel Prize winner Alexis Carrel was one of the prominent pioneers of tissue culture. Working at the Rockefeller Institute in New York City, Carrel achieved worldwide publicity for his accomplishment in keeping a chicken heart beating in vitro for over a decade. With its extension in the development of organ culture by Honor Bridget Fell and Robinson in 1929, tissue culture laid the foundation for modern biomedical techniques ranging from in vitro fertilization to organ transplantation, including the animal to human transplantations whose press furor greeted me on my arrival in London (Russell).[5]

As it has developed, then, tissue culture exists at the nexus of two boundaries currently under renegotiation: the boundary of the species and the boundary of the human life span. Two forms of growth are typically investigated in tissue grown in vitro: organized growth and unorganized growth. Researchers have accessed them—with iconic overdetermination—by culturing the embryo and the cancer cell, potent images of life and death.[6] Since its earliest years, this Janus-faced technique has drawn on—and catalyzed—fantasies extending beyond the discrete set of cells in the laboratory culture. In lectures, scientific papers, popular science articles, and fiction, people weigh the potential of tissue culture for the assault on aging, the prolongation of life, and, indeed, the fantasy of immortality. Tissue culture calls into question the definition of the individual, the boundaries of the body, the relations among species, and the authority of medical science. Moreover, it also challenges the conventional or accepted structure of the human life span.[7] Cultured tissues, whether from animals or human beings, play a crucial part in the broader twentieth-century project of reshaping the human life, in biomedicine (through interventions in conception, gestation, birth, aging, and death) and in literature and popular culture.

In the past decade, scholars have begun to hypothesize that the life course functions as a social institution, changing as other institutions have changed in response to Western processes of modernization.[8] Many now argue that the life span is "a discursive or imagined production, symbolic of a culture's beliefs about living and aging," and they understand the arc from birth to death not as merely natural but as the product of a culture's beliefs about human life and the shape of the human story (Katz, "Imagining" 61). However, such a constructivist position has its limits. Refuting "a view of the life course in which culture is granted the overarching power to mold nature in any form it chooses," these scholars hold instead that "human beings share with other species an embodied existence inevitably involving birth, growth, maturation and death" (Featherstone and Hepworth 375).

The biotechnology of tissue culture participates in this broader twentieth-century reconceptualization and reconstruction of the human life span. As I demonstrate in the longer project from which this essay is drawn, this reshaping of life is being accomplished, simultaneously, in biomedical science and in literature and popular culture, in ways I want to exemplify briefly. One kind of material and cultural reconfiguration of life is produced by biomedical practices, so that people are conceived differently, born differently, and age and die differently. The new technique of cloning, or more precisely nuclear fusion, which has tissue culture as its foundation, has already provoked a rethinking of the notion of aging on the animal level that is certain to travel—even if the technology does not—to the human realm.[9] Another restructuring of human life is accomplished by fiction, science fiction, popular science writing, and journalism, all part of a powerful set of inscription technologies which shape human beings to meet the needs of society. Through the vehicles of image, genre, character, and plot, literary and para-literary texts help to set the boundary conditions for the human condition. Such representations help us to understand our experience as individuals, including our experience of the body. This process of identity construction shapes both the symbolic and the material realms, as new kinds of life stories both catalyze and confirm new beginnings and endings.[10] A third kind of restructuring occurs at the boundary of the scientific and the literary, in the medical narratives that interweave materiality and representation, objectivity and interpretation (Hunter). "The journey from birth to death," Michael Holquist has observed, "serves as a biographical rhythm that entrains information into narrative so that it may best be processed as meaning by men and women who are born and die" (21). Such an implicit structure lies behind even the case history and the illness narrative, so that diagnostic facts arrive already shaped by representation.[11] When the journey from birth to death is rerouted, lengthened, or curtailed, meaning too is changed. In each of these different settings (the scientific, the literary, and their intersection in various forms of medical writing), the practical and symbolic resources of creatures that border on the human (animals and human embryos and fetuses, as well as tissues cultured from them) are used to recast that birth to death journey and thus redefine the human. This robs human beings of some old certainties and enables us to imagine new options.

If we analyze the (re)construction of the life course, as carried out in symbolic process and in scientific practice, we can perhaps recover the fantasies, responses, fears, and even the sense of possibilities, that may have been obscured or erased as we accepted without scrutiny this massive technoscientific and cultural reshaping of life. Moreover, when we transgress the disciplinary boundaries that limit knowledge, and study these acts of reconstruction in relation to each other, we can also illuminate literary and scientific practices. In what follows, I will trace some of the resonances of the new biotechnology of tissue culture (including its specifically literary resonances) within this larger modern and postmodern project of reshaping the course of human life. My focus is one specific site: the Strangeways Research Laboratory in Cambridge, England.

Tissue culture came to Great Britain when Thomas Strangeways Pigg Strangeways made the technique the sole focus of the laboratory he founded (Fell, "Tissue" 1). Initially a research hospital investigating rheumatoid arthritis when it was established in 1905, by the teens and early twenties the laboratory abandoned the clinical medical component, when Strangeways determined to focus on the microphysiology of disease, as revealed by this new technique he had learned from Alexis Carrel.[12] Tissue culture continued to be central to Strangeways Laboratory research throughout its founder's life. As the *Lancet* recounted in Strangeways's obituary, "By a curious irony the last lecture which he ever delivered was on Death and Immortality, and he illustrated it by a preparation bearing marvelous witness to his technical skill—a living culture of tissue made from a sausage purchased in the town" ("T. S. P. Strangeways" 56). With his death in 1926, the Research Hospital became the Strangeways Research Laboratory, funded by the Medical Research Council of Great Britain. Its directorship was assumed by twenty-eight-year-old Dr. Honor Bridget Fell, who, after completing a Ph.D. in zoology at Edinburgh University, had studied tissue culture at Cambridge with Thomas Strangeways (Strangeways; Hall).

The papers of the Strangeways Research Laboratory—in the 1920s and 1930s the only laboratory in the United Kingdom to concentrate exclusively on tissue culture—provide a flavor of the scientific, technical, cultural, and social issues this relatively new biotechnology raised.[13] Scientists dedicated themselves to exploring "the tissue-culture point of view," a phrase Dr. Honor Fell coined to refer to the scientific and popular mind-set bred by the micropractices of this novel technique. Three genres of documents all register the impact of this new biotechnology on our understanding of the beginning and the end of life: (1) a series of lectures on tissue culture that Honor Fell delivered, (2) press coverage of Strangeways's experiments, and (3) poems Strangeways researchers wrote in response to their experimental work. I approach these documents with two questions in mind. How do they reflect the broad cultural and scientific impact of tissue culture within and beyond the laboratory? Is there a genre-linked difference in the way that impact is registered and articulated? As we begin to investigate the tissue-culture point of view, we should hold in our minds a line from the nursery rhyme that has provided the opening epigraph of this essay: "I do not like thee, Dr. Fell." While this poem originated nearly three centuries earlier, C. H. Waddington's daughter recalls reciting it as a child and assuming that it referred to her father's laboratory colleague. As her memory suggests, the point of view bred by the practices of tissue culture inspired both fascination and unease (McDuff).

A series of lectures Dr. Honor Fell called "Tissue Culture," which she delivered between 1936 and 1938 to vocational and postgraduate medical students, conveys the impact of the new technology on contemporary biomedical practices and cultural attitudes toward life and death. Dr. Fell is explicit about "the tissue-culture point of view," and her lectures set the context for the reception of the prototypical Strangeways tissue-culture experiments, both within and beyond the laboratory.

1. The tissue-culture point of view raises questions about—and catalyzes reconceptions of—the boundaries of death and life. "The tissue to be cultivated," Dr. Fell specifies dryly, "should be taken from the animal immediately after death. This, however, is not essential and tissue has been known to grow quite well in vitro when taken from the body as much as a week after death or considerably longer if the body has been kept in cold storage" (Fell, "Tissue" 3). Perhaps in order to increase the drama—because this lecture is to vocational students—Dr. Fell shifts registers as she explains the implications of this fact. She moves from the animal (for it is exclusively animal tissue that is being cultured at Strangeways) to the human: "From this we can see that when a doctor pronounces a patient

'dead' he is only using the word 'death' in a restricted sense" (3). Working to make her material accessible to a non-professional audience, Dr. Fell is clear about the broader implications of tissue culture, both dramatizing and even popularizing them. "There is one very interesting and important result of this capacity for unorganized growth. . . . [I]t makes *certain* tissues (at least) *potentially immortal*" (5; emphasis in original).[14]

2. The tissue-culture point of view has a conceptual and methodological scaling effect that links the human to the animal, the whole organism to the cell, and the mature to the embryonic.[15] "Scaling," or recursive symmetry, is the repetition of a pattern across different dimensions and in a graduated series.[16] Dr. Fell demonstrates this scaling event when, linking the microscopic with the human dimension, she observes, "cells in culture can be used as experimental animals. This is so true that we can even vivisect them by a kind of microsurgery" ("Technique" 10). The tone here is less popular and more technical, perhaps because she is addressing postgraduates rather than vocational students, and the language, which might provoke discomfort in the general public, invokes an insider audience comfortable with experimental procedures. But the invocation of vivisection also gestures to the world beyond the laboratory: Dr. Fell's inclusion of this point in a discussion of the "value of tissue culture in research" suggests that the modern trend toward miniaturization in experimental science may be, in part, a response to the late-nineteenth- and early twentieth-century outcry against vivisection, so powerfully embodied in fiction in H. G. Wells's "The Island of Dr. Moreau," published only a decade earlier.

Dr. Fell's assessment of the impact of tissue culture illustrates one sort of scaling effect: the classic scientific strategy of drawing analogies between human and animal. Indeed, because the cells that she cultures are both embryonic and mature, her scaling effect incorporates two dimensions: the evolutionary and the (human) developmental. Strangeways cultures both embryonic and mature cells, and the cells thus cultured can stand in, Dr. Fell observes, for laboratory animals.[17] She cites this as one of the advantages of the tissue-culture method: "We are . . . able to use cells as experimental animals and subject them to the influence of various experimental agents or to changes in their environment" ("Technique" 6).

3. The tissue-culture point of view resituates the "body of science" from the realm of the static, graphic, and dead to that of the dynamic, photographic, and living. This shift is accomplished through visual images which resemble surveillance photography or unposed and spontaneous snapshots. The remarkable thing about tissue culture, Dr. Fell explains, is the fact that "[c]ells growing out of the body in this way can be watched and photographed whilst going about their ordinary business." The technique has changed the study of cell biology, she observes: "Originally histology was one of the most static of sciences: tissue culture has made it dynamic" ("Tissue" 1).[18]

4. The tissue-culture point of view appeals to scientists as a particularly glamorous or romantic technique, an appeal that tends to enroll more practitioners. Dr. Fell admits to her insider audience that this can be one of the drawbacks to tissue-culture research: "Tissue culture often suffers from its admirers. There is something rather romantic about the idea of taking living cells out of the body and watching them living and moving in a glass vessel, like a boy watching captive tadpoles in a jar. And this has led imaginative people to express most extravagant claims about tissue culture, which our actual experience in the laboratory quite fails to justify" ("Technique" 5).[19] Tissue culture may be used when another method would work just as well or even better, she observes, due perhaps "to a subconscious desire to use a spectacular method in preference to a simple one" ("Technique" 11). Yet if Dr. Fell admits that the associations clustering around tissue culture and the subconscious desires of researchers may lead to the overuse and overestimation of this research technique, she also risks contributing to that problem in her lecture to the vocational students, when she frames the implications of the technique broadly, without the qualifications that she is careful to draw in this lecture to postgraduate medical students.

5. The tissue-culture point of view is premised on decontextualization and abstraction. Unexamined analogies from the cellular to the human, while a tempting part of the tissue-culture point of view, should be resisted, Dr. Fell argues. "[R]esults obtained from experiments and observations made on cells growing under the abnormal conditions obtaining in vitro are very possibly not applicable to cells living in their normal environment in the body, because cells in vitro may be physiologically modified and therefore not comparable with normal cells in vivo" ("Technique" 8). By its procedure of isolating tissues in a culture medium, the tissue-culture method reduces the complex interplay of variables shaping any cell in vivo, and thus any findings should be received with caution. "One of the greatest advantages of tissue culture vis., the simplification of the environmental conditions of the cells, is at the same time one of its greatest limitations" ("Technique" 11). Dr. Fell's observation anticipates the position of some contemporary feminist critics of science, which has focused on the methodological limitations of laboratory science in general, in particular its strategy of contextual narrowing.[20]

If we consider how the prototypical Strangeways experiments were represented (and received) by the scientists within the laboratory and the non-scientists beyond it, we can see how the tissue-culture point of view generated scientific, technical, cultural, and social issues specific to this biotechnology. Three kinds of experiments and one specific research methodology typify the laboratory's approach to studying tissue culture in the 1930s, embodying what Dr. Fell called the "tissue-culture point of view."[21]

Embryo culture, or the study of the developmental center of mammalian and avian embryos by grafting in vitro, was best exemplified at Strangeways by the work of C. H. Waddington. He provided "the first demonstration of the activity of organization centres in mammals," by grafting pieces of the primitive streak into the neural centers of rabbit and chick embryos ("Strangeways Research" n. pag.).[22] As Waddington described his work in *Listener* in 1936:

> I found that the very young [chick] embryo would stay alive for a time if one put it on the surface of a particular sort of nutritious jelly. It starts on the jelly as a plain flat disc of cells, and it stays alive for about two days, and develops the beginning of a brain and eyes and ears and a heart which has begun to beat. It has an organiser, too. . . . In fact, probably all backboned animals have organisers in their eggs. Presumably man has. You yourself owe your brain to the working of your organiser. (812)

Waddington's explanatory passage demonstrates the scaling effect here, characteristic of the tissue-culture point of view, as he draws a parallel between embryo and mature organism, animal and human. His research also embodied the double investigative focus of tissue-culture research, both on the sources of life and the causes of death. As he observed in a letter to *Nature* in 1935, "I was . . . originally responsible for the suggestion that work on embryonic organizers may throw some light on the induction of malignant growth" (606).

Organ culture, or the growth of embryonic body parts in vitro, was a major part of the laboratory's activities. Dr. Fell saw this as the center of her career.[23] Researchers at Strangeways worked on the culture of a range of organs, including ears, mammary glands, ovaries, salivary glands, the pancreas, and hair and teeth.[24] Among the experiments, one in particular garnered publicity: the culture of embryonic rat's teeth, by Miss S. Glasstone. Press coverage emphasized the prominent role women played in the development of this new technique. As the *New York Herald* reported:

> The first recorded instance of "growing teeth" outside the body is contained in the annual report of the Medical Research Council issued here [London, 12 March 1936]. The achievement stands to the credit of the Strangeways Research Laboratory at Cambridge, and is regarded as an outstanding illustration of the power of the new technique of "tissue culture." . . . A woman, Dr. Honor B. Fell, is at the head of the small band of scientists responsible for the work, and Miss. S. Glasstone has been responsible for the laboratory's latest success. . . . ("Rat's"; Spear/Strangeways 1936)

With women scientists often active at its helm, the field of tissue culture seems to have been characterized by heightened gender awareness and gender play. This may explain the intriguing boundary crossing evident in a mock ceremony held at Strangeways Research Laboratory to mark the opening of the new X-ray department on 4 March 1937. The ceremony featured the visit of "Lady LeBec" to present the key of the new department to Dr. Fell. Three photographs of Lady LeBec survive in the Strangeways archive. A tall man in a white hospital gown who bears a striking resemblance to Strangeways researcher Lancelot Hogben, Lady LeBec wears a wig, a flower-bedecked hat, and much rouge and lipstick.[25] When the scientists were male, the act of growing tissue in special cultures may have seemed a humorous appropriation of a feminine, even maternal, function, while with women scientists it may have seemed worthy of note because of its divergence from the norm, as a method of "unnatural" growth.[26]

Finally, Strangeways scientists engaged in cancer research, studying unorganized cell growth and the biological effects of X-rays.[27] This research area, too, exhibits the tissue-culture point of view in its reconceptualization of life and death. While researchers working on embryo culture addressed the beginning of life, those working with wound tissue, culturing entire organs, or studying the effects of radiation on cancer cells turned their attention to the latter end of the life course, working on projects that would make significant contributions to the treatment of cancer and wartime injuries.[28]

When reported in the popular press, these three representative experiments in tissue culture provided the romanticized and technologically glamorous spectacle of which Dr. Fell warned in her lectures. Two examples will give the flavor. In 1937, in the London *Daily Mirror*'s column, "Science This Week," Harrison Hardy reported on Petar N. Martinovitch's experiments in ovary culture in vitro:

> The possibility that at some remote future time babies may be grown in bottles is an idea which cautious scientific men leave to the writer of fantasies. Though bottle babies may never be a fact, science moves slowly in their general direction. Mr. Petar N. Martinovitch, a biologist working in the Strangeways Research Laboratory, Cambridge, is growing germ-glands on watch glasses. From newborn mice and unborn rats he cuts out the sexual glands and plants them in clots of mixed fowl plasma (blood extract) and fowl embryo extract. *Female glands remained healthy after twenty-two days, male glands after seventeen days.* Mr. Martinovitch, being a serious biologist, does not talk about bottle babies, but says the results may "provide a new experimental approach to problems of sex physiology." (n. pag.; emphasis in original)

The article reflects a familiar tension between scientific and popular discourses—the careful assessment of the scientist and the futurological speculations of a "writer of fantasies"—while also engaging in a remarkable amount of translation and education on the methods of tissue culture.

Another article, Norah Burke's "Could You *Love* a Chemical Baby? For That's What Science Looks Like It's Producing Next" appearing the following year in the tabloid *Tit-Bits,* dispensed with the scientific education in favor of a futurological and fantastical presentation. Claiming that the worldwide alarm about a falling birth rate may be solved "in a few years' time by the production of living babies from scientists' laboratories," this article provides a lurid survey of tissue culture. It highlights the accomplishments of a number of researchers: an unnamed British woman doctor who "has succeeded in growing teeth in a test tube"; Alexis Carrel, who "has kept human hearts alive without their bodies, and bodies alive without their hearts"; Gregory Pincus, "renowned biologist [who] has 'created' living rabbits"); and "a Russian eye-specialist [who] has kept a human eye alive for six days in an air-tight jar in an ice chest." More is to come, the article implies: "Already other parts of the human body have been grown in test-tubes" (3). Here again, the popular press emphasizes the prominence of women in the Strangeways administrative and research pool. Remarking on the "unusual specialty of the Strangeways Research Laboratory at Cambridge, where the new technique of 'tissue culture' is carried out," the article notes "Dr. Honor B. Fell, a woman, is at the head of the work, and another woman, Miss S. Glasstone, has been responsible for the test-tube teeth" (3).

Commenting that "tissue culture has stupendous possibilities," Burke goes on to speculate: "It may not be long before life is possible without breathing! Test-tube human beings may be produced fully grown, by having oxygen injected directly into their blood until the lungs are ready to work" (3). The passage anticipates Jep Powell's short story "The Synthetic Woman," to appear two years later in the pioneer science fiction magazine *Amazing Stories,* in which a woman is synthesized chemically and grown to full physical maturity in the laboratory, with the help of injections of "oxydyne." Yet if Burke's article anticipates the science fiction genre, it also illustrates the characteristic gender tensions produced by the tissue-culture point of view:

> What are the gigantic realms of discovery and of human life in the future to which these experiments may lead us? . . . It may be that the human species will produce test-tube babies, or full-grown human beings, entirely chemically.
> *What sort of creatures will these be? All the rabbits "created" by Dr. Pincus were female. Will the test-tube human beings be unisexual, or sexless? Will the human race become like the social insects with three genders—male, female and neuter (or workers)? Will these sexless, soulless creatures of chemistry conquer the true human beings? Will our size be violently altered? Remember that chicken heart that went on growing and growing.* (3; emphasis in original)

While these newspaper stories demonstrate the broad popular interest catalyzed by the experiments in embryo and organ culture carried on at Strangeways, a different sort of publicity earlier greeted the most significant research technique developed there: the "Canti method," as it was dubbed by the popular press. This was a method of cinema photomicography developed by Dr. Ronald George Canti, a researcher who worked on "normal and malignant cells and their response to radiation." Canti was "the man who first turned the cinema camera in this country into an instrument of medical research," as the London *Evening Standard* put it in his obituary (Barker 2). Dr. Canti, who had studied with Alexis Carrel at Rockefeller Institute in New York, found himself frustrated—so the story goes—by the slow, clumsy techniques for culture filming used by Alessandro Fabbri, of Carrel's lab. In response, Canti devised an apparatus for photographing cell cultures in vitro, using what we would now call time-lapse photography (Foxon; Barker). Canti's films, with titles such as "Cultivation of the Whole Chick Embryo Femur *in vitro*," "Beating and fibrillation in the chick embryo heart *in vitro*," and "Investigation of the Normal and Parthenogenetic Division of the Rabbit Ovum," made in collaboration with C. H. Waddington, Honor Fell, and Gregory Pincus, captured the popular imagination.[29] The films were shown at 10 Downing Street, on television, and at scientific congresses throughout Europe.

The Canti technique played a central role in producing the tissue-culture point of view, shifting science from the static and graphic to the dynamic and photographic. As a press report on the public screening of one of his films reveals, Canti's films had great rhetorical power, making scientific research appealing to the public. They may also have embodied some interesting gender dynamics, enacting and thus satisfying a fantasy of patrilineal control at the microscopic level. In June 1921, in its coverage of one of Canti's film screenings, *The World* reported:

> How often have you heard a proud mother exclaim: "I can see my boy grow."
>
> Of course she cannot see her boy grow. She sees him every day . . . and she cannot see him grow any more than she or anybody else can see a plant grow. A rank weed may spring up with wondrous speed, but no human being can see it grow. . . .
>
> If she could take one picture of her son each day and then show them in the succession in which they were taken and at the rate of sixteen a second, she could see her son's growth and so could everyone who looked at the pictures. *But she and they would see him grow much faster than he does really; so fast that if his growth were as rapid as appears on the screen he would be able to hitch his toy wagon to a star long before he was as old as Carrel's chick heart.*" ("Scientists" n. pag.; emphasis added)

As this coverage indicates, Canti's films harnessed a perennial human obsession (human growth, emblematized in the growth of one's children) to a specific set of scientific questions characteristic of tissue-culture research: the growth of normal and abnormal cells. They also built on the recent desire to have not still photographs but a cinematic record of growth.

Further, the films drew on the public outcry caused by Alexis Carrel's accomplishment in the long-term culture of the chicken heart. Dr. Fell's tissue-culture lectures, with which I began my discussion of the Strangeways documents, also used Carrel's experiment to capture the implications of tissue culture. However, for Fell the effect is a disruption of normal size and mass, not of time: "[A] strain of heart fibroblast has been kept alive by Ebeling and Carrel for over 20 years. It has been calculated that if all the possible descendants of this patriarchal culture were alive today, the total bulk of tissue would vastly exceed the volume of the Earth" ("Tissue" 5). Significantly, both the *World* columnist and Dr. Fell conceptualize the long-lived tissue as being male, whether a mother's beloved son or a patriarchal culture with its descendants.[30]

The coverage in *The World* in 1936 not only linked cellular and embryonic growth to the development from child to adult but also related technological development to the wished-for human "development" that would lead to immortality. In 1938, a film collaboration between R. G. Canti and C. H. Waddington was given a public showing that catalyzed both sets of hopes:

> The time may come when, by means of television, classes all over the country may be able to watch laboratory experiments more difficult and intricate than could be carried out under ordinary school conditions. In this category come the "Artificial Immortality" tests to be conducted before the television cameras in the evening transmission . . . when Mr. C. H. Waddington, of the Strangeways Laboratory, Cambridge, will show how tissues can be kept alive indefinitely. ("Seer" n. pag.)

The embryo culture to be televised could not, of course, grow fast enough for television. The new televisual technology needed the help of its more established sibling, and "a film [was] transmitted which was taken during a previous experiment at the rate of one picture a minute. When these [were] projected at the normal rate of twenty-four frames a second, the culture [could] be seen growing" ("Seer" n. pag.). In effect, viewers of the film experienced two new cultures growing parallel to each other, as the film's portrait of nutrient-bathed animal tissues was transmitted by the relatively new technology of television. As press coverage revealed, this novel technology fed both the futurological curiosity of the general public and the stodgy scopophilia of the more cautious scientists.

> What we are missing by having to wait for television may be gauged from the following advance description of an Alexandra Palace studio item for to-night under the title of "Artificial Immortality" in the "Experiments in

Science" series. How a small piece of animal flesh can be kept alive indefinitely will be demonstrated in an unusual television feature. . . . Science is less interested in achieving the "artificial immortality" than in obtaining an opportunity by these means, of watching cells while they are alive. In the living body, it is practically impossible to watch them closely. Still, why shouldn't our biologists have a night out, now and again? ("What We Miss!" n. pag.)

If the public responded to Strangeways's experimentation with a mixture of excitement, mockery, and alarm at the ways it destabilized human identity, the life span, and human gender relations, how did the participating scientists view their activities? The official Strangeways position was one of a carefully limited statement of goals and implications: we are only interested in achieving information about the behavior of certain cells, the problems of sex physiology, et cetera. But we see something different if we examine communications not from the laboratory to the "outside world" but among scientists within the laboratory. Three poems composed by researchers at Strangeways, whose principal interests were the three central foci of Strangeways research (embryonic development, organ culture, and processes of disorganized or cancerous growth), reveal the internal responses to the tissue-culture point of view.[31]

We can begin with a poem the embryologist C. H. Waddington composed in January 1937. This poem concerns the organizer, the element in the embryo that controls its development for the first fourteen days. As Waddington explained the concept in the same year in *Listener:*

The actual experiments I do nearly all involve grafting little bits of the egg from one place to another. At this stage the egg has begun to develop but has hardly got any organs at all. It is still a round ball, and it is just beginning to fold in from the surface a sort of tube which will later turn into the gut or intestine. The region round about the edge of this tube, where it folds in from the surface, is the most important part of the egg. It is not only the place where the first organ, which is the gut, begins to appear, but it also controls the formation of all the other organs as well. ("Scientist" 811–12)

Waddington's poem explores the principle of the organizer from the inside, considering, as it were, how the egg views its changed potential. Although focused on that most microscopic of protagonists, the fertilized ovum, this poem participates in a genre of visual and verbal fetal representations that moves from the microcosmic to the macrocosmic, portraying the fetus as the product of social as well as scientific processes.[32] Waddington was comfortable making the leap from biology to politics, with its implied social constructivist viewpoint; in a later autobiographical essay, he went even farther, arguing that his particular scientific bent was shaped by an underlying metaphysic.[33] The poem begins with a celebration of the laziness of the fertilized egg:

Happy the egg, the bubble blastula,
does nothing much, but all the same,
each part as good as every other
or would be if it could,
but better its situation;
the opportunist roof, the sodden floor,
_____ with a space between:
a capital notion, but a lazy affair.

Moving from the collectivist happiness of the "bubble blastula", the egg develops until "every separate part is tied / to particular performance but still within itself is free / to organise its own affair," and the fertilized ovum now seems to be autonomous, rights bearing subject of the liberal civil state. In its movement from consciousness of individual freedom to functional specialization, Waddington's poem seems a miniature recap of the development of the Enlightenment individual from collectivity into class hierarchy, economic segmentation, and professionalization:

Feeling, thinking, feeding come
by limiting this freedom
of each being any,
now each for all
performs its special function.

The developing blastocyst moves from collective identity to individual functional identity and (through that) to agency. Identifying with the developing embryo, the poem affirms its change as progress.[34]

Now every separate part is tied
to particular performance
but still within itself is free
to organise its own affair

Finally, the functionally specialized embryo has attained not only economic and civic autonomy but psychic agency: "a special place among the few . . . exchanged / the possibility to be / for the act to do."

Two more poems, written on the same day in January 1938 by Strangeways researchers Petar Martinovitch, who was growing ovaries in vitro, and Arthur Hughes, whose work was on a study of vascularization in the embryonic chick, extend this identification with the objects of scientific study. Hughes instigated the exchange of poetry, writing this poem describing Martinovitch's work culturing mammalian "testicles and ovaries":

Testicles and ovaries
Explanted in a row
Grown by Martinovitch
in vitro.

The comic strategy of the Hughes poem puts the process of tissue culture in context, both institutional and personal. We seem to "pan" the laboratory looking for evidence of growth, our vision moving from the rows of testicles and ovary explants in vitro to the other sites of development: volumes in the library; sexualized nurses at the hospital (who wonder whether they "should know / About sexual maturation / In vitro"); and researchers who look forward to Martinovitch's next breakthrough ("Colleagues in the tea room / Wait for him below / To announce the new advances / In vitro"). Finally, the poem rounds on itself, suggesting that tissue culture reveals something not only about the cells but about their researcher. Martinovitch is another such tricky life form in culture, hard to get to meals, prompting his landladies to ponder "if they couldn't nourish him / in vitro." Humorously flouting the scientist's public posture of careful epistemological containment, the poem illustrates the scaling effect crucial to the tissue-culture point of view, linking the cellular to human, the laboratory experiment to its context, and sexual development to scientific and social development. Moreover, as it takes Martinovitch step-by-step out of the laboratory, it also exemplifies the tissue-culture strategy of abstraction and decontextualization.

Martinovitch's poem, "A Reply," lacks the metric smoothness of Hughes's effort, but it too articulates the tissue-culture point of view the Strangeways Research Laboratory produced.

At the Strangeways lab a certain lab
Sits on the bank (always left) of a blood vessel and follows
Myriads of little creatures
Of similar birth, but different features
Carried by a stream of swift motion,
With powerful sweep and great commotion—
to their unknown destiny.

The poem reflects the double orientation toward life and death characteristic of Strangeways research. In jerky rhythms, the poem identifies Arthur Hughes's research goal: through studying the circulation of the blood, to discover the "effect, if not the cause of—obliteration." The poem also indicates Hughes's method, in which the cinema played a central part; for Hughes was "one of the pioneers in adapting phase-contrast cinematography to the study of cellular processes" (Fell, "Cell" 22, 19). Finally, the poem's description of Hughes's work takes into account not only its scientific results but its broader popular implications. In contrast to the careful rejection of any talk of immortality when Strangeways scientists responded to public questions, the poem suggests that these scientists are hoping for professional scientific immortality: the perpetual survival of the narrative they create as scientists.

But, to make the story immortal,
The show must be filmed:
With cameras of special make
With a little twist and a little shake
With the trick A and the trick B
And behold! What do we see?
We see exactly what we have related before,
And there is no use of seeing more.

There is marked cynicism in Martinovitch's position here. The poem implies that the new scientific imaging the Canti method makes possible does not advance scientific knowledge ("the story of circulation is an old information"), it merely improves its communication to the public: "never before— / This lad does not hesitate to tell,— / Has it been followed so well."

The ambiguity in those lines is worth pausing over. Do they emphasize Hughes's egotism in not hesitating to tell of his success? And who is it who is described as following the narrative so well: the camera that pays such close attention to the scientific processes whose story it tells or the lay audience that is able to follow the old story of circulation better than ever before? The poem pauses, in the ambiguous gap between scientific self-aggrandizement and improved public education and communication, to consider the effect on scientific practice of a new set of rules: the disciplinary paradigms of the cinema.

The motto of the undertaking
Of motion pictures making
Is: You must tell a tale!
So, of late, an attempt has been made
By chromosome X and his mate,
(the latter in its non-existant [sic] state),
To imitate
Mickey Mouse and his gait.

Martinovitch's poetic point is worth repeating. A shift has recently occurred toward narrative in the sciences, he suggests, due to the increasing influence of the performance paradigms of cinema.[35] Not only have the objects of scientific study become subjects, they have taken on the qualities of one celebrated and highly commodified subject: Mickey Mouse. With its allusion to *Steamboat Willie,* the famous inaugural cartoon that follows Captain Mickey in his steamboat down the Mississippi River, Martinovitch's poem grants science the power to generate a fresh kind of value: not only epistemological but commercial. This new value is premised not on originality but on (cinematic) replication.[36] We have thus shifted from the microscopic to the human, from the river of blood in an embryonic chick to the larger than (human) life scale of the Mississippi.

Locus of agency has also relocated. In the course of the poem, we have moved from the lad, whose concerns and goals are the focus of the first two stanzas, to the cinematographer, who "must" film the story and whose camera twists, shakes, and tricks preoccupy the third stanza, to the viewer and producer of the film, whose skill at reception and projection are the subject of stanzas four and five, and, finally, in stanza six, to the anthropomorphic chromosomes X and Y ("his" mate, curiously), who try to imitate another anthropomorphic creature: the cinema icon Mickey Mouse. The poem closes with a summary of the effect of the turn to cinematography on the production and consumption of scientific knowledge: "If you like to see the show,— / You need not spend your dough,— / Come, and you will be told, / Precisely, what you wish to know." Again, the ambiguities are multiple. Strangeways's research was funded by nine organizations— including the Medical Research Council, the Royal Society, the British Empire Cancer Campaign, the Rockefeller Foundation, and the Wellcome Trust—and one wonders how much this prominent list of institutional sponsors colors the line "You need not spend *your* dough." And then there is the curious matter of the concluding couplet: "Come, and you will be told, / Precisely, what you wish to know." What does "precisely" modify— the way of telling or the content of what is told? Will the audience be told precisely, because this is science speaking, after all? Or does the cinema's entertainment mission drive the transfer of knowledge, so that the audience for this little film will be told "precisely, what [they] wish to know [and no more, no less]"? The passage links scientific popularization, cinematography, and what some of the Strangeways researchers believed to be a particularly American form of intellectual pandering.[37]

What are the implications of the tissue-culture point of view as explored in these three genres of documents? Honor B. Fell's lectures suggest that scientists are not always so monolithically concerned with control, management, and alienation of the embodied being—whether human or animal—as has at times been argued by the journalists who were their contemporaries, the cultural critics of science who are our contemporaries, and even ourselves.[38] The para-scientific writings on tissue culture of Dr. Fell make clear that there was a fairly high level of self-consciousness about the distortions of tissue culture, its excisions, and the links between (rather than replacement by) cells, organs, animals, and the human body. Fell's lectures also make it clear that Strangeways researchers understood the turn to tissue culture as an alternative to animal vivisection. As she put it:

"cells in culture can be used as experimental animals. . . . [W]e can even vivisect them by a kind of microsurgery" (Fell, "Technique" 10). This is in vivid contrast to the "virtual test tube *in vivo*": the rabbit-ear chamber that Eliot Clarke and Elinor Clarke developed in the United States, which—as Lisa Cartwright has shown—rather than abstracting and decontextualizing tissue, held the rabbit captive to the viewing cylinder surgically inserted into its sensitive, translucent ear tissue.[39] The writings on tissue culture reveal a tendency to identify with the tissue cultures as subjects rather than objects of study. In several documents, Strangeways scientists conceptualize cells as having an agency and autonomy. Fell portrays cells "going about their ordinary business" while the tissue culturist watches and photographs them; Waddington tells the story of earliest embryonic development as a condensed survey of the growth of the functionally specialized Enlightenment individual; Hughes draws parallels between the problems of feeding gonads in tissue culture and the problems of feeding the sexually and scientifically developing scientist; and Martinovitch imagines the chromosomes filmed under the microscope as performers akin to Mickey Mouse. This work suggests that there may have been ethical or psychological as well as practical reasons for preferring this method of investigation through decontextualization.[40] Finally, the poems disclose a more cynical, pragmatic, and prescient response to cinema photomicography than the credulous embrace of the technoscientific gaze we have recently been led to expect.

What do we learn by comparing these various reception sites for the complex new technique of tissue culture? We have seen that distinctly different, and valuable, insights can occur when the practices of science are confronted in the discursive medium of literature. Poetry puts in process a set of positions for identification that enables scientists to think against the grain of their scientific work. They can identify with the corpuscles they follow through the bloodstream, the tissues they culture, and even the developing blastula. In addition, the generic demands of poetry for a concise movement to articulate, and resolve, tension provide the incentive for counter-disciplinary truth telling. Waddington identifies across gender to speak from the position of the ovum, articulating a nostalgia for the undemanding passive position ("to be"), if ultimately affirming the move toward agency ("to do"); Hughes gives us a scientist who—although he nurtures and disciplines the sexuality and development of other species— seems in need of both discipline and nurture; and Martinovitch admits that the Canti method does not so much uncover truth as it enrolls allies, in the Latourean sense. We learn, in short, the ambivalences inherent in the tissue-culture point of view.

And these ambivalences bring us back to the central ambivalence: the simultaneously productive and problematic ambiguity of Honor Fell's phrase. While in her lectures she purported to represent "the tissue-culture point of view," that was what neither she nor her researchers could do. Rather, Fell articulated the perspective of the researchers engaged in culturing tissues: the human/scientific/instrumental/agentic point of view. Despite their claims to speak for the ovum, the blastulum, the cultured gonad, or the chromosome, the Strangeways researchers have no access to the point of view of the culture itself. The point of view they articulate is that of the tissue culturist. The ambiguity at the heart of Dr. Fell's phrase is both problematic, because when unacknowledged it obscures the scientists' epistemological limitations, and productive, because if acknowledged it can encourage scientists to draw on imagination as an aid to epistemology.

As our culture increasingly explores the biotechnologies founded on tissue culture, from animal-human organ transplants to reproductive technologies and fetal tissue transplants, the perspective provided by examining the problems and potentials embodied in the tissue-culture point of view will be crucial. It can enable us to access those counter-identifications that can reveal the underside of glamorous new biotechnologies, and it can provide us with some healthy skepticism about the powers of cinema photomicography—the "Canti method"—and the contemporary visualization and imaging technologies that are its legacy. Ultimately, awareness of the tissue-culture point of view may give us a more complex set of responses, both emotionally and intellectually, to medical crises such as the one that took me to London: my friend's slow dying of lung cancer. It will certainly produce a powerful respect for the basic laboratory skill of tissue culture: a competence that if used right makes it possible to diagnose a specific type of cancer and if not mastered—as in my friend's case—will produce misdiagnosis and inadequate treatment.[41] More than that, it may slow the rush to xenotransplantation as a feasible solution to the problem of diseased organs, in the realization that medicine first needs to deal with the epistemological, ethical, and technical issues foundational to the technique since its inception. Finally, mindfulness of the tissue-culture point of view can provide a richer perspective on the issues at stake in the complex play of boundary transgressions between human and animal, pre-birth and post-partum, life and death comprising the twentieth-century project of reshaping the human life span.

NOTES

This essay is in memory of Ian Crispin.

1. The decision would disappoint "thousands of people awaiting transplants .... It will also be a great disappointment to the Cambridge-based company, Imutran, the world leader in so-called xenotransplantation, which might now lose its pre-eminence to American, Japanese or Italian rivals" (Laurance).

2. There was a press outcry in August 1996, when nearly four thousand human embryos were destroyed in order to comply with the Authority's five-year storage deadline (Goldbeck-Wood).

3. A good survey of the uses of tissue culture is Philip R. White's *The Cultivation of Animal and Plant Cells.* As a hypothetical technique, tissue culture dates back at least to 1812, when physiologist Le Gallois observed that, "'if one could substitute for the heart a kind of injection . . . of arterial blood, either natural or artificially made . . . one would succeed easily in maintaining alive indefinitely any part of the body whatsoever'" (qtd. in Carrel and Lindbergh 621–22).

4. Harrison's full paper on the subject was published in 1910, and, according to J. A. Witkowski, the tissue-culture technique spread through Europe in the 1910s and 1920s. See also Fell, "Tissue" 622; Carrel and Lindbergh 622.

5. The cases of xenotransplantation performed to date cover a range of organ and tissue-culture practices. Solid organs such as the heart, kidney, liver, and lungs have been transplanted from chimpanzees, pigs, baboons, and sheep into humans. Other animal-human tissue transplants have included the pancreatic islets of Langerhans from fetal sheep, neural tissue from fetal pigs, and baboon bone marrow. Skin, corneas, and blood from other species, while not currently being transplanted, are all being studied as potential xenotransplantation possibilities (Great Britain).

6. "Tissue cultivated in vitro shows two forms of growth—sometimes called *unorganised growth* and *organised growth* respectively" (Fell, "Tissue" 3). This passage comes from one of a series of lectures on tissue culture Dr. Fell delivered between 1936 and 1938.

7. As the Chairman's Foreword to *Animal Tissue into Humans,* the report by the Advisory Group on the Ethics of Xenotransplantation, explains, "Those who need tissue for transplant, whether it be a solid organ such as a heart [or a lung] . . . or cells such as pancreatic islets . . . may look to xenotransplantation as, at last, the answer to their needs" (Kennedy vii).

8. See Featherstone and Wernick; Katz, *Disciplining.* The social sciences are also involved in this reconceptualization of the life span. For example, "experimental gerontology . . . tries to manipulate natural aging to extend the life-span of experimental animals and eventually that of man" (Medvedev 10).

9. The cloning of Dolly has catalyzed a reevaluation of how age is measured: "Our 7-month-old lamb actually has a 6-year, 7-month-old nucleus in all her cells. It's going to be interesting to see what happens with the aging of this animal," notes Grahame Bulfield, director of Roslin Institute ("Ewe" 132).

10. The relationship between literature and science is reciprocal rather than unidirectionally mimetic because literature functions not only to consolidate scientific fields and assimilate scientific discoveries but frequently to set the epistemological agenda for science. For an extended analysis of the two-way traffic between literature and science in the construction of the new field of reproductive technology, see Squier, *Babies.*

11. For analyses of the role of representation in shaping "facts" in medicine and bioethics, see Donnelly; Chambers; and Poirier et al.

12. As E. D. Strangeways (the scientist's widow) later put it, he dedicated himself to "the study of living cells by means of tissue culture *in vitro* and *in vivo,* a field in which he was one of the chief pioneers" (Strangeways 12).

13. "So far as I know, your lab is the only one in this country devoted entirely to tissue-culture" (Hancox).

14. It is the unorganized nature of the growth, not whether the tissues are embryonic or mature, that provides the tissue with its potential immortality. Both

embryonic tissue in unorganized growth and cancerous tissue result in the similar process of outgrowth, or as Waddington put it, "the induction of a cancer can be homologized with the induction of an embryonic tissue" ("Substances" 111). The continually growing tissue can be subdivided, with new cultures placed in fresh culture medium, and for certain tissues the process can continue, Fell observed, "indefinitely." Fell's definition is helpful here: "tissue undergoing unorganised growth *loses its normal histological structure.* For example, kidney tubules undergoing organised growth merely spread out over the glass of the culture vessel as an indifferent sheet of epithelium . . . and completely loses [*sic*] the anatomical features of the original tubules. This loss of histological structure is sometimes known as *dedifferentiation*" (emphasis in original) ("Tissue" 3).

15. The phenomenon of scaling was identified by Kenneth Wilson, Nobelist for physics in 1982, and later elaborated upon in nonlinear mathematics by Benoit Mandelbrot. "Scaling, as Mandelbrot uses the term, does not imply that the form is the *same* for scales of different lengths, only that the degree of 'irregularity and/or fragmentation is identical at all scales' " (Hayles 166).

16. As N. Katherine Hayles has described it in her study of nonlinear dynamics in literature and science, "Chaos theory looks for scaling factors and follows the behavior of the system as iterative formulae change incrementally" (170).

17. Honor B. Fell's chapter "Cell Biology" in the official *History of the Strangeways Research Laboratory* reveals that the following different kinds of cells were cultured there: chick embryonic cells, mammalian sternum cells, ear, mammary gland, ovary, salivary gland, pancreas, hair and teeth cells (animal of origin unspecified), human thyroid cells, fetal rat skin cells, chicken skin cells, rabbit red blood cells, rat cancer cells (rhabdomyosarcomata), human fetal lung cells, rat salivary gland cells, human cancer cells, and many varieties of bacterial cells.

18. In the passage from his textbook on tissue culture that I have used as my second epigraph, White echoes this stress on the dramatic and performative aspects of the technique: "Every student can not only see these things but have the thrill of preparing them for himself" (vii).

19. Her image recalls an illustration from Charles Kingsley's *The Water Babies* in which two scientists gaze at a water baby in a jar. This image, I argue in *Babies in Bottles,* shaped the scientific maturation of zoologist Julian Huxley, for whom it invoked the scientific drive to control the processes of life and death.

20. Specifically, feminists have critiqued the tendency of scientists to argue from experimental findings to the macrosocial context, without taking into account the processes of social construction that—as sociologists and anthropologists of science document—are an integral part of the construction of scientific facts. See Harding; Longino; Spanier; Keller; Latour and Woolgar.

21. As *Nature* described it in 1937, "the Strangeways Laboratory of Cambridge . . . has become one of the most distinguished places in the scientific world for the prosecution of morphogenetic studies." The passage continued, "Here the technical methods, tissue culture *in vitro,* tissue grafting, the growth of large explants under a variety of conditions, embryological experimentation, the registration of results by the Canti-cinema method . . . have been used abundantly. . . . Dr. Fell and the Strangeways Laboratory have placed British anatomists deeply in their debt" ("Morphogenetic" 1036).

22. This is one of several documents that exist as clippings in the Strangeways Research Laboratory files of the Contemporary Medical Archives Centre, Wellcome Institute for the History of Medicine, London. While I have given the most complete citation possible, page numbers were not noted on the clipping and thus are not available. The primitive streak—"a linear thickening, visible with minimal

magnification, that lies in the head-to-tail axis of the embryo-to-be"—is the first marker of developmental singleness, according to embryologist Clifford Grobstein (26). Among the researchers involved in embryo culture, according to H. B. Fell, were: Mr. Michael Abercrombie, Dr. Aron Moscona, Dr. J. Grover, Dr. B. McLoughlin, Dr. L. Weiss, Dr. P. D. F. Murray, Dr. A. Hughes, and Dr. Fitton Jackson (Fell, "Cell" 21–22).

23. As Honor Fell put it in an unpublished paper, "'probably my main contribution to science has been the development and application to biomedical research of the organ culture technique'" (qtd. in Vaughan 239). The paper was titled "Evaluation of Research," and Vaughan speculates it was written in 1982.

24. Dr. Fell summarized the work in this area undertaken at Strangeways in her contribution to the laboratory's *History*: "Progressive development of many different organs was obtained: ear rudiments (H. B. Fell), mammary gland (H. Hardy), ovary (P. N. Martinovitch), salivary gland (E. Borghese), pancreas (J. Chen), hair (D. H. Strangeways and later H. Hardy), and teeth (S. Glasstone)" (Fell, "Cell" 21). As the *British Medical Journal* summarized the Strangeways Research Laboratory report of 1936, "Miss S. Glasstone has shown that whole tooth-germs from rat and rabbit embryos explanted before cusp formation had begun [to] form cusps *in vitro*" ("Strangeways Research" n. pag.).

25. "Lady LeBec" appeared in their midst just as Dr. Fell had read a telegram praising her as "a remarkable person in many ways. Not only is she justly celebrated for her princely generosity and sparkling intelligence, but she is also reputed to be one of the loveliest women in England" (Spear/Strangeways 1937).

26. The gender transgression involved in Honor Fell's work surfaced in headlines such as "Woman Scientist Cultivates Life in Bottles," *Daily Express* (London), 16 March 1936, an article by the special correspondent, who was probably Charlotte Haldane, science writer, novelist, and wife of the eminent geneticist J. B. S. Haldane.

27. As Dame Janet Vaughan has observed, "the earliest and pioneering work on radiobiology originated in the Strangeways. . . . It has been said 'British Radiobiology stems from the Strangeways and there must be hundreds of people who owe their education on cells and radiation to the Strangeways School'" (249).

28. During World War II, Strangeways researchers worked extensively on the problem of wound healing. They also carried out a series of experiments focusing on "the metabolism of cell cultures grown under different experimental conditions," under the auspices of the British Empire Cancer Campaign. However, the most prominent cancer-related work in the Strangeways Research Laboratory was probably the work in radiobiology, which was made the focus of weeklong courses in radiobiology the laboratory mounted between 1947 and 1949.

29. The normally staid *Lancet* enthused about the photomicrographic apparatus, describing it as "a marvel of ingenuity," and praised Canti for "bringing the behavior of tissue cultures within the range of ordinary vision" (Cinematograph).

30. We can trace a lineage (and it *is* patriarchal) back from these male images at the other end of the tissue-culture micrograph to Anton van Leeuwenhoeck's discovery of tiny animals in sperm and marvel at what Lisa Cartwright has called the "masculine fascination with the exaggerated image of minuscule bodily organisms" (84).

31. These poems are held in File PP/FGS/C18–19 in the Contemporary Medical Archives Centre, Wellcome Institute for the History of Medicine, London.

32. For a discussion of two literary examples of the genre, see Squier, "Fetal Voices." For visual examples of the same genre, see Henderson; and Newman.

33. "I should like to argue that a scientist's metaphysical beliefs are not mere

epiphenomena, but have a definite and ascertainable influence on the work he produces" (Waddington "Practical" 123). Waddington's daughter also reminisces about her father, "He liked 'all over' paintings with no particular focus (Sam Francis, Jackson Pollock) and the idea of diversity, and [he] always told me how much A. N. Whitehead influenced him with his idea of different nexuses of events, and fundamentally distinct (but overlapping) points of view with no one better than any other" (McDuff).

34. Waddington's daughter observes: "My father did believe in progress—in fact he wrote a book, *The Ethical Animal,* which was (as far as I remember) an attempt to formulate what progress is from a scientific (evolutionist) point of view, though he may not have remained with that belief" (McDuff).

35. A parallel exists to the powerful role narrative has played in medicine. As visualization technologies increasingly dominate modern medical practice, we may find a conflict between two distinct and different narrative models, the earlier drawn from literature and culture and the more recent influenced by film. See Hunter.

36. The shift from original to replicant as privileged object of science has been inflected in fascinatingly different ways by contemporary feminist theorists. Donna J. Haraway has written movingly in celebration of a new life built not on origin stories but on replication with a difference, while Vandana Shiva sees the movement to patent intellectual and genetic property (in order to replicate it) as part of the centuries-old enclosure movement, which engages in "piracy from the mothers and grandmothers of the Third World."

37. My thanks to Dusa McDuff for this observation.

38. See Lesley A. Hall's invaluable article, "The Strangeways Research Laboratory: Archives in the Contemporary Medical Archives Centre."

39. "The Clarks literally retooled their animals' bodies and habits to make them conform to experimental procedures . . . . [I]n an attempt to render a part of the rabbit body more easily observable under the microscope, they physically implanted an instrument for observing growth within the living ear" (Cartwright 100).

40. This evidence anticipates—although due to its in-built hierarchy does not finally attain—contemporary representations of the multiple, nested human subject. As Dorion Sagan has observed, "The body is not one self but a fiction of a self built from a mass of interacting selves" (370).

41. A troubling intersection exists here between my examination into the Strangeways Research Laboratory's role in the development of the technique of tissue culture and the most celebrated line of cultured tissue cells, the HeLa cell line. Henrietta Lacks's carcinoma was misdiagnosed (epidermoid carcinoma of the cervix rather than the actual, far more malignant, adenocarcenoma), leading to an untimely death in her case as in the case of my friend, Ian Crispin. See Hannah Landecker, this volume.

WORKS CITED

Barker, Dudley. "Cameraman of the Medical World: Filmed Cells through a Microscope: Lord Horder's Tribute to Dr. Canti." *Evening Standard* [London] 9 Jan. 1936: 2.

Burke, Norah. "Could You Love a Chemical Baby? For That's What Science Looks Like It's Producing Next." *Tit-Bits* 16 Apr. 1938: 3.

Carrel, Alexis, and Charles A. Lindbergh. "The Culture of Whole Organs." *Science* 81 (1935): 621–23.

Cartwright, Lisa. *Screening the Body: Tracing Medicine's Visual Culture.* Minneapolis: University of Minnesota Press, 1995.

Chambers, Tod. "Dax Redacted: The Economics of Truth in Bioethics." *Journal of Medicine and Philosophy* 23 (1996): 287–302.

Cinematograph Work. Strangeways Research Laboratory Scientific 1932 Cinematograph Work, R. G. Canti, SA/SRL 23/ H.2. Contemporary Medical Archives Centre. Wellcome Institute for the History of Medicine, London.

Donnelly, William J. "Taking Suffering Seriously: A New Role for the Medical Case History." *Academic Medicine* 71 (1996): 730–37.

"Ewe Again? Cloning from Adult DNA." *Science News* 1 Mar. 1997: 132.

Featherstone, Mike, and Mike Hepworth. "The Mask of Ageing and the Postmodern Life Course." *The Body: Social Process and Cultural Theory.* Ed. Mike Featherstone, Mike Hepworth, and Brian Turner. London: Sage, 1991. 371–89.

Featherstone, Mike, and Andrew Wernick, eds. *Images of Aging: Cultural Representations of Later Life.* New York: Routledge, 1995.

Fell, Honor Bridget. "Cell Biology." *History* 19–33.

———. "The Technique of Culture and Its Value in Research." Postgraduate School of Medicine. 3 Mar. 1937. Strangeways Research Laboratory File PP/ HBF/E12. Contemporary Medical Archives Centre. Wellcome Institute for the History of Medicine, London. N. pag.

———. "Tissue Culture." Postgraduate School of Medicine. 8 Jan. 1936. Strangeways Research Laboratory File PP/HBF/E12. Contemporary Medical Archives Centre. Wellcome Institute for the History of Medicine. N. pag.

Foxon, G. E. H. "Early Biological Film—the Work of R. G. Canti." *University Vision: The Journal of the British Universities Film Council* 15 (1976).

Goldbeck-Wood, Sandra. "Europe Is Divided on Embryo Regulations." *British Medical Journal* 313 (1996): 512.

Great Britain. Advisory Group on the Ethics of Xenotransplantation. *Animal Tissue into Humans: A Report by the Advisory Group on the Ethics of Xenotransplantation.* London: HMSO, 1997.

Grobstein, Clifford. *Science and the Unborn: Choosing Human Futures.* New York: Basic, 1988.

Hall, Lesley A. "The Strangeways Research Laboratory: Archives in the Contemporary Medical Archives Centre." *Medical History* 40 (1996): 231–38.

Hancox, N. M. Letter to Honor Bridget Fell. 25 Nov. 1932.

Haraway, Donna J. *Simians, Cyborgs, and Women: The Reinvention of Nature.* New York: Routledge, 1991.

Harding, Sandra. *The Science Question in Feminism.* Ithaca: Cornell University Press, 1986.

Hardy, Harrison. "Any Tin in the Sun?" *Daily Mirror* [London] 20 Mar. 1937: n. pag.

Hayles, N. Katherine. *Chaos Bound: Orderly Disorder in Contemporary Literature and Science.* Ithaca: Cornell University Press, 1990.

Henderson, Andrea. "Doll-Machines and Butcher-Shop Meat: Models of Childbirth in the Early Stages of Industrial Capitalism." *Genders* 12 (1991): 100–119.

History of the Strangeways Research Laboratory (Formerly Cambridge Research Hospital), 1912–1962. Strangeways Research Laboratory File SA/SRL/J.3. Contemporary Medical Archives Centre. Wellcome Institute for the History of Medicine, London.

Holquist, Michael. "From Body-Talk to Biography: The Chronobiological Bases of Narrative." *Yale Journal of Criticism* 3 (1989): 1–35.

Hunter, Kathryn Montgomery. *Doctors' Stories: The Narrative Structure of Medical Knowledge.* Princeton: Princeton University Press, 1991.

Katz, Stephen. *Disciplining Old Age: The Formation of Gerontological Knowledge.* Charlottesville: University Press of Virginia, 1996.

———. "Imagining the Life-Span." Featherstone and Wernick 61–75.

Keller, Evelyn Fox. *Secrets of Life, Secrets of Death: Essays on Language, Gender, and Science.* New York: Routledge, 1992.

Kennedy, Ian. "Chairman's Foreword." Great Britain vii–viii.

Kingsley, Charles. *The Water Babies.* New York: Garland, 1976.

Latour, Bruno, and Steve Woolgar. *Laboratory Life: The Construction of Scientific Facts.* Princeton: Princeton University Press, 1979.

Laurance, Jeremy. "Transplant of Pig Hearts to Be Banned: Blow for Patients Awaiting Surgery." *London Times* 16 Jan. 1997: 1.

Longino, Helen. *Science as Social Knowledge: Values and Objectivity in Scientific Inquiry.* Princeton: Princeton University Press, 1990.

McDuff, Dusa. Personal communication with the author, 1996.

Medvedev, Zhores Alexandrowitsch. "The Structural Basis of Aging." *Life Span Extension: Consequences and Open Questions.* Ed. Frederic C. Ludwig. New York: Springer, 1991. 9–17.

"Morphogenetic Factors of Bone." *Nature* 19 June 1937: 1036–37.

Newman, Karen. *Fetal Positions: Individualism, Science, Visuality.* Stanford: Stanford University Press, 1996.

Poirier, Suzanne, et al. "Charting the Chart: An Exercise in Interpretation(s)." *Literature and Medicine* 11 (1992): 1–22.

Powell, Jep. "The Synthetic Woman." *Amazing Stories* Sept. 1940: 100–124.

"Rat's Teeth Grown in Laboratory in Cell-Growth Study by British." *New York Herald* 13 Mar. 1936: n. pag.

Russell, K. "Tissue Culture—A Brief Historical Review." *Clio Medica* 4 (1969): 109–19.

Sagan, Dorion. "Metamazoa: Biology and Multiplicity." *Incorporations.* Ed. Jonathan Crary and Sanford Kwinter. Cambridge: MIT Press, 1992. 362–85.

"Scientists May Now Watch Living Connective Tissue Reproduce Its Ultimate Cells." *The World* 12 June 1921, sec. 2: n. pag.

"The Seer." *World Radio* [London] 11 Feb. 1938: n. pag.

Shiva, Vandana. Lecture. The Pennsylvania State University. 21 Apr. 1997.

Spanier, Bonnie B. *Im/partial Science: Gender Ideology in Molecular Biology.* Bloomington: Indiana University Press, 1995.

Spear/Strangeways 1936. Strangeways Research Laboratory File PP/FGS/C.17, item 2816. Contemporary Medical Archives Centre. Wellcome Institute for the History of Medicine, London.

Spear/Strangeways 1937. Strangeways Research Laboratory File PP/FGS/C.18. Contemporary Medical Archives Centre. Wellcome Institute for the History of Medicine, London.

Squier, Susan M. *Babies in Bottles: Twentieth-Century Visions of Reproductive Technology.* New Brunswick: Rutgers University Press, 1994.

———. "Fetal Voices: Speaking for the Margins Within." *Tulsa Studies in Women's Literature* 10.1 (1991): 17–30.

Strangeways, E. D. "1905–1926." *History* 7–12.

"Strangeways Research Laboratory." *British Medical Journal* (1936): n. pag.

"T. S. P. Strangeways." Obituary. *Lancet* 1 Jan. 1927: 56.

Vaughan, Dame Janet, FRS. "Honor Bridget Fell." *Biographical Memoirs of Fellows of the Royal Society.* Vol. 33. London: The Royal Society, 1987. 237–59.

Waddington, C. H. Letter. *Nature* 135 (1935): 606.

————. "The Practical Consequences of Metaphysical Belief on a Biologist's Work: An Autobiographical Note." *Towards a Theoretical Biology 2: Sketches.* Edinburgh: Edinburgh University Press, 1969. 123–32.

————. "Scientist at Work: What Controls the Development of Animals?" *Listener* 23 Oct. 1936: 811–12.

————. "Substances Promoting Cell Growth." *British Medical Journal* 28 (1936): n. pag.

Wells, H. G. "The Island of Dr. Moreau, Part I." *Amazing Stories* Oct. 1926: 636–55, 671–72.

————. "The Island of Dr. Moreau, Part II." *Amazing Stories* Nov. 1926: 702–23.

"What We Miss!" *Birmingham Mail* 16 Feb. 1938: n. pag.

White, Philip R. *The Cultivation of Animal and Plant Cells.* New York: Ronald, 1954.

Witkowski, J. A. "Alexis Carrel and the Mysticism of Tissue Culture." *Medical History* 23 (1979): 279–96.

"Woman Scientist Cultivates Life in Bottles." *Daily Express* 16 Mar. 1936: n. pag.

Chapter 2

Immortality, In Vitro

A History of the HeLa Cell Line

HANNAH LANDECKER

> A tissue is evidently an enduring thing. Its functional and structural conditions become modified from moment to moment. Time is really the fourth dimension of living organisms. It enters as a part into the constitution of a tissue. Cell colonies, or organs, are events which progressively unfold themselves. They must be studied like history.
>
> —Alexis Carrel, "The New Cytology"

> The double is neither living nor dead: designed to supplement the living, to perfect it, to make it immortal like the Creator, it is always "the harbinger of death." It disguises, by its perfection, the presence of death. By creating what he hopes are immortal doubles, man tries to conceal the fact that death is always already present in life. The feeling of uncanniness that arises from the double stems from the fact that it cannot but evoke what man tries in vain to forget.
>
> —Sarah Kofman, *Freud and Fiction*

In 1951, a piece of cancerous cervical tissue was cut from a woman named Henrietta Lacks. Lacks died eight months later of cancer. Live cells from the biopsy were grown in test tubes, supplied with nutrient medium, and kept at body temperature in an incubator. Named HeLa, from the first two letters of Lacks's first and last names, and called an immortal cell line, descendants of these original cells continue to grow and divide in laboratories around the world. Proliferating with these glass-bound populations of cells are narratives of their life and origin.

The cells live and the woman does not. They somehow stand for her and she for them; otherwise, this pair of circumstances would not present itself as a paradox, much less one that has generated such fascinated attention from 1951 to today. That one party in this relation should be alive and the other dead creates a dramatic tension which continues to generate scientific papers, newspaper and magazine articles, and television documentaries. The resolution of the paradox in these narratives is always the same: the woman and the cells are immortal, the woman through the cells' life and the cells through the woman's death. It is a personification of the woman who died that gains immortality, while the woman's death is necessary to elevate the cells from unremarkable life—maintained in laboratories for over forty-six years—to immortal life.

It is not surprising in itself that HeLa should be personified. Cell lines are made to stand in for persons in the first place; they function in the laboratory as proxy theaters of experimentation for intact living bodies. The visualization of cellular processes by placing living cells under glass, where they are accessible to microscope and camera, has become part of what we understand to be the "life" of the body.[1] As sites of manufacturing—of viruses or proteins or antibodies—cell lines are the tools of the industry whose product is human health. Their identification as "living" and "human" entities cannot fall from them, because it is this origin that gives them commercial and scientific value as producers of biological substances for use by humans and their validity as research sites of human biology.

Given that these living technologies are thus necessarily understood to be human, I wish to ask more specifically how the material existence of the cell line redefines the designations used to describe that existence. To this end, I trace the history of the cell line and its personifications. Lacks's story is simultaneously what happened to a person and her body and a narrative vehicle through which journalists and scientists have imagined and witnessed the possibilities for lives and bodies constantly being changed by the rapid development of these "technologies of living substance" made from human tissues.[2] Lacks's photograph graces many of the accounts; the cell line bears fragments of her name; the cells bear various proportions of the genetic material of which her body was composed when it was alive, the body that was the source of cells whose varied descendants continue to live and reproduce in laboratories all over the world. However, the meaning of this material lineage is repeatedly being renegotiated in the changing personifications of the cell line.

More than an exercise in cataloguing variations on a theme, this history demonstrates how the personifications of the cell line shift alongside the development of differing experimental roles in biology, medicine, and biotechnology. The physical matter, technical practice, and economic significance of growing cells in vitro—tissue culture—generated new knowledge about and fresh meanings for the concepts of human, alive, and immortal. These are both shaped by and interact with wider cultural narratives, from modern medicine's triumph over polio in the 1950s to anxieties over race and purity in the 1960s and 1970s to, most recently, a recasting of the story in economic terms.

In a sense, I mimic the narratives I am analyzing, by structuring the following history of the HeLa cells around the changing definitions of in vitro immortality over the course of the twentieth century. This is meant, in the end, to serve as a critique rather than a retelling; the final point of this essay is to highlight that the death of the person who was Henrietta Lacks has been obscured by the personification of her cells as an immortal entity.

IMMORTALITY IN THE HISTORY OF TISSUE CULTURE

HeLa cells were called immortal within a year of their cultivation in vitro. The only way to understand what seems a rather rapid jump to conclusions is to place the establishment of this line in the context of the history of tissue culture. This essay is not the place to recount the history of tissue culture; instead, I choose to take up a single strand of its development in the United States.[3] The work of Alexis Carrel at Rockefeller Institute in New York City from 1910 to 1938 is important in the context of this essay, because it was Carrel who first proposed the concept and the supporting technology of indefinite life of tissues in vitro. He drew his initial inspiration from Ross G. Harrison, an embryologist, who had shown in 1907 that he could keep a piece of embryonic frog neural tissue alive long enough to watch a single nerve fiber growing out from it. Harrison thus demonstrated how valuable information could be gleaned from the isolation and maintenance of a living system in which the "behavior of certain cells could be observed when removed from the bewildering conditions . . . within the embryonic body" (6).

However, it was Carrel who first tried to grow human cells in vitro. He did not aim to keep tissues alive long enough to observe some aspect of their behavior; rather, his goal was their "indefinite" or "permanent" life. Drawing on the philosophy of Henri Bergson, Carrel developed a theory of corporeal life in which physiological processes were the "substratum of duration." Time was recorded only when the metabolic products of these processes were allowed to remain around the tissues being grown in vitro:

> If these metabolites are removed at short intervals and the composition of the medium is kept constant, the cell colonies remain indefinitely in the same state of activity. They do not record time qualitatively. In fact, they are immortal. (Carrel, "Physiological Time" 620)

This theory translated into meticulous technical practices such as washing ("rejuvenating") the cells every few days, the design of special glassware for these procedures, and strict protocols to ensure asepsis. By 1912, Carrel declared the "permanent life of tissues outside of the organism" an issue only of more perfect technique, and he claimed to have established a culture of embryonic chicken heart fibroblasts. These cells would live in vitro for thirty-four years, when they were discarded two years after Carrel's death (Carrel, "Permanent Life"; Ebeling).

From its beginnings, the living cell in culture—in particular the human cell—has been an uncanny object. Tissue culture was developed using living matter cut from fetal cadavers and tumors. The living qualities of these cells, acted out in isolation from the organism and visible to the observer— contracting, beating, forming synaptic nets, proliferating, migrating—are what make them useful for biology. Their isolation gives them the character of autonomous life—which is especially evident in the early fascination of Carrel and other tissue culturists with applying new techniques of time-lapse cinematography to the study of cells in culture. These silent films of cells enlarged to screen size made visible the movement and division of entities previously seen only in the fixed, stained state of classical histology, and they enhanced the perception of an autonomous sphere of life. "Tissue and blood cells are always in the process of becoming," wrote Carrel. "They do not show their true physiognomy . . . under the microscope. . . . [C]ells appear on the film as mobile as a flame" ("New Cytology" 337).

From this early history comes not only the sense of a kind of life extracted from the confines of the animal or human body but an enduring connection with magic and sorcery. One textbook of cell biology states that "until the early 1970s, tissue culture was something of a blend of science and witchcraft," which refers both to the understanding of successful tissue culture as an "art" that had to be learned in a hands-on apprenticeship and the aesthetic setting of the early tissue culture laboratories (Alberts et al. 161). Carrel and those he trained staffed their laboratories with technicians dressed in black robes and hoods, ostensibly to minimize reflections which might interfere with the delicate operations, while the air was kept moist with "witches' cauldrons" of steam (Witowski 281). Although contemporaries and historians have blamed Carrel for making tissue culture out to be more difficult—and more occult—than it really was, thus scaring off potential practitioners, to my mind there is no doubt that his sense of the possible was extremely important to the establishment of in vitro immortality as a desired scientific object.

Getting tissue or cells to live in glass has also been the venture to get them to live indefinitely outside of the animal body. What good, after all, is a technology that only lives for a matter of days or months? Much better is one that will—if fed and maintained—reproduce itself and serve as a constant medium for repeatable experiments. However, the permanent life of cells outside of the body did not turn out to be an easily achievable goal, and HeLa was established only after years of effort with other cells, animal and human.

THE ESTABLISHMENT OF THE HELA CELL LINE

In the laboratory of George Gey and Margaret Gey, at Johns Hopkins University Hospital, the ongoing attempt to establish cell lines from human tissues intersected with another research program to determine the relationship between two types of cervical cancer. The first was a non-invasive form of cancer involving only the epithelial surface of the cervix. The second was invasive carcinoma involving the deeper basal layers and leading to metastasis. Although it is now understood that the former is a precursor of the latter, this was still unresolved in 1951. George Gey had been recruited to the project to grow cervical cancer cells in the hope that their life under glass would reveal something about their action in the body. It was into this context that Henrietta Lacks entered when she sought treatment for intermenstrual bleeding. After initial uncertainty on the part of the Johns Hopkins' treating physicians as to the nature of the lesion they saw on her cervix, they took a biopsy of the lesion and made the diagnosis of cervical cancer.[4] Without her knowledge or permission, Lacks became part of the cervical cancer research project when a piece of the biopsy material was sent to the Gey laboratory.

In 1951, when it became clear that HeLa cells were going to continue growing and dividing unperturbed by their artificial environment, it did not take long before the label of "immortality" was applied to them and their role as a cell line quickly overshadowed any part in the cervical cancer study. George Gey distributed samples of HeLa to his colleagues around the world. Because—as one tissue culturist put it—"HeLa cells can be grown by almost anyone capable of trypsinizing cells and transferring them from one tube to another," their cultivation quickly became a widespread practice (Bang 534).[5]

Gey never attempted to patent or otherwise limit the distribution of HeLa cells, clearly not anticipating the chain letter effect of sending out cultures which were then grown up, split into parts, and sent on to others. Almost immediately, a company called Microbial Associates, Inc., began growing HeLa cells for commercial sale. In 1954, Gey expressed dismay over the number of laboratories working on HeLa in a letter to a colleague. Gey's correspondent, Charles Pomerat, reacted to this statement with some amusement:

> With regard to your statement . . . of disapproval for a wide exploration of the HeLa strain, I don't see how you can hope to inhibit progress in this direction since you released the strain so widely that it now can be purchased commercially. This is a little bit like requesting people not to work on the golden hamster![6]

Indeed, by this time, HeLa cells were being mass-produced as part of a push for the rapid evaluation of the polio vaccine, which Jonas Salk developed in 1952. HeLa cells were chosen as the "host" cell for measuring the amounts of antibodies the poliovirus antigen produced. The Tuskegee Institute, a historically black college in Alabama, was appointed by the National Foundation for Infantile Paralysis to be the locus of production. A laboratory was committed to the sole purpose of producing as many as twenty thousand tube cultures of HeLa per week (Brown and Henderson).

George Gey had as little control over the story that he released into the public domain as he had over the cells. Because of intense national interest in the subject of polio, the HeLa cell line came quickly to the attention of journalists. Gey did not want to release Lacks's name, and so he informed an interested writer from the National Foundation for Infantile Paralysis that he could not see fit to reveal the name of the cells' donor. However, it was clear in the reply to this refusal that the writer had asked Gey's permission only as a formality; he had already learned Lacks's name.

> An intrinsic part of this story would be to describe how these cells, originally obtained from Henrietta Lakes [*sic*], are being grown and used for the benefit of mankind. Here is a situation where cancer cells—potential destroyers of human life—have been channeled by medical science to a new, beneficent course. . . .

"Incidentally," the writer smugly concluded, "the identity of the patient is already a matter of public record inasmuch as newspaper reports have completely identified the individual."[7] Another journalist writing for *Colliers* in 1954 was more discreet, referring to "an unsung heroine of medicine named Helen L." (Davidson 79).[8] Helen L. was characterized in this piece as a young Baltimore housewife whose unfortunate early death turned her into an "unsung heroine" because of the HeLa cells' research role. Her death and her immortality were uttered in the same sentence: "Mrs. L. has attained a degree of immortality she never dreamed of when she was alive, and her living tissue may yet play a role in conquering many diseases in addition to the cancer which killed her" (Davidson 80).

In this version of immortality, the cells were understood to be a piece of Henrietta Lacks that went on growing and living, encased in a test tube instead of a body. The cells were seen as universal human cells. They served as a substrate in the design of a polio vaccine that was to be applied to millions of people. They were used to produce standardized nutrient media for use in culturing all kinds of cells. They were utilized to figure out methods for growing other cells and how to produce large numbers of them. They were referred to as the "golden hamster" of cell biologists, and their concomitant personification was in the form of an angelic figure, an immortalized young Baltimore housewife, thrust into a kind of eternal life of which such a woman would never dream.

To understand how death becomes a footnote to immortality in this narrative, I will do a close reading of one of the retellings of HeLa's origin story. When George Gey died (of cancer) in 1970, his colleagues wrote a peculiar memorial tribute to him in the journal *Obstetrics and Gynecology*, entitled "After Office Hours: The HeLa Cell and a Reappraisal of Its Origin." They wrote that the original biopsy secured for the patient, Henrietta Lacks (fig. 2) as HeLa, an immortality which has now reached 20 years. Will she live forever if nurtured by the hands of future workers? Even now, Henrietta Lacks, first as Henrietta and then as HeLa, has a combined age of 51 years. (H. W. Jones et al. 945)

Beside this statement is figure 2—a photograph of a young woman, smiling into the camera, hands on hips. Underneath the photograph, the caption reads "Henrietta Lacks (HeLa)," as if the photograph of the woman held the image of the incipient cell line, as if the woman *was* the cell line, that according to these gynecologists, "if allowed to grow uninhibited under optimal cultural conditions, would have taken over the world by this time."[9]

The reappraisal promised by the tribute's title was another look at the original biopsy slides of Henrietta Lacks. Upon reexamining the slides, Gey's colleagues wrote that what had been originally diagnosed as epidermoid carcinoma of the cervix in its early stages was actually an adenocarcinoma, a rarer, more aggressive form of cancer involving a different kind of cell. While some readers might expect this admission to cause reconsideration of the treatment of the patient, who died within eight months of diagnosis—in particular, as one of the authors was a physician responsible for this patient's diagnosis and treatment—this is not the case. They wrote:

> while it is necessary to record that the first continuous cell strain is not of epidermoid carcinoma of the cervix . . . the exact histopathologic nature of HeLa is but a footnote to the abiding genius of George Gey. (946)

Thus the woman is paradoxically made immortal by the engine of her death, in the form of a biopsy used to diagnose her ailment (inaccurately) that becomes research material without her permission, to end up as a footnote to the abiding genius of the scientist.

The question of whose immortality is involved in the establishment of cell lines is further accentuated by the predilection some scientists showed for trying to establish immortal cell lines from pieces of their own bodies or the bodies of close relatives. In 1961, Leonard Hayflick of Wistar Institute established a cell line from his newborn daughter's amniotic sac. The amnion, which grows from the fetal tissues, was of the same genetic makeup as Leonard Hayflick's daughter, and because it carried his genes, it was literally his "daughter cell" (Hayflick, "Establishment" 608).[10] He named the cell line WISH, an acronym which stood for Wistar Institute and Susan Hayflick, his daughter.

In 1966, Monroe Vincent was diagnosed as having a benign tumor of the prostate. He promptly attempted to grow some of the cells taken from his prostate, and he established the cell line MA160, named not after himself but after the biomedical supply company Microbial Associates, Inc., in which he was a partner. HeLa was itself scientific progeny—letters to

George Gey referring to HeLa called the cell line "your precious baby."[11] Gey is fondly remembered for hand-delivering the cultures personally to other scientists: "'He would put his glass tubes containing the cells in his shirt pocket, use his body heat to keep them warm, and then fly to another city and hand them to a fellow scientist.'"[12]

FROM BENEFICENCE TO NOTORIETY

This benign version of immortality came to an abrupt end in the 1960s. First, new scientific work that studied aging through cell culture revealed that only cancerous cells had the ability to keep dividing indefinitely. This drew a sharper line of definition between normal and cancerous cells. Carrel's famous immortal chicken heart cell culture had supposedly been composed of normal cells, in which "permanent life" had been induced by removing them from the body and manipulating their environment, but in 1961 this was shown to have been something of a fraud, when Hayflick demonstrated that normal somatic cells in culture consistently divide for a set number of generations and then all die at once. Cells reproduce by replicating their DNA and dividing into two daughter cells. When a whole population of cells goes through division, it is said to double. Hayflick showed that cells taken from human fetal tissue will always undergo about fifty doublings before dying. Even if the culture is frozen, and no matter how long it stays frozen, when thawed it will pick up where it left off and in total complete approximately fifty doublings. Cells taken from adults consistently go through about thirty doublings (Hayflick, "Biology").[13] What is more, the finite number of doublings is species-specific. Chicken cell culture will go through thirty-five doublings at the most, not thirty-five years worth of doublings. Thus it seemed impossible that Carrel's culture could have been composed of normal chicken cells.

Hayflick concluded that the chick embryo extract preparation used as nutrient medium for Carrel's cultures provided new viable embryonic cells at each feeding. Others have hypothesized that Carrel's proximity to Peyton Rous at Rockefeller Institute led to the normal somatic chicken cells being infected with Rous sarcoma virus and thus rendered capable of the same kind of unlimited division seen in cancerous cells (for example, see Harris 45). The cause for the famous culture's immortality could not be investigated as it was thrown away in 1946.

More important was the stark distinction drawn between normal body cells and cancer cells with this work. Intrinsic to this distinction was the finite life span of populations of normal cells, a limit only cancer cells could break. Normal somatic cells were euploid, that is, contained a normal number of chromosomes. Cancer cells were aneuploid, showing abnormal chromosome numbers. Immortality was not available to normal, euploid cells except through freezing. They could be "transformed" with a virus or mutagen, but then they became aneuploid and behaved like cancer cells.[14] Immortality was thus a characteristic solely of cancerous, aneuploid cells, and it was one of the traits that made cancer a menacing and mortal disease of the body.

Malignancy and cancer were already associated with uncontrolled cell proliferation and metasticization, but it was not until after 1966 that HeLa cells were understood or described in these terms. Certainly it was recognized that HeLa cells came from the cancerous tissue that caused Henrietta Lacks's death, but the emphasis had been on their control by scientists, their harnessing as producers of knowledge in the victorious battles against polio, and the less successful but still hopeful attempt to understand and contain cancer. This sense of control came to an abrupt end with the second and more profound disruption of a benign image of in vitro immortality.

This disturbance was the announcement that HeLa cells had contaminated and overgrown many of the other immortal human cell cultures established in the 1950s and 1960s. Because one human cell looks much like another, only cross-species contamination—which could be seen by counting chromosomes—had up to this time been identified in cell culture. This changed with the introduction of techniques of genetic identification. At the Second Decennial Review Conference on Cell Tissue and Organ Culture in September 1967, geneticist Stanley Gartler announced that he had profiled eighteen different human cell lines and judged them all to have been contaminated and overtaken by HeLa cells.

Gartler had tested the eighteen lines electrophoretically for a set of enzymes known to be genetically polymorphous, that is, to differ slightly among different people. All eighteen cell lines contained exactly the same enzyme profiles, indicating that they were all the same rather than eighteen distinct human cell types. All eighteen had the same profile as the HeLa cell. The key piece of evidence in this study was the profile for a particular enzyme called G6PD (glucose-6-phosphate dehydrogenase), which is a factor in red blood cell metabolism. Gartler stood up in front of an audience of tissue culturists and said:

> The G6PD variants that concern us are the A (fast) and B (slow) types. The A type has been found only in Negroes. . . . The results of our G6PD analyses of these supposedly 18 independently derived human cell lines are that all have the A band. . . . I have not been able to ascertain the supposed racial origin of all 18 lines; it is known, however, that at least some of these are from Caucasians, and that at least one, HeLa, is from a Negro. (Gartler, "Genetic" 173)

The terminology of cell culture was already dense with the connotations of lineage, culture, proliferation, population, contamination, and, most recently, malignancy. With the delivery of this paper, Gartler used these terms in a scientific explanation which marked the contaminating cell line as black and the contaminated lines as white.

At this moment, the narratives surrounding the HeLa cell changed dramatically. Prior to Gartler's work in 1966, race had not entered into the discussions of either HeLa cells or their donor, Henrietta Lacks. In fact, Gartler had to write to George Gey early in 1966 to ask about Lacks's race.

> I am interested in the racial origin of the person from whom your HeLa cell line was initiated. I have checked a number of the early papers describing the development of the HeLa cell line but *have not been able to find any information pertaining to the race of the donor.*[15] (emphasis added)

After 1966, the race of the donor was central to the scientific evidence of cell culture contamination, and metaphors and stereotypes of race framed scientific and journalistic accounts of the cell line.

The following analysis traces this transformation of scientific and popular rhetoric in detail. It is not sufficient to assert that one discourse of contamination merged with one of miscegenation, as if this were the inevitable course of events. It was by no means inevitable; I argue that Gartler emphasized the least sound piece of his repertoire of evidence of contamination, an error which was then promulgated by those who tested, extended, and reported on Gartler's work. This particular course of events reveals much about the functioning of concepts of biological race in American biology in the late 1960s and 1970s.

First, it is necessary to support my assertion that Gartler emphasized the weakest part of his evidence of HeLa contamination of other cell lines, keeping empirically stronger arguments in the background as supporting evidence. Gartler did not have to explain his results in the manner he did, highlighting the G6PD typing. When he tested each of the cell lines for particular enzymes, he was looking for variations in structure between the same genes in different humans. The resulting enzymes differ slightly in amino acid sequence, a difference which can then be visualized as lines on a gel. In addition to G6PD, Gartler also used three other sets of polymorphic electrophoretic variants. Each of these three variants occurs in differing proportions across the world's population, much like blood types, and could not be categorized as being specific to any one population. In Gartler's own words, the statistical likelihood of all eighteen cell lines carrying the same profiles for these systems of polymorphisms was statistically "absurdly low," regardless of the purported race of the various donors of the cells (Gartler, "Apparent" 750). In other words, Gartler's evidence of contamination would have been conclusive even without any reference to racial difference.

Any eighteen people, in particular eighteen unrelated people from the diverse population of the United States, would show different enough enzyme profiles to be distinct from each other. The same enzyme profile for all eighteen would be strong evidence of contamination of all of them by the same cell line. Thus an explanation based on human genetic variation, which lurked in the background of Gartler's paper as supporting evidence, was empirically more sound than that which he chose to highlight, the G6PD types and their supposed racial correlations.

The presence of G6PD type A or B in a cell culture was correlated in his account with the racial categorization of the patient donor as black or white, a categorization presumably noted by the physician-scientist on the basis of the patient's self-designation or the physician's visual assessment. The correlation implied an equivalence of the two kinds of racial categorizations—by G6PD variation and visual judgment. Gartler suggested that an apparently white donor could not have been the origin of a cell line expressing G6PD type A, which is not true at all. This weakness was noticed by the founder of one of the cell lines said to be contaminated by HeLa. As this was a commercially marketed cell line, contamination was a threat to its economic value. Almost immediately, Monroe Vincent—whose cell line was made from his prostate—published a denial that MA160 was HeLa-contaminated. He hypothesized that the G6PD type A in MA160 could be inherited from potential remote "Negro ancestry" in his lineage (Fraley et al. 541).

Even with this weakness, G6PD continued to be the main marker in testing for HeLa contamination of cell lines. The consequence of this emphasis was the essentialization of cells as "black" or "white"—an error possible only in the prevailing confusion about concepts of biological race. An essential black/white difference was simpler for Gartler to explain and easier for his audience to understand as "proof" than would an explanation based on human genetic variation. This fact indicates that the concept of population had not yet managed to displace the concept of race for these biologists. Gartler had been working since the early 1960s with a cell culture that carried a mutation at the G6PD locus, and as a geneticist he was well acquainted with the voluminous literature concerning the incidence of G6PD variation in populations around the world. However, his audience of cell biologists was not trained in population genetics, nor were these cell biologists familiar with its methods. Their responses gave G6PD type A a simplified, essentialized status as a "black gene," as if the register of race went from skin to cell to enzyme to gene. This marker, taken from the context of population genetics and used as an identifying test for contamination in cell culture, lost all the subtleties and complications of a gene frequency within a population and became, instead, an absolute indicator of difference.[16]

Once the lines between cell types were demarcated in this fashion, the crossing of these boundaries meant that the scientific community experienced two simultaneous disruptions. First was the disruption of a complacent sense of control. If Gartler was right, these scientists had mistakenly been doing experiments on cells that they thought were breast cancer or colon cancer or amnion cells but were in fact all HeLa cells. This was a threat to the integrity and value of past work, an imputation of carelessness in their technical practice, and the sudden switch from HeLa as a founding success to HeLa as the source of catastrophe. Second, it was the disturbance of a previously unarticulated presumption of race. There had been no "information pertaining to the race of the donor" to this point, and in its role as a breakthrough, a standard, a universal, HeLa and its concomitant personification as an angelic and immortal Henrietta Lacks were unmarked and assumed to be white.

At the same time, the synecdoche between cell and person functioned to make the cell populations of petri dishes analogous to populations of people. The scientists moved readily between the language of cells in culture to that of people in culture. One respondent to Gartler's paper stood up to remind the audience that cross-species contamination occurred easily in tissue culture, a statement which exemplifies the facile slide from "human cell" to "human":

> We all remember clearly a number of years ago—maybe 5—when this contamination business began and everybody was very defensive: the L cells contaminating the rabbits, the HeLa cells contaminating the mouse. . . . Now, here comes the HeLa: human contaminating human.[17] (Hsu 191)

Human contaminating human, explained in terms of racial difference, meant an immediate introduction of the metaphors of miscegenation. The immediacy of this response is better understood within the larger context of American history.

The late 1960s saw the arguing of the landmark United States Supreme Court case of *Loving v. Virginia*, ruling in 1967 that the Virginia miscegenation law was unconstitutional. As in science, the validity and utility of racial categories were being challenged. This ruling was followed by a general move on the part of most American states to repeal statutes that defined racial categories, usually by blood proportion (Pascoe 67–68). This included the "one drop of blood" rule which defined a person as black if they had so much as a single drop of black blood. Miscegenation laws, present from the 1660s to the 1960s, asserted an absolute interdiction of sexual or marital crossing of the racial border. However, the existence of the one drop rule, and its aim to demarcate black from white absolutely, admitting no middle ground, indicated that this border was crossed all the time. Originating in American slavery, when the master's rape of the female slave was an "open secret," the one drop of black blood criterion and miscegenation laws worked to deny any kinship across racial boundaries (JanMohammed).

That these boundaries were still anxiously regarded—and that kinship across them was still being denied—is evident in the audience discussion after Gartler delivered his paper. There was a great deal of defensive rejection of his conclusion of widespread HeLa contamination from these members of the tissue culture community, some of whom had founded the cell lines Gartler was identifying as contaminated. Leonard Hayflick's WISH— the cell line made from his daughter's amniotic sac—was one of the cell lines Gartler identified as carrying the genetic marker he said was "found only in Negroes" ("Genetic" 173). Hayflick, apparently a white man, is reported to have stood up during the discussion following Gartler's paper and said "'I have just telephoned my wife, who assured me that my worst fears are unfounded'" (qtd. in Gold 30). Themes of miscegenation and pollution, the fear of impregnation of the white woman by the black man, and doubt in the genealogy of the scientists thus came to the fore within minutes of Gartler's conclusions.

After Gartler made his argument about HeLa contamination, the description of what happened to cells in culture was structured by these metaphors of miscegenation. Scientists passed on this explanation to journalists, who used this narrative to tell the HeLa story to a larger audience. The scientists also read the journalists' accounts, footnoting them in their scientific papers. The warnings about the danger of HeLa contamination, for example, played up a "one drop of HeLa" theme: "If a non-HeLa culture is contaminated by even a single HeLa cell, that cell culture is doomed. In no time at all, usually unnoticed, HeLa cells will proliferate and take over the culture" (Culliton 1059). One drop was enough.

The racial metaphors altered but did not completely change the way tissue culture had been understood up to this point. Even with the earlier famous chicken heart cell culture, there was a consistent obsession with hypothetical calculations of the total volume of cells the immortal culture produced; with HeLa, these were calculations of a swamping of a white population by a black one. Ross Harrison had mused in 1927 that had it been possible to allow all of the cells in Alexis Carrel's chicken heart culture to multiply, it would "now greatly outweigh the terrestrial globe," while in 1937 P. Lecomte De Noüy envisioned the same set of cells reaching a volume "more than thirteen quatrillion times bigger then the sun" (Harrison 18; De Noüy 104).[18] This "mathematical calculation" abruptly became a threat, a literal "fear of a black planet" in the case of HeLa. The calculation was of a fleshliness that not only outweighed the globe but threatened to take it over: "HeLa, with a generation time of about 24 hours, if allowed to grow under optimal cultural conditions, would have taken over the world by this time" (H. W. Jones et al. 947). The calculation of the putative volume of the culture when "allowed" to multiply freely was not just of a cell culture but of how much Henrietta Lacks would weigh now, if all her cells were to be put back together—an "incredible amount" (Curtis).

Gartler's findings and methodology were taken up by Walter Nelson-Rees, director of a cell culture laboratory at the University of California, Berkeley, who was charged by the National Cancer Institute with keeping stocks of standard reference cells. Starting in 1974, Nelson-Rees began publishing lists of cell lines he judged to be HeLa-contaminated—an alarmingly high number. Contamination proved to be widespread. It is impossible to estimate how much research was invalidated by the findings that the researchers were mistakenly working on the wrong type of cell. Contaminated cell lines included a set of six cell cultures given to American scientists by Russian scientists under a biomedical information exchange that Nixon and Brezhnev negotiated in 1972 (Nelson-Rees et al. 751).

High profile incidents such as these, the emphasis on the "provenance" of cell lines (one of Nelson-Rees's favorite terms), the consistent use of the G6PD marker system, and Nelson-Rees's penchant for personifying HeLa cells all contributed to a revived interest in the figure of Henrietta Lacks in the 1970s and into the 1980s. The inability of scientists to explain why HeLa contaminated other cultures, but rarely the other way around, fed into a characterization of the cells as voracious, aggressive, and malicious.

A large number of articles about HeLa and Henrietta Lacks appeared in magazines and newspapers from *Science* to *Rolling Stone* between 1974 and 1977. Unlike the writers in the 1950s, these authors were not interested in the figure of the self-sacrificing housewife. Although cell cultures were being identified by this time by karyological studies—the appearance of their chromosomes—and a number of other systems of genetic polymorphism not characterized as specific to black or white populations, cell identity was still being explained primarily through the G6PD system. HeLa cells were depicted as having a distinct, identifiable biological race due to their particular genetic structure. Michael Rogers, reporting in the *Detroit Free Press,* explained this to his readers by writing, "In life, the HeLa source had been black and female. Even as a single layer of cells in a tissue culture laboratory, she remains so" (D4).

This identity as black and female was combined with a character described variously as "vigorous," "aggressive," "surreptitious," "a monster among the Pyrex," "indefatigable," "undeflatable," "renegade," "catastrophic," and "luxuriant." The narrative of reproduction out of control was linked with promiscuity through references to the cell's wild proliferative tendencies and its "colorful" laboratory life. Rogers reported that he first heard about Henrietta Lacks through graffiti on the wall of the "men's room of a San Francisco medical school library" (D1). Nelson-Rees, the self-appointed watchdog of HeLa contamination for the cell culture community, was fond of talking about the appearances of "our lady friend." When describing the letters Nelson-Rees wrote to his fellow biologists when he suspected they were working with HeLa-contaminated cell lines, another journalist wrote, "It was like a note from the school nurse informing the parents that little Darlene had VD." Problems of contamination of cell lines were described as the scientific community's "dirty little family secret" (Gold 63, 64, 72).

In the personification of Henrietta Lacks as promiscuous and lascivious, in the characterization of HeLa cells as akin to venereal disease, in the facile linking of race and contamination, and in what I call the "one drop of HeLa" narratives of miscegenation and hybridity, I see the long reach of what Hortense Spillers has called the "American grammar" of race, that "the ruling episteme that releases the dynamics of naming and valuation remains grounded in the originating metaphors of captivity and mutilation" (68). The distressing literality of the excision, cultivation, exchange, mutation, and sale of a living, reproducing fragment of a black woman's cervix—without her or her family's knowledge or permission—evokes Spillers's theorization of the "captive body," as that which has been severed from its motive will and active desire. The captive body is "culturally unmade" by becoming "a thing for" the captor, removed and renamed (67).

I would argue further, that the racialized rhetoric of contamination is not something apart from the immortality narratives that I discussed earlier. With Gartler's 1967 bombshell to the tissue culture community, the desired scientific progeny—the immortal cell line—the "precious baby," exhibited more autonomy than was expected or wanted of it and the promise turned to menace. The baby transformed into a monster, supplanting and destroying more legitimate scientific progeny, such as the tokens of political good will on the part of Russian scientists, the WISH line, and Monroe Vincent's prostate cells. Because this transformation was detected and narrated in terms of racial difference, the already menacing aspect of malignant immortality became inextricably wound with a threat to scientific order and a set of racial and sexual metaphors of contamination and miscegenation.

These anxieties provoked by the HeLa cells' repetitive appearances are underpinned by the more general transgressions of the object of the immortal cell line across understandings of the life and death of intact bodies. Tissue culture was designed as a microcosm of the human body; it is the double, made for the visualization of disease processes and as the stuff of experimental practice. In some cases, it is made from the bodies of the experimenters. This double, as Sarah Kofman observes, is "neither living nor dead"; it is "designed to supplement the living, to perfect it, to make it immortal" (148)—but faced with this double, there is a shock, a lack of recognition.

CONCLUSION

The two forms of immortality that I have gone through here—the beneficent immortal chicken heart, wonder fathered by modern science, cells in a test-tube body form of immortality, and the racialized, malignant, out-of-control immortality have both functioned to deflect attention from what mobilized them in the first place—Henrietta Lacks's death. This is evident in the treatment of her apparent misdiagnosis as a footnote to scientific genius. It is also manifest in the volatile, threatening personification arising from the contamination narrative.

This effacement of death has not dropped away with the 1990s version of the HeLa story. Rather, the immortalized Henrietta Lacks has taken on a distinctly economic cast. In media accounts, she has become a figure of economic exploitation, with a contemporary right to sue for compensation. Lacks has become personified as the holder of an investment account, where the original capital was those first biopsy cells (for example, see Stepney). These should have had a dollar value from the beginning, because look what they would be worth today, after all these years in the investment account that is the burgeoning biotechnology industry. Lacks's family is cast into the role of the rightful heirs to the proceeds of this "investment" who cannot collect, because nobody ever patented the cells and thus it is difficult to pin down either past or present profit, or any one party who could benefit from the commercial exchange of HeLa cells and all their products and permutations.

Race reenters the story here as demarcating lines of economic power and privilege. As one of George Gey's colleagues commented to him in 1954, it was "out of the goodness" of Gey's heart that HeLa cells, only three years after their establishment, had become "general scientific property."[19] As a black woman from a black family, Henrietta Lacks walked into a clinic at Johns Hopkins University Hospital, where there was no institutional, ethical, or legal framework to ensure that she or her family was in a position to execute any kind of decision—out of the goodness of their hearts or otherwise—as to the fate of the cells.

Her family and friends, long left out of the story, are now being interviewed as key players in a drama where Henrietta Lacks's cells became important tools of modern medicine without her or her family's permission. With contemporary awareness that significant tools of modern medicine are also valuable commodities, endless reproduction and worldwide distribution remain part of a story of Lacks's immortality, but the metaphors have become those of the growth of capital and those of miscegenation and contamination have retreated into the background of the story.

An analysis of the significance of this story to contemporary discussions of the body as property, the implications of laws which allow for patenting of living organisms and materials, and other cases of immortal cell lines made from human tissues remains to be written. In anticipation of these concerns, this essay has attempted yet one more return to the origin of the HeLa cell line. What this return had indicated is that at the establishment of this cell line exists a moment of irredeemable silence. Lacks's illness and treatment in 1951 at Johns Hopkins took place in an institutional, cultural, and scientific setting that had no room in it for heroic agency or any other expression of personal will on her part, even the simple act of donation. Recognition of this absence places the establishment of the first immortal human cell line and the science of tissue culture within the long and troubled history of human experimentation (see Lederer).

The relationship between immortal cell lines and human experimentation has been obscured by a "false and misleading plenitude" of personification.[20] This functions to animate the cells with an autonomous will, as though they were beneficent or malevolent independent of the scientific apparatus and constant tending that maintains their "life" in the laboratory. The narrative of immortality—beneficent, malignant, or monetary—masks the death at its origin. Although it is difficult to say whether an accurate diagnosis of adenocarcinoma would have helped in Lacks's treatment, the total absence of questioning of the circumstances and adequacy of her medical treatment—even with the clearly stated admission of diagnostic error published in 1971—indicates the power of the concepts of immortality produced by the life of these cells.

The HeLa cell line, even though referred to as an individual entity with a clear physical relation to an individual named Henrietta Lacks, does not exist in a single place, is not a tiny vial containing an ever-living cell. The cells grow and divide, are split up to make new cultures, and are distributed to live and divide further in laboratories around the world. Various generations of them remain in stasis, stored in the freezers of cell culture banks. The original biopsy was a piece of tissue, not a single cell, and there exist many subtypes of the HeLa cell line. HeLa cells have been used for years as one part in hybridization procedures which merge two kinds of cells to make a new and distinct cell line. To speak of the HeLa cell line is to speak of a distributed, heterogenous thing which is always growing, multiplying, and changing.

HeLa thus serves as its own metaphor. It is not a story to which there is a single conclusion. Immortality, the uncanny double, and the cultural, scientific, and individual effects of ideas of biological race have existed in an intricate reciprocity with the matter and practice of the science of tissue culture in this history. The resulting personification of HeLa simultaneously captures and erases human experience of this twentieth-century biomedical reach toward the technical alleviation of aging and death.

NOTES

This project was originally inspired and guided by Barbara Johnson in the teaching of *Persons and Things*. Thanks to Evelynn Hammonds and Christopher Kelty for invaluable critical suggestions. Audiences and readers from the MIT STS program, the Harvard Life Sciences Working Group, the Cornell STS Program, and the participants of Biotechnology, Culture, and the Body have contributed much in the way of critical advice and suggestions for source material. This work was supported in part by a Doctoral Fellowship from the Social Science and Humanities Research Council of Canada, grant number 752–96–0493.

1. Cellular "life" and the synecdoche between cell and body do not remain stable over time, and they are part of a larger history of the twentieth-century life sciences than I can address here. For an in-depth analysis of the rhetorics of "life," see Doyle.

2. The phrase "technologies of living substance," which I believe is an appropriate description of tissue culture, is taken from a letter Jacques Loeb (1859–1924) wrote to Ernst Mach in 1890 about Loeb's ambitions for a biology that would manipulate, transform, and control living matter (Pauly 4). See also Rabinow for a discussion of a more recently established cell line.

3. For more detail on the history of tissue culture, see Susan M. Squier, this volume.

4. Before any biopsy was taken, Lacks's physicians sent her to be tested for syphilis, and this detail should be viewed in the context of American medical history. James H. Jones has written about the characterization of American blacks as a "notoriously syphilis-soaked race" by a white medical establishment and the role of this perception in the founding of the Tuskegee syphilis experiments to track the course of the disease in untreated black males. Elizabeth Fee, in an analysis of Baltimore in the first half of the twentieth century, writes that the situation there was similar to the South; venereal diseases among the black population were seen "as both evidence and consequence of their promiscuity, sexual indulgence, and immorality" (182). In the context of this history, Lacks's being sent to the syphilis clinic reveals to some extent the doctors' perception of the patient; furthermore, when this event reappears as part of the narrative of the origin of the HeLa cell, the cell line is personified with metaphors of promiscuity and contagion.

5. Trypsinizing means shaking the cells apart by using the digestive enzyme trypsin.

6. Charles Pomerat, letter to George Gey, 5 March 1954, George Gey Papers, Alan Mason Chesney Medical Archives, Baltimore. All letters cited are quoted with permission from the George Gey Papers.

7. Ronald H. Berg, letter to George Gey, 24 November 1953. Berg was referring to a story that had appeared in the *Minneapolis Star* in 1954 and gave Lacks's name, although it is unclear who released the name.

8. After this point, there is a proliferation of pseudonyms made from the HeLa letters: Helen Lane and Helen Larson are two of these, probably spread by Gey, who was startled to find that Lacks's real name had leaked out without his knowledge and thought that a fictitious name would serve equally well.

9. As was the case with the biopsy cells, this photograph is used in this and many other publications, such as a 1973 textbook of medical genetics, without any indication that permission was sought or given for its use, either from Lacks or her family.

10. When a cell divides, the two resulting cells are referred to as daughter cells.

11. C. M. Pomerat, letter to George Gey, 19 February 1954.

12. Dr. George Gey, Jr. (qtd. in Kelly A1).

13. Thanks to Anne Fausto-Sterling for pointing out the importance of aging research to this story.

14. "Transformation" is the technical term used to describe an event such as a

mutation, chromosomal rearrangement, or viral infection after which cells in culture grow to higher densities, in several layers rather than a monolayer, and cause tumors when injected into animals. This event can either occur spontaneously or be induced.

15. Stanley Gartler, letter to George Gey, 16 March 1966.

16. Donna J. Haraway has described the concepts of race and population in biology circa 1950–1970 as follows: "Occasionally still a convenient notion, 'race' was generally a misleading term for a population. The frequency of interesting genes, like those coding for immunological markers on blood cells or for different oxygen-carrying hemoglobins, might well differ more for individuals within a population than between populations. Or they might not; the question was an empirical one and demanded an explanation that included consideration of random drift, adaptational complexes, and the history of gene exchange" (343).

17. T. C. Hsu then equivocated by adding, "-if it is HeLa," reflecting the series of skeptical questions the alarmed audience directed at Gartler.

18. Thanks to Evelyn Fox Keller for this source.

19. Charles Pomerat, letter to George Gey, 5 March 1954.

20. "We return to those empty spaces that have been masked by omission or concealed in a false and misleading plenitude" (Foucault 135).

WORKS CITED

Alberts, Bruce, et al. *Molecular Biology of the Cell.* 2nd ed. New York: Garland, 1989.
Bang, Frederick. "History of Tissue Culture at Johns Hopkins." *Bulletin of the History of Medicine* 51 (1977): 516–37.
Brown, Russell, and James Henderson. "The Mass Production and Distribution of HeLa Cells at Tuskegee Institute, 1953–1955." *Journal of the History of Medicine* 38 (1983): 415–31.
Carrel, Alexis. "The New Cytology." *Science* 73 (1931): 331–45.
———. "On the Permanent Life of Tissues outside of the Organism." *Journal of Experimental Medicine* 15 (1912): 516–28.
———. "Physiological Time." *Science* 74 (1929): 618–21.
Culliton, Barbara J. "HeLa Cells: Contaminating Cultures around the World." *Science* 184 (1974): 1058–59.
Curtis, Adam. Interview. BBC Radio. 14 Apr. 1997.
Davidson, Bill. "Probing the Secret of Life." *Colliers* 14 May 1954: 78–83.
De Noüy, P. Lecomte. *Biological Time.* New York: Macmillan, 1937.
Doyle, Richard. *On Beyond Living: Rhetorical Transformations of the Life Sciences.* Stanford: Stanford University Press, 1997.
Ebeling, Albert H. "Dr. Carrel's Immortal Chicken Heart." *Scientific American* Jan. 1942: 22–24.
Fee, Elizabeth. "Venereal Disease: The Wages of Sin?" *Passion and Power: Sexuality in History.* Ed. Kathy Peiss and Christina Simmons. Philadelphia: Temple University Press, 1989. 178–98.
Foucault, Michel. "What Is an Author?" *Language, Counter-Memory, Practice.* Ed. Donald Bouchard. Trans. Donald Bouchard and Sherry Simon. Ithaca: Cornell University Press, 1977. 113–38.
Fraley, Elwin E., et al. "Spontaneous *in vitro* Neoplastic Transformation of Adult Human Prostatic Epithelium." *Science* 170 (1970): 540–42.

Gartler, Stanley. "Apparent HeLa Cell Contamination of Human Heteroploid Cell Lines." *Nature* 217 (1968): 750–51.

———. "Genetic Markers as Tracers in Cell Culture." *Second Decennial Review Conference on Cell Tissue and Organ Culture, September 1967*. National Cancer Institute Monograph 26: 167–90.

Gold, Michael. *A Conspiracy of Cells: One Woman's Immortal Legacy and the Medical Scandal It Caused.* Albany: State University of New York Press, 1985.

Haraway, Donna J. "Universal Donors in a Vampire Culture: It's All in the Family: Biological Kinship Categories in the Twentieth-Century United States." *Uncommon Ground: Toward Reinventing Nature.* Ed. William Cronon. New York: Norton, 1995. 321–66.

Harris, Henry. *The Cells of the Body: A History of Somatic Cell Genetics.* Cold Spring Harbor: Cold Spring Harbor Laboratory Press, 1995.

Harrison, Ross G. "On the Status and Significance of Tissue Culture." *Arch. exp. Zellforsch* 6 (1927): 4–27.

Hayflick, Leonard. "Biology of Human Aging." *American Journal of the Medical Sciences* 265 (1973): 433–45.

———. "Establishment of a Line (WISH) of Human Amnion Cells in Continuous Cultivation." *Experimental Cell Research* 28 (1962): 14–20.

Hsu, T. C. "Discussion of Genetic Markers in Cell Culture." *Second Decennial Review Conference on Cell Tissue and Organ Culture, September 1967*. National Cancer Institute Monograph 26: 191–95.

JanMohammed, Abdul R. "Sexuality on/of the Racial Border: Foucault, Wright, and the Articulation of 'Racialized Sexuality.'" *Discourses of Sexuality: From Aristotle to Aids.* Ed. Donna Stanton. Ann Arbor: University of Michigan Press, 1992. 94–116.

Jones, Howard W., et al. "After Office Hours: The HeLa Cell and a Reappraisal of Its Origin." *Obstetrics and Gynecology* 38 (1977): 945–49.

Jones, James H. *Bad Blood: The Tuskegee Syphilis Experiment.* 2nd ed. New York: Free, 1993.

Kelly, Jacques. "Her Cells Made Her Immortal." *Baltimore Sun* 18 Mar. 1997: A1 +.

Kofman, Sarah. *Freud and Fiction.* Trans. Sarah Wykes. Oxford: Polity, 1991.

Lederer, Susan. *Subjected to Science: Human Experimentation in America before the Second World War.* Baltimore: Johns Hopkins University Press, 1995.

Nelson-Rees, Walter, et al. "HeLa-like Marker Chromosomes and Type-A Variant Glucose-6-Phosphate Dehydrogenase Isoenzyme in Human Cell Cultures Producing Mason-Pfizer Monkey Virus-like Particles." *Journal of the National Cancer Institute* 53 (1974): 751–57.

Pascoe, Peggy. "Miscegenation Law, Court Cases, and Ideologies of Race in Twentieth-Century America." *The Journal of American History* 53 (1996): 44–69.

Pauly, Phillip. *Controlling Life: Jacques Loeb and the Engineering Ideal in Biology.* Berkeley: University of California Press, 1990.

Rabinow, Paul. "Severing the Ties: Fragmentation and Dignity in Late Modernity." *Essays on the Anthropology of Reason.* Princeton: Princeton University Press, 1996. 129–61.

Rogers, Michael. "The HeLa Strain." *Detroit Free Press* 21 Mar. 1976: D1 +.

Spillers, Hortense. "Mama's Baby, Papa's Maybe: An American Grammar Book." *Diacritics* 17.2 (1987): 65–81.

Stepney, Rob. "Immortal, Divisible: Henrietta Lacks Died 40 Years Ago but Her Cells Live On and Multiply." *Independent* 13 Mar. 1994: 50.

Witowski, J. A. "Alexis Carrel and the Mysticism of Tissue Culture." *Medical History* 23 (1979): 279–96.

Part II: Maternity in Question

Chapter 3

"From Generation to Generation"

Imagining Connectedness in the Age of Reproductive Technologies

THOMAS W. LAQUEUR

Strange, a strange thing is the common blood ["splankhon" =
that which springs forth] we spring from—
—Aeschylus, *Seven against Thebes*

Suppose I had one hundred percent access to the facts, and one
hundred percent knowledge of the laws of nature. None of this
would tell me whether a surrogate mother should keep her baby.
—Richard Feyman, in response to a request
to join a "science court," 1988

I want to begin with an extraordinarily tangled and complex set of claims
made by a gestational surrogate living in northern California.[1] In response
to the question, "Did you have any difficulty giving up the baby you had
carried for nine months," Ms. A. offers the following reply:

"You know, I think it just sits in your mind the whole time that it's not my
baby. I'm just letting it use my body. I'm just growing it for someone else
who can't do it." (qtd. in Roberts)

Three things interest me about these sentences. While the baby sits in
Ms. A.'s womb, "it"—presumably the thought "not my baby"—sits in her
mind. The manifest physical fact of gestation is opposed by an idea. (Actually, she hides a referent at still another level; "your" mind could either
mean "my" mind or more probably "one's," the normative mind.) Second, the difficulty of discovering more precisely what "it" refers to suggests
the enormous amount of cultural work being done in this single sentence.
"It" has no grammatical antecedent; it functions to prefigure the phrase
"that it's not my baby," but this leaves us to wonder whether she regards
this as a fact, a belief, as something she subscribed to when signing the
surrogacy contract. In short, what are the grounds for Ms. A. believing that
what sits in her mind is so easily abstracted from what sits in her womb?

Finally, there is the ancient metaphor of the mother as gardener to which I will return shortly: "I'm just growing it for someone else." But she is making an even more modest claim: Ms. A. asks no more than to be considered a tenant farmer.

Surrogacy contracts generally require that the surrogate mother agree to have an abortion if the child is found to be defective or a selective reduction if more than one fetus is growing. The deal is for one, and only one, normal child. In this case, Ms. A. is personally opposed to abortion except in cases of rape; that is, she would not abort a child for any of the reasons stipulated in the contract. And yet, she claims to be perfectly agreeable to having an abortion should the conditions stipulated in the contract pertain. The reason for this apparent inconsistency is, once again, that just because the baby is growing in her body and the procedure would therefore have to be performed on/in her does not mean that she is having the abortion or that it is her child.

The "body" in any ordinary language sense of the word seems irrelevant. Abortion should be a matter of choice for any woman, Ms. A. says, only she is not the relevant woman with respect to this abortion and this fetus.

> "'I do not have any responsibility to make that decision for them. And although I would not terminate my own pregnancy, they're terminating their own pregnancy. *Just because it's in my body does not have anything to do with me.* That's how I stand on it.'" (qtd. in Roberts; emphasis added)

"Just because it's in my body does not have anything to do with me" is a stunning, even shocking, claim. It seems almost a parody of alienated labor, a blatant case of false consciousness, a real-life instance of the sort of feminist dystopia imagined in Margaret Atwood's *The Handmaid's Tale,* an illustration of the fundamental immorality of various reproductive technologies, surrogacy generally, gestational surrogacy in particular, or of the various relevant legal and financial arrangements they entail. Maybe. But I want to take Ms. A.'s claim—unusual as the circumstances under which it was made or thought might be—as an entry into exposing how fundamentally fraught is the search for the grounds of connectedness. If being in her body has nothing to do with her, one might well wonder what does.

One answer is as old as Aeschylus's *Eumenides.* While the technology that impregnated Ms. A. is of very recent vintage, her analysis echoes Apollo's famous exculpatory speech (lines 656–73). No, the god declares, Orestes could not have murdered his mother if by that is meant a person to whom he had blood bond because neither he nor anyone else has a mother in that sense: "the mother is no parent of that which is called her child, but only nurse of the new planted seed that grows," "a stranger"

(26–28). Ms. A.'s explanation, on this account, is not based on a belief about real blood: she is as bound to the child within her through blood, flesh, and all manner of vessels as is any other woman who became pregnant in more usual circumstances. If common sense does not make this self-evident—which of course it does—the sonogram and other imaging technologies make it easier than ever before to imagine the mother's connections to the child within. Admittedly, as Rosalind Pollack Petchesky has shown, they also allow us to imagine the contrary—a fetus as an independent being with temporary residence in the womb—which is why viewers of the anti-abortion film *The Silent Scream* can sympathize with the supposed plight of the tiny embryo that fills the screen floating bravely in a sea of amniotic fluid.

Ms. A.'s fetus has nothing to do with her, on this account, because her standard for connectedness is the one Freud associates with the bond between the father and child: abstract not physical: "*der Triumph der Geistigkeit über Sinnlichkeit,*" the new abstract Hebrew God over the old gods available to the senses (Freud, *Mann* 16, 221).[2] This is not quite fair because Ms. A. probably has in mind some modern genetic account of connections in which fleshliness does not have much purchase but in which there is still a material basis for the bond between mother and child, the bond of a plan of schemata. But even admitting this, Ms. A. is putting an enormous amount of reliance on that tiny fraction of all the base pairs in the DNA of her germ cell that distinguishes it from the DNA of all other humans. If we share 95 percent of our code with monkeys and a higher proportion with our fellow humans, what remains to differentiate us from others of our species is very small indeed. In short, Ms. A. has managed somehow to isolate "her" code and to construe its absence as the grounds for the child not being hers. She would not be having an abortion because this genetically foreign fetus seems to her as alienable as those anodyne body parts or fluids which we give up without a thought. (Except, of course, when some bit of blood or a few cells from a biopsy turn out to have commercial value, but that is another story.) Ms. A.'s reasoning may not quite be Freud's "conquest (*Sieg*) of intellectuality over perceptibility," but it comes close.

And moreover, Ms. A.'s perspective rejects what has been the dominant image of maternal connection from antiquity to the present. An enormous amount of cultural work—perhaps a whole system of sacrifice from Abraham and Isaac on—has gone into creating agnatic kinship against the "natural" connections of the flesh (Jay). And still today that is what matters in English law. The old common law notion that giving birth is what makes a woman a mother is the basis for current law governing in more technologically advanced times: "the woman who is carrying or has carried a child . . . and no other woman" is the mother however the embryo or the sperm and egg got there. Ms. Mary Beth Whitehead's supporters in her contest with

Mr. Stern argued *not* on the grounds of her equal genetic connection to Baby M. but to her superior physical tie. She carried the child; he only contributed his sperm.[3]

I have, so far, concentrated on the unusual case of Ms. A. who—like Orestes—denies that real, actual, here and now flesh and blood creates cultural flesh and blood, the consanguineous relationship with a child that seems to be the basis for kinship, for motherhood. In the history of father-hood, in contrast, real-life flesh and blood has always been of dubious significance. It seems to me far from clear whether blood, as in "of my blood," ever literally meant blood. Of course semen was taken to be the finest concoction of blood and therefore somehow passed its essence from one generation to another. But in no classical account of procreation did any actual blood pass from father to child through the mother. At best, and the Aristotelian tradition would deny this, some representation of blood, some froth, entered into the "conceptus." On Aristotle's account, nothing of the father's blood is physically passed between generations, just as no part of the shoemaker—other than his skill as artificer— becomes part of the shoe. Thus when, for example, Bolingbroke addresses his father John of Gaunt as "thou, the earthly author of my blood," or when the Duchess of Gloucester says to Gaunt that he was "one of seven vials of his (father's) sacred blood," or when Buckingham says to Richard, in a play that is all about power and bloodlines, that he, Richard the hunchback, is the true heir of his father—the "right idea of your father, both in your form and nobleness of mind"—the claim is not about material blood. It is about blood as legitimate kinship, as lineage, as partaking in the flesh of the ancestors in a way that invokes corporeality at the same time as it elides it.[4]

Thus real blood can be imagined to count for nothing and the merest *sanguinis pneuma* of the right sort can mean the world. Perhaps this observation does nothing more than reinvent the very old distinction between kinship and genealogy, between a cultural system of meaning and biological map, an ordering—if one wants to consider only the maternal line—of mitochondrial DNA. But if this is so, I make the observation in order to collapse a distinction which simultaneously demands our continued allegiance. The recognition of oneself in another seems, at one and the same time, absurdly corporeal and profoundly cultural. And so, the distinction between blood and not blood ties, the real and the not quite so real flesh that matters and flesh that does not, will crumble in much the same way two sex construals of male and female crumble. Concurrently, a conflation of these categories—whether of sex or of connection between genera-tions—cries out for once again separating them. Blood—the inner connec-tion—like sex, will turn out to be a remarkably unstable, local category.

Beginning with the founding texts of classical antiquity, kinship is pro-jected onto the anatomy and physiology of connectedness. In Sophocles'

Antigone, Teiresias speaks of a child coming from his *father's* innards, his "splankima" (1066–67). It comes not just from his loins but from his innards, from his "womb." (The term of course also refers to the mother's womb, for which Liddle and Scott's Greek-English lexicon cites Pindar, "Olympu" 6.43.) And in Aeschylus's *Seven against Thebes,* the English translation of the notion that children are born from the "splankhon" common to both mother and father bespeaks the elusive quality of what it is from which we come. "Strange, a strange thing is the blood we spring from" (1031–32).

And strange it still is—in sperm banks and ovum brokerages, in fertility clinics but also among adopted children and lesbian mothers, among people struggling to establish heterodox kinship ties. This strangeness, however, is not born of technology. It is at the root of kinship; heterodox ways of making families—high, low, or no technology—only expose it for what it is, as if our own, Western, structures were exposed to the scrutiny anthropologists more usually reserve for other people. In the great scheme of things it is also tempting to dismiss the ethnography, indeed the politics of reproductive technology more generally, as a sideshow of American politics put on by a prurient and sensationalistic press. The number of cases of disputed custody of children born of so-called surrogate mothers—the "traditional" or genetic or the far less common gestational or host surrogate variety—is tiny: as of 1993 a mere sixteen out of an estimated five thousand surrogacy contracts had been disputed; almost all of the at least two hundred "surrogate births" a year in California go unnoticed by the public. There have been no litigated cases involving the claims of ovum donors against contracting parents, and over thirty thousand instances of donor insemination a year pass by routinely.[5] Adoption is similarly widespread and children pass from biological to adoptive parents regularly. Child custody disputes between lesbians are more frequent but one might well understand them as but one aspect of the far more general issues of child custody upon dissolution of the legal or de facto relationships into which the children were born. (Insofar as they raise special issues—the primacy of the biological over the "other" mother—it is because of the legal limbo of homosexual marriage, adoption, and partnership contracts more generally. Advanced technology is irrelevant to the problem of rights and kinship.[6])

Yet the *Los Angeles Times* gave greater prominence on its front page to the Orange County, California, superior court's decision that the genetic parents were to receive full custody of the child conceived from their egg and sperm and that the "gestational surrogate mother" had no claims than the newspaper gave to President George Bush's veto of the Civil Rights Bill. Baby M. captivated the nation in 1988; the lesbian in Oakland who held up a turkey baster in court when asked to identify the father of her child was a local celebrity. Baby Jessica, the child caught between legal jurisdictions in Michigan and Iowa and between the only parents she had known and her real parents was one of the biggest media events of 1993.

I want to speculate in this essay on why these numerically insignificant cases have become representative anecdotes—to use Kenneth Burke's phrase—of much deeper and politically charged issues. They are, I will suggest, moments in a cultural drama that go well beyond the immediate issues and speak to far broader questions: of identity—which in the United States often means ethnicity; of what it is to be a person; of the place of the body in imagining a history of the self; of the recognition of one's self in others; and of community. The representative power of these unusual cases is evidence for a crisis of kinship—the structures that connect society to the natural world and generations to each other—that is refracted by new technologies but is in no sense caused by them.

Traditional kinship structures in the postindustrial West are increasingly shaky, but I mean here something more fundamental: a weakening of the epistemological and ontological foundations of identity, of the subject, and of human connectedness. The distinction between nature and culture is profoundly troubled and the flaws in a naturalistic epistemology—evident since David Hume but averted until recently by a tissue of social and cultural fictions—are now too glaring to be ignored. In short, our belief in the power of nature to dictate social relations is equivocal at best and is tested to its limits at the time when molecular biology seems to support a new reductionism.

Technology is of relatively little importance in all of this, and the biopolitics of family, abortion, race, and sexual preference are critical. Clearly, technology is not entirely irrelevant: gestational surrogacy is possible only because of in vitro fertilization; HLA compatibility—and in the near future probably, DNA—testing as evidence for both maternity and paternity depend on advances in immunology and molecular biology. But in most of the cases I will be discussing—*Baby M.* being the most famous—the technology of fertilization is that first described in the medical literature by the eminent eighteenth-century John Hunter who collected sperm from an opening in the side of a patient's penis—the result of a congenitally deformed urethra—and injected it into the vagina of the man's wife (see Poynter). Artificial insemination of animals has been practiced for centuries and probably since Egyptian antiquity.

The capacity to imagine donor insemination is well established already in a famous sixteenth-century text: "They interfere with the process of normal copulation and conception by obtaining human semen and themselves transferring it." Or, "they are able to store the semen safely, so that its vital heat is not lost; or even that it can not evaporate so easily" (Kramer and Sprenger 22). The "they" in question are, of course, succubae who transfer the human semen to incubi who in turn impregnate unsuspecting or, worse, suspecting women. Conversely, manifestly novel technologies are construed in accordance with old cultural narratives. Thus, for example, it is assumed that when semen of the appropriate status enters a woman's body it stakes a claim there; like an explorer in *terra nullus*, it takes possession and assumes for its genitor the appropriate rights and obligations. The right kind of semen somehow incorporates a child into culture and the wrong kind excludes. Moreover, a child without a legal father—without a claim to status-conferring semen—is, in early modern English jurisprudence, a *"natural* child," one outside of human conventions. Because of these foundations of patriarchy, a great deal of legal and institutional effort goes into depatriarchalizing semen in situations in which either the donor or the recipient affirmatively desires sperm free of its cultural baggage. Enormous energies go into maintaining the anonymity of sperm donors; nameless sperm begets nameless children. Under pressure of the adopted children's movement and certain women's organizations, considerable legal ingenuity has recently gone into devising a regime under which total anonymity might be lifted once a child reaches majority, without at the same time invoking a whole panoply of paternal rights and obligations. But this only underscores the power of the underlying assumptions.[7]

The case for ovum donation is entirely different. Because there is no historical model for women's reproductive substances venturing forth, that is because motherhood has almost exclusively been construed as "bearing" and not as begetting, there is no long collective memory to overcome. There are few assumptions about ova. When I visited an ovum brokerage near Berkeley, California, I was ushered into the sort of nondescript but pleasant office that I associate with prosperous suburban stock or insurance brokerages and shown catalogs containing detailed descriptions and pictures of potential donors by the score.[8] The name, age, address, and any number of winsome pictures are offered for each woman. When I remarked that they all seemed so attractive, I was told that I had seen nothing yet: Jennifer still awaited me. Her vita was kept out of the book because she would have skewed the market. Everyone would want the eggs of a Stanford honors graduate, a championship athlete, who had also worked as a model. While sperm banks and physician-mediated insemination use a variety of means to assure absolute anonymity and while much anguish on all sides is generated when this policy is relaxed, everything about ovum donation is out in the open.

When I asked for an explanation for the unequal treatment of two genetically equal gametes, I was told that anonymity simply was not an issue. Patrons bought eggs, and then they owned them. And while this is probably true, it is the case only because there is no countervailing assumption that ova make claims or entail obligations. Ova do not colonize; they do not reproduce the blood of the ancestors. The one breech in this story is that before women are accepted into the ovum donor program I visited, they have to demonstrate the appropriate psychological profile on the Minnesota Multiphasic Preference Test (MMPT). That is, they have to show themselves to be sufficiently "masculine" to be willing to give up their eggs without a fuss, namely, to treat ova as if they were sperm at the moment of ejaculation. (There seems to be a shortage of Jewish ova, not because Jewish women are unwilling to participate in donor programs but because their MMPT profiles either show them to be too "feminine"—possessive of their germ seed—or otherwise unsuitable.)

While modern assays make possible specific new determinations of "biological" parentage and hence changes in evidentiary codes and while reproductive technologies have expanded well beyond the usual, artisanal, heterosexual, penis in vagina method of conception, the underlying, competing models of what it is to be a parent have changed little since the Greeks: to be a parent is to contribute the plan of the child, to pass along the essence of a lineage as well as of a genus, or, to be a parent is to act the role of the parent, to perform its emotional and physical labors ranging from the provision of sustenance to gestation. Plan versus labor, roughly speaking.

The notion of surrogacy, as American courts have repeatedly noted, goes back at least to Abraham and Sarah, when she suggested that he "should go unto my maid that I may obtain children by her" (Gen. 16.2). This turns out to be an interesting story for my purposes: it shows that surrogacy is not primarily a reproductive "technology" but rather a cultural splitting of motherhood. It also raises the question of why, if patriarchal descent is what matters, Ismael (the son of Hagar) should not be the equal of Isaac, the son of Sarah. (The later biblical texts, which, according to Nancy B. Jay, take patriarchy for granted, see no problem here; the earlier "E" text requires Isaac's near sacrifice to establish his unique connection with the ancestors through the father). In other words, unusual reproductive strategies demand that a culture be clear about the structures of paternity and lineage which it otherwise takes for granted. They do of course provide the occasion—as in the case of Isaac and Ismael—to put pressure on a culture's commitment to this or that strategy of producing connectedness. We, for example—unlike old-regime France—find adoption to be acceptable but reproductive technology makes us ask how far will we go: how about adopting a "prefabbed" zygote—or eventually even embryo—left over from someone else's in vitro fertilization? Were adoption difficult or ruled out entirely, this dilemma would not arise. If we, as a culture, were clearer about what could and could not be legitimately or ethically or legally owned or sold or purchased, the case of *Baby M.* would not have arisen. In this landmark case, the "technology" is hundreds if not thousands of years old; the novelty is in the social relations of production and reproduction which we are willing to accept.

Or conversely, so-called new reproductive technology and alternative forms of parenting are problematic because of new cultural possibilities and the destruction of old cultural verities, both largely or entirely independent of science. These verities have long been thought of as somehow natural, which inclines us to regard changes in our capacity to manipulate nature—shifts in technology—as perforce affecting them as well. But this is not the case. In thoroughbred horse racing, for example, where the meaning of filiation is not in dispute, neither old techniques—artificial insemination—nor new technologies—embryo transfer—have caused a stir. Of course they can be used to improve the bloodlines or performance of non-thoroughbreds, but within the sport they are irrelevant because an unchallenged, unambiguous rule—section 5, rule 1, paragraph D—defines the circumstances of which mare and stallion count as the dam and sire of any given foal: "Any foal that is the product of either Artificial Insemination or Embryo Transfer . . . is not eligible for registration." No turkey basters or other props—"A foal must be the result of a stallion's natural service with a brood mare (which is the physical mounting of a brood mare by a stallion"; no surrogacy—"a natural gestation must take place in and delivery from the body of the brood mare in which the foal was conceived." (There is a slight role for human assistance: ejaculate produced during "cover" may "immediately be placed in the uterus of the brood mare" [The Jockey Club].[9])

This is not a rule that derives from nature or from technological possibilities. Genetics matters desperately but genes could be—and are in almost all other breed associations—passed along under less stringent conditions. One particular clan—The Jockey Club—has in place a generally accepted kinship system that is untouched by the sorts of questions that trouble our less secure tribe.

I do not want to suggest that only culture matters. Of course every society develops complex kinship rules—"a coherent system of symbols and meanings" in David M. Schneider's terms—but even as we make our own history we do not do it as we please: "nature" has been the gold standard of inter-generational connectedness through the ages (Schneider 8). This tension between kinship as culture and kinship as nature is everywhere. Consider the great eighteenth-century English jurist William Blackstone on the subject. On the one hand, he argues that no specific laws are required to force parents to care for their children because:

> providence has done it more effectively than any laws by implanting in the breast of every parent that insuperable degree of affection . . . which neither deformity of person or mind of the parent or wickedness, ingratitude and rebellion of children can overcome. (435)

To Blackstone and his contemporaries, for fathers or especially mothers to act otherwise was to act "un-naturally." (Leave aside that this is not universally true, as Nancy Scheper-Hughes has shown in her recent book about situations in which mothers do not naturally develop an affection for their children.) But natural affection counts for absolutely nothing when apportioning power; then, the jurist invokes the naturalness of patriarchy: "a mother as such is entitled to no power, but only to reverence and respect," except of course when the mother stands outside of patriarchy, that is, mothers of illegitimate children who have custody and are responsible for their maintenance until they are seven.

Of course, we might scrutinize more closely what jurists mean by "natural" or "inherent" in a post-Lockean age. The puzzle is precisely that it is so hard to imagine what it means that "there is a right *inherent* in the father, recognized by positive law, and to no degree dependent on the discretion of the Lord Chancellor, to act as a guardian of his children?" When Lord Hardwicke ruled in the mid-eighteenth century that "the father is the *natural* guardian of the sons during their minority" or Lord Eldon in the early nineteenth century that there is such a thing as the "legal natural right of the father," the quality of this nature eludes us. Indeed, Blackstone informs us that "the empire of the father continues even after his death"; the Father, like the King, seems to have two bodies, one of which is eternal. But then, in the early 1840s, a contemporary legal text tells us, the Legislature adopted "the opinion that these rights (of the mother) fall short of what *natural* feeling and public policy demanded." And so began the revolution in family law which between 1800 and 1900 turned ancient assumptions on their head: the mother, not the father, was now assumed, by nature, to be the appropriate guardian for a child.[10]

The naturalness of family relations—and of ascriptive ordering more generally—has long been under attack. One might argue that the increasing reference to the biology of sexual difference—to corporeal instrumentality—in liberal political theory from Thomas Hobbes through Jean-Jacques Rousseau and well into the nineteenth century was a response to the collapse of a familial order grounded in a higher nature, that is, in revelation or metaphysics. It was thus argued that breast-feeding was in the nature of women as mothers, so much so that the capacity to give milk became the essential feature of the phylogenetic class to which we belong (Schiebinger 382). Conversely, *not* breast-feeding was regarded as unnatural, resulting in so-called diseases of civilization. (Rousseau's *Emile* is the classic text here.[11]) Thus when one sort of nature—the "nature" of a divine hierarchical order—crumbled, another nature—this time, biology—was mobilized to take its place as the ultimate ground for all sorts of distinctions and connections.

The normative collapse of a supposedly natural family in popular culture occurred well before reproductive technologies added new twists to an old story. The motif of the single mother, which caused such stir when our then Vice President Dan Quayle attacked Murphy Brown of the eponymously named network television show, goes back at least to the soap operas of the mid 1960s: *As the World Turns,* and to more recent prime-time programs: Mary Jo in *Designing Women,* Carla in *Cheers,* Maddie Hayes in *Moonlighting,* and Susannah in *Thirtysomething* are all unwed mothers. And of course one could go back to the popular culture of the nineteenth century, that is, the novel: Hedy Sorel in George Eliot's *Adam Bede* and Oliver Twist's mother. (The difference is that in the twentieth-century fiction, single mothers stand a chance at surviving.)

In addition, there is a genre of television shows about alternative paternity. In *My Two Dads,* a teenage girl is raised by two men who resolutely refuse to learn which one is the "real" father. After a particular quarrel, they take a so-called paternity test, but when the female Jewish judge who lives upstairs comes in to announce the results, they decline to hear them. In another show, a girl and her mother live on earth while the space creature who fathered her appears periodically as a crystalline triangle.

An increasing divorce rate, which since 1945 has resulted in legal separation exceeding death for the first time as the cause of the end of marriage, has clearly created the context for frequent evaluations of what constitutes parenthood (see Philips; Stacey). And, more constructively, public demands for the recognition of gay and lesbian relationships and the explicitly new familial relationships which result from these unions has enormously expanded the boundaries of "motherhood" and "fatherhood."

Into this space, vacated, or seemingly vacated, by a normative, putatively natural family order enter new practices, new ways of making communities or new roles for old practices. Major issues such as abortion fit in here. The debate is between two mutually uncomprehending sides. The one, called pro-choice, might be construed as an aspect of freeing a woman's body from nature, that is, from the natural course of pregnancy. It assumes that even if motherhood is natural, terminating pregnancy is not an unnatural act but rather an act of disposing over one's body. The other side, called pro-life, aims to create through rhetoric (apostrophe, for example) and through science an image of the fetus that makes it recognizably human and hence protected by natural law and, of course, also positive law. Humanness is pushed further and further back into the nidus of life, such that the fertilized egg—the completed plan waiting for its fulfillment—is by nature, according to this view, something like a person. It is in this context that the ruling of a Tennessee court which held that a husband could prevent his estranged wife from using a fertilized egg belonging to the two of them is considered a victory for the pro-choice side. That is, in this case the court regarded a zygote as having no rights, contrary to the view that the "pre-born" do enjoy some or all of the rights of the already born.

I want in the rest of this essay to concentrate on how the crisis of nature and what is imagined through it is represented in two sorts of legal cases, venues which focus deep anxieties and cleavages in American society: first surrogacy and then same-sex partners with children. (The latter, I should note here, is part of a larger debate about adoption, both in relation to the secrecy that has traditionally surrounded birth parents and as an alternative for women to reproductive technology as a way of having children when other ways are not available.)

Surrogacy burst onto the American scene with the *Baby M.* case, one of the biggest news stories of the late 1980s. (There had by then been at least one thousand uncontested cases of so called traditional surrogacy, according to one account [Fleming 87].) The facts are simple. Mr. Stern, a child of holocaust survivors, desperately wanted to assure the survival of a family, which was nearly destroyed, by having a "natural child." His wife, Dr. Stern, is a biochemist who either could not, or chose not to, have children because she suffered from the early stages of multiple sclerosis. Through a broker, the Sterns contracted with Ms. Mary Beth Whitehead, a relatively poor working-class woman, whose marriage was acknowledged to be shaky, to bear a child conceived in her body with Mr. Stern's sperm and to relinquish any claim to this child once it was born. Dr. Stern would then adopt the baby and it would "belong" to the contracting couple. For these efforts, Ms. Whitehead was to receive $10,000. After the baby's birth, Ms. Whitehead refused to give up the child, and the Sterns went to court to enforce what their lawyer construed as an ordinary contract.

The lower New Jersey court agreed with the Stern's lawyer and upheld the deal: $10,000 in consideration for the pregnancy, that is, the work, resulting in the baby. (Anti-slavery laws prohibit the actual sale of children.) The New Jersey Supreme Court reversed this judgment. It held that both Mr. Stern and Ms. Whitehead were the parents, that Mr. Stern should be given custody on "best interests of the child" grounds, and that Ms. Whitehead should have rights of visitation as might any party in a divorce settlement.[12]

I will return to the verdict shortly, but I want first to sketch the outlines of the debate which the case engendered. In the first place, it brought into sharp focus anxieties about class and thus also about the body. A relatively rich man had contracted with a relatively poor woman, a sort of synecdoche for the gendered distribution of wealth and income and indeed for the routine condition of much of American economic life. In the context of the President Ronald Reagan revolution, the question of what one could sell and for how much loomed large. When the California legislature debated surrogacy legislation, this issue was made explicit: "There is something repugnant," said a legislator, "about offering $50,000 to a woman to do something she would not want to do, and it is also repugnant not to adequately compensate her." In the morally less suspicious circumstances of everyday economic life, it is, of course, the body that is sold in the labor market; in many, many cases, people do what they would not ordinarily want to do in return for money. Surrogacy—like prostitution in the nineteenth century—may reveal in naked detail these uncomfortable moral dilemmas of capitalism, but they are scarcely new.

More interesting for the purposes of this collection, however, are debates that focus on women's bodies and the claims that might be grounded in them. Advocates of Ms. Whitehead argued that the trial court judge had discounted her labor—her *in utero* mothering, her intimate physical connection to the baby. The judge had recognized that Ms. Whitehead had borne the child for nine months, but in his ruling he had shown that he regarded the mother simply as a vessel. As in Atwood's *Handmaid's Tale* or in the debates over a Florida indictment of a pregnant woman who was charged with giving drugs to a minor, the judge, Ms. Whitehead's proponents asserted, had radically misconstrued the moral claims of concreteness.[13]

Ms. Whitehead's supporters, of course, wanted to reverse this valuation of nature. So, Mr. Stern's plan, that is, his sperm, was mocked; Ms. Whitehead's intentions, that is, her initial and considered resolve to give up the baby, were discounted. Instead, the claim was of a resoundingly Lockean sort: the physical labor of gestation and birth made the baby Ms. Whitehead's. Dr. Stern's plans—her hopes for a baby, her preparing the nursery—figured not at all. In a sense, the New Jersey Supreme Court handed a resounding victory to what I think is an ancient and historically ambivalent notion of motherhood and its foundation in nature. None of Ms. Whitehead's advocates, as far as I know, argued for her claims to the baby on the grounds of her genetic connection, that is, on the basis that it was her egg. They simply reversed the Aristotelian valences and reasoned that her rights derived from everything having to do with physical connectedness. Needless to say, Ms. Whitehead's proponents tended to be identified with an essentialist as opposed to a performative or constructivist account of the category "woman." So, Mr. Stern's rights were founded on the old-fashioned Aristotelian grounds that he provided the plan or design, although this contribution was discounted by Ms. Whitehead's more strident defenders. Ms. Whitehead's rights derived from her material contribution to gestation. And Dr. Stern had no rights.

Of course, "plan" does not always have such purchase. If Mr. Whitehead—Mary Beth's husband—had made claims on the baby, the result might have been different. Mr. Stern's rights might well have run afoul of the old doctrine, most recently upheld in the case of *Michael H. v. Gerald D.*, that a child born in wedlock is presumptively the "natural" child of the father, even if acknowledged biological facts are to the contrary. (In this case, the court held that there was no protected liberty interest for the adulterous "biological" father and he had no claims on his progeny against the claims of the lawful husband of its mother.) The rules of kinship in action.[14]

Given modern technologies, it was only a matter of time before the

Orange County case of *Johnson v. Calvert*, finally decided by the California Supreme Court, would provide in reality what many thought was the relevant thought experiment in the *Baby M.* case. (The questions in that experiment are not new, although this time it is a woman's and not a man's "idea" or plan for the child versus another woman's labor or material invested in that child that are in conflict as the foundation for connectedness.) The *Johnson* case puts enormous pressure, in my view, on the essentialist arguments from nature made for Ms. Whitehead's motherhood. Its facts are as follows: Crispina Calvert, a woman of Filipino descent, had her uterus removed to resolve an apparently intractable problem with fibroids; her ovaries remained intact. She and her husband, Mark, of English descent, began to consider surrogacy, and when Anna Johnson, a black woman and sometime welfare mother, heard of their plight from a coworker and offered to bear a child for them, a deal was struck. Crispina and Mark Calvert signed a contract on 15 January 1990 with Anna Johnson, in which the latter agreed to bear a child conceived in vitro from the Calvert's egg and sperm in return for $10,000 and medical benefits. The money was to be paid in installments, the last of which was to be due six weeks after the birth.[15]

On 19 January a zygote was implanted and Johnson's pregnancy began. By June, that is, three months before the due date, various annoyances, class resentments, and perhaps covert racial tensions precipitated a crisis. Johnson felt that the Calverts had not done enough to secure an insurance policy for her and that they had been insufficiently supportive during a bout of premature labor; the Calverts were upset that Johnson had not revealed that she had suffered several miscarriages during prior pregnancies. In a letter to Mark and Crispina in July, Anna Johnson demanded early payment of what was still owed to her because she needed to move and did not have the money to pay the deposits for a new apartment and telephone connection:

> I am unable to return to work until the delivery of this baby some income is limited. I do not get enough disability to make a two month rent deposit plus, the security deposit & have the telephone reconnected. I don't think you'd want your child jeopardized by living out in the street. I have looked out for this child's well being thus far, is it asking too much to look after ours. (Johnson)

It is unclear whether the pregnancy had caused Johnson's unemployment; there is no question about the differential resources of the protagonists in the ensuing drama. After this plea for money, Johnson issued a demand and then a threat that sound not much different from any that might result from a business deal gone sour:

> There's only two months left & once this baby is born, my hands are free of this deal. But see, this situation can go two ways. One, you can pay me the

entire sum early so I won't have to live in the street, or two you can forget about helping me out but, calling it a breach of contract *& not get the baby*! I don't want to get this nasty, not coming this far, but you'd want some help too, if you had no where to go & have to worry about yourself and your own child *& the child of someone else!!!* Help me find another place and get settled in before your baby's born. (emphasis in original)

And then the threat:

This is the last letter you will get from me. The next letter you will receive will be from my lawyers, unless I hear from you by return mail at the end of the week—7/28/90.

Both sides then went to court to establish parentage of the *in utero* baby. Despite her clear, emphatic reference to "the child of someone else," Johnson played the only card she could. She and her supporters argued that, on the grounds that Apollo used against the mother of Orestes, the child was hers. She had done the labor of producing it. Work is explicitly what is sold—alienated—in a surrogacy contract. "How much am I paid for my *time and energy*" (emphasis added) asks the question in an information booklet for prospective surrogate mothers. And work is what does not historically entitle one to the product of one's labors, either in manufacturing or in motherhood. But Johnson's advocates turned this on its head; work is what counts. As a psychiatrist who is a consultant for the National Coalition against Surrogacy put it:

She is the mother because she gestated and gave birth to this child. What makes her a mother is her emotional and physical *work* in nurturing the fetus and the way in which her body builds the baby. It brings oxygen. It takes away waste. It protects the baby from bacteria and external harm. (emphasis added)

There is no room in this definition of motherhood for any historical claims of connectedness, no ground onto which one might project the world of the ancestors. Motherhood, as in the *Baby M.* case, is understood as a concrete condition of the here and now.

The trial judge in this case, affirmed by the California Supreme Court in substance although not in legal reasoning, disagreed. Physical labor did not create parentage, he ruled, and relied instead on a new, gender neutral valuation of plan—of efficient cause in the Aristotelian sense—as conferring the rights of both fatherhood and motherhood. (True, Aristotle would not have countenanced two efficient causes, but . . . . An amicus brief to the California Supreme Court suggested that the baby might have two mothers but no one accepted this not outlandish argument.) So, the baby became one person and not another: Christopher Michael, and not, as Johnson had called him, Matthew. Nothing Johnson could have done would, in all likelihood, have changed the filial decision. But she talked as if she too would discount physical labor, that is, the traditional foundation of motherhood, in favor of something far less substantial. Like Ms. A., with whom I began this essay, Anna Johnson revealed that love and the feeling of bondedness could really be transferred—I use the term in its psychoanalytic sense—only onto something that represented historical continuity, presumably onto those parts of the genetic pool that belonged to her. In an interview to the press, Johnson admitted that "if it had been my egg, it would have made a real difference. But with it [in vitro fertilization] there is no connection to me. There's been a detachment from the baby from day one." Of course, the phrase "if it had been my egg" may well be only a way of saying "if I thought that I could keep the baby," so that the detachment of which she speaks is to be understood as the merciful adjustment of the sensibility to a brutal reality and not to the absence of genetic connection. But still, this absence seems to make the task of taming projective identification much easier.

Genes, the trial judge noted, "set in motion human development." The bonding of birth is temporary; the connection of genes is for the long haul: "Heredity can provide a basis of connection between two individuals for the duration of their lives. . . . We want to know who came before us and who's coming after." So, ironically, the collapse of the patriarchal model in which the history of the person is conveyed by the father resulted in the first case in human history in which the woman who gave birth to a child was judged not to be its mother. In California, this conclusion is the outcome of an explicitly liberal definition of parentage embodied in the 1975 Uniform Parentage Act.[16] One of the central purposes of this act was to abolish legal distinctions between legitimate and illegitimate children. It listed the usual ways of establishing parenthood and added to the list new, gender-neutral scientific tests: "the provisions of this part, applicable to the father and child relationship, shall be applicable to the mother." To be the natural father means the same thing as to be the natural mother, namely, to be the source of half of the child's genetic material.

The appellate court therefore read Section 7003.1 of the Uniform Parentage Act—"Between a child and a natural mother it (the parent-child relationship) may be established by proof of her giving birth to the child"—as inapplicable to Anna Johnson.[17] She is not the natural mother, in the gender neutral sense in which Mark Calvert is the natural father and Crispina Calvert the natural mother. That is, "Anna was excluded, as clearly shown by the report of the director of the parentage testing laboratory received into evidence as exhibit 1 by stipulation of both parts." In short, the director's report "flatly excluded Anna: 'not the mother' of the baby boy" (234 CA App. 3rd 1557, 1569).

So, we now have the politically peculiar situation in which the National Council of Bishops, various fundamentalist Christian groups, the American

Civil Liberties Union, and the National Association of Women support the value of physical labor while surrogacy agencies and supposedly class-biased judges champion a way of imagining the history of the self as proceeding equally from both parents. The California Supreme Court, I should note, did not reason along the grounds I have just outlined; it construed the case as a contract action. It ruled for the Calverts and against Johnson on the grounds of their respective intent; while both women could argue on biological grounds for being the mother of the child, only Crispina had "intended to bring about the birth of a child that she intended to raise as her own." The decisive power of intention which six of the seven justices advocated strikes me as confirming a Humean view that there are no "natural connections" in the world and that we literally make the ties that bind, whether between generations or between any other discrete things or events in the world.

I do not, of course, claim to have begun to have exhausted what is really at stake in these cases. Questions that hover around money— of what can and cannot be sold, disputes that go beyond surrogacy to the sale of body parts on the one hand and prostitution and pornography on the other— are central and I have largely ignored them. They are ancient. But I have given enough evidence to suggest how vexed the meaning of "flesh of my flesh" has always been and how modern technologies rehearse old ambiguities in new contexts.

So far, emotions, community, and identity politics have played little role in this essay, and I want now to turn to these matters in the context of a discussion of lesbian motherhood. In one sense, the notion of a lesbian community based in families is a triumph of culture over nature. A lesbian relationship is not the "natural" configuration in which to produce children. But at the same time, children are regarded within lesbian relationships, as they are more generally, as the foundation for imagining an existence of self and community through time. Biology is appropriated in the interest of a group which through political struggle and cultural development has come into its own: "Now that we are out of the closet and feeling better about ourselves, we can have kids. We're worthy of having families, of the respect of having families."[18] Producing families of this sort requires in most cases a reconfiguration of the claims of sperm. Indeed, the legal foundations have been laid for cleansing it of patriarchy. I asked one woman at a sperm bank why she was going to the considerable expense—8 cc of sperm which has been frozen in buffered glycerol and guaranteed for 180 days costs $116—to buy something she could probably get for free from friends. "It's really cheaper to pay for it," she said. "Then it has no history." In short, the cash nexus does for fatherhood what it did for the ascriptive relations of the old regime; money, now as then, is the great social solvent.[19]

An entire practice has evolved to further depersonalize the sperm and make it the property of the would-be couple. In cases of informal, that is, non-professional donation, a "go between" transports the sperm, often in a baby food or marinated artichoke jar, with their large openings and minimum depth. One regular likes to put the jar in a sock, which she carries between her legs as she drives. (This is all recounted in Debra Chasnoff's Academy Award–winning film.) In general, the further, the more mediated is the relationship between sperm donor and the recipients, the more attenuated the claims of fatherhood and the stronger the claims of the mother(s). (In California, the law is clear only in cases of anonymous donation to sperm banks; in practice, a physician can "de-paternalize" sperm by brokering its passage from donor to user.) The point is that, in principle, emotion and cultural cohesion are meant to take the place of biology in lesbian parenthood. The DNA of sperm is systematically discounted.

But like the repressed, biology returns. In the first place, fathers reassert their rights if the sperm has been insufficiently cleansed of paternity. A gay man who had initially renounced his claims but who had an ongoing relationship with the lesbian couple changed his mind when he professes to have looked into his "son" Sean's eyes and seen "the whole history of my family there." The court supported his projection.[20]

Second, in cases of dissolution the law more or less mandates an unequal struggle between biology and the emotions in deciding the claims of motherhood. (If lesbian or gay marriages were legally sanctioned, the "nonbiological" parent could adopt the child to which it was not connected by "blood" and thereby enjoy equal rights with the biological parent. Thus, in these, as in disputed cases arising out of so-called reproductive technologies, the problem is not created by the reproductive act itself but by the kinship muddle in which it transpired.) Michelle G. and Nancy S. started living together in August 1969.[21] In June 1980, their daughter, K., was born; four years later they had a son, S. Michelle inseminated Nancy with a turkey baster so that she might conceive both children; Michelle is listed on both of their birth certificates as the father. (In many lesbian couples, both parents are called "mother" by their children; in others, the inseminating parent is listed as the father and gets father's day cards even if not actually called "Dad." I do not have the ethnographic information to know the frequency of various forms of address.) In any case, both children were given the "father's," that is, Michelle's last name.

In January 1985, Michelle and Nancy's "marriage" ended, and they agreed to a complicated visitation arrangement under which each child would spend five days at one parent's house and then five at the other's but would have four days together at one venue. Three years later, Nancy sought to change these terms so that both children would spend 100 percent time together, 50 percent time with each parent. Michelle objected, and Nancy went to court to have herself declared the sole custodian of both children. In a dispute that rocked the San Francisco lesbian community, the case finally ended with a complete victory for the "biological" mother. Whatever intentions or arrangements the two might have had, the court held that the natural mother should have full custody rights and that the "emotional" parent had a claim only if it was not against the interest of a competent natural mother.

There have been two views of this result. The initial is outrage. In a brave new world based on building alternative sorts of bonds and connections, it is curious, to say the least, to assert the claims of biology to the exclusion of culture. Indeed, on the one hand, Nancy's position seemed to some entirely antithetical to what the gay and lesbian community stood for. On the other hand, there are other voices speaking for biology, however it might be construed. In the first place, battered women's groups, in most but not all cases representing partners in heterosexual relationships, vehemently oppose the extension of parental rights because each broadening of the definition of parents allows an abusive partner, not biologically related to a child, grounds for further contact and legal maneuvering.

Second, there has been since 1982 a strong adopted children's movement that is asserting the right to know the truth about both biological fatherhood and motherhood. Its arguments make an emotionally wrenching case for the continuing power of the "imaginaire" of blood and only blood. Against the view that "all options should be open" (including anonymous sperm donation and anonymous giving up of children for adoption), that we are happily "expanding and recreating the meaning of family," stands a real fundamentalism.[22] A coalition, begun by one of the children whom Korean mothers sent to the United States by the thousands for adoption, labels the erasure of blood as the "legalized kidnaping of one's history and heritage."

> Voices: Biology is "the basic fact of our existence"; we "don't choose to be cut off from half of our heritage"; we have "full rights of heritage"; "from a child's point of view," without knowledge of who your parents are, "you don't feel you are real. You are floating around"; "You're missing half your life."

I end with another voice—that of an adopted child—perhaps not as puzzling as that with which I began, but still mysterious:

> "There are places in each person's mind for the father whose genes are theirs; there is place in each person's mind for the mother who bore them into the world and whose genes are theirs. Neither ideology or law will remove that place, that void."[23]

Despite the palpable sense of geography, of a physical place, the critical word is "mind." "It just sits in your mind," as Ms. A., with whom I began this essay, says about the child that is in her womb but is not hers. I emphasize mind, that is, culture because my point has been to insist upon a long and fraught history of connectedness largely independent of technology. We have imagined already, and resolved, almost all of the possibilities which new reproductive techniques have brought our way. This is not to diminish their importance; they make manifest with increasing intensity the strains in our structures of kinship. And alternative social relations, absent any technology or any beyond the millennia-old artificial insemination—adoptions, lesbian parenting—do likewise. It is possible that some technologies—the storing up of frozen zygotes and their use decades or centuries after the death of those who contributed the gametes—will so disrupt our ideas of generational succession that they will seriously challenge our sense of kinship. But so far blood and flesh, mother and father, and sire and dam are as clear or as muddled as they have always been.

NOTES

1. A gestational surrogate is a woman who has contracted to bear a child conceived from the sperm of the contracting man and the egg of the contracting, or some other, woman. She is to be distinguished from the so-called traditional surrogate—Ms. Mary Beth Whitehead in the *Baby M.* case, for example—who contracts to bear a child conceived from her own egg. The material that follows comes from an interview conducted by Elizabeth Roberts, currently a graduate student in anthropology at the University of California, Berkeley. It appears in her prize-winning undergraduate anthropology student honors thesis which I helped to supervise. Roberts has gathered what I think is some of the best ethnographic material in existence on the question of imagining connectedness.

2. James Strachey in the *Standard Edition* translates this passage as "the victory of intellectuality over sensuality" (23, 114), but *"Sinnlichkeit"* here has the sense of "perceptibility," the point being that what matters is the connection made through ideas rather than flesh, through that which is recognized by the senses.

3. See also Human Fertilisation and Embryology Act, which Great Britain passed in 1990.

4. On classical theories of conception and the literature cited there (*Richard II* 1.3.69–70 and 1.2.13; *Richard III* 3.7.13–14), see Laqueur, *Making Sex* ch. 2.

5. It is extremely difficult to get reliable figures and these have been gathered from a variety of clinics and informed lawyers. At the time of the *Baby M.* case there had been more than five hundred surrogate-mother arrangements carried out with "few apparent snags" (Gest 60). The *New York Times* thought there had been one thousand (see In re Baby M.). The domain of reproductive medicine has increased dramatically, and I suspect that the numbers in each category have risen considerably as well. Certainly the number of artificial inseminations are in the tens of thousands a year. The fact that *Baby M.* was the first big public case does not of course mean that prior surrogacy contracts were all happily met. The National Coalition against Surrogacy, for example, collected a number of cases of "abuses in surrogacy" that transpired between 1980 and 1987. These are reprinted in the

mimeograph *Surrogacy Information Packet* issued by the Foundation on Economic Trends. Litigation is currently on a new frontier—the status of "leftover" zygotes from in vitro fertilizations, for example. The questions are (a) how they are to be treated in relation to ordinary adoptions, and (b) how, if at all, do the clinics circumvent restrictions against baby selling. This last point would be moot if clinics performed their services "at cost." Other questions include the disposition of zygotes belonging to couples who divorced or died after a successful in vitro procedure.

6. See, for example, Rassam, who suggests that the *Johnson v. Calvert* gestational surrogacy litigation and a recent lesbian custody case in New York—the one steeped in technology, the other not—should both be decided on the basis of a "dynamic feminist approach to defining motherhood." One need not agree with Rassam's hermeneutic prescription to appreciate her conclusion that technology is not the deciding factor.

7. O'Hanlan posits that lesbians procure anonymous sperm from a sperm bank rather than from known individuals so as to prevent the possibility of paternity suits. Seligson discusses the controversial and much attacked policy of the Sperm Bank of California in dealing directly with mothers, or parents generally, and not through doctors. This eliminates one further, culture-erasing, mediation. A more radical innovation of this sperm bank is that it reveals the identity of donors—those who have explicitly given their permission—when the offspring reaches eighteen. I spent time in this clinic which has been attacked in some quarters for this breach of total anonymity and in others—the adopted children's movement—for not going far enough and eliminating anonymous donation entirely.

8. These agencies find women who, for a fee of between $2,500 and $3,000 (perhaps more now; my fieldwork was undertaken in 1994), will take large amounts of hormones that will result in the production of many eggs. Market forces—the dearth of Jewish or Japanese ovum donors, for example—determine the payments. It is the pictures of these women and their stories that one finds in the books I discuss. The eggs are "harvested" trans-vaginally and fertilized in vitro. The resulting zygote is then implanted either in the woman who bought the eggs or in a third woman, a gestational surrogate. The sperm usually comes from a woman's partner, but there are cases in which someone with an intact uterus, malfunctioning ovaries, and no partner wants to bear a child. In such cases, sperm can of course be obtained from a sperm bank.

The particular office I visited also arranges for gestational surrogacy, but the two sides of its business are generally kept separate. That is, either a couple provides their egg and sperm and hires a surrogate or they buy eggs and the woman of the couple carries the fetus. I was told that the market in northern California would not bear the costs, about $70,000 for a three-way deal in which eggs that woman A bought from woman B are implanted in woman C. In Los Angeles, this division of labor is apparently more common.

9. Other breeds once had this rule and gradually gave it up. The Jockey Club has stuck with tradition not out of any commitment to traditional kinship but for economic reasons. Artificial insemination allows semen to be diluted down to the minimum of five hundred million spermatozoa per insemination, so that instead of covering the 70–90 mares each season that is usual for a stallion he can cover 150–200. Artificial insemination has thus led to a glut on the market for foals and hence lower prices for breeds which allow it. I am also grateful to Professor Craig Wood, an expert in equine sciences in the College of Agriculture, University of Kentucky, for his help on this matter.

10. See on these points Macpherson 61–67, 142.

11. See Fildes, *Breasts;* and on the meaning of the non-nursing mother, see Fildes, *Wet Nursing.*

12. In re Baby M.

13. See, for example, Chesler, in which she explicitly reverses the old hierarchy by arguing for Whitehead's right to the baby based on the "fact" of mother-hood—in other words, her material contribution—while "Bill" has no claim because he provided only an "idea." For Chesler, mother and uterus seem to be the same thing, although without the usual dystopian resonances that this trope usually suggests. See also Corea, *Man-made, The Mother Machine.*

14. *Michael H. v. Gerald D.*

15. The story was widely recounted in the press, and I gleaned various small details from a variety of newspaper articles, but there is a full summary of the facts in the California Supreme Court judgment, 851 P. 2d 776, May 1993. The appellate court decision, because its legal reasoning differed from that of the supreme court, cites facts and provides considerable documentary material not available elsewhere: 234 CA App. 3rd 1557. See Rose's account of this case as indicative of the continuing strength of the old authorship paradigm—now gender neutral—of parenthood.

16. In Britain, the Warnock Commission recommended, and Parliament enacted into law, the rule that the gestational mother—"the woman who is carrying or has carried a child as a result of the placing in her of an embryo or of sperm and eggs, and no other woman"—is *the* mother and that the other mother and father are on equal terms with each other in competing with the gestational mother for custody, visitation, and rights (Human Fertilisation).

17. I leave aside here the sort of question a legal realist might pose: would the judge have arrived at this interpretation if Anna Johnson had been white, richer, or better educated? One has the distinct feeling that somehow a legal interpretation would have to be found to keep a white child with a potential middle-class home from being assigned a poor black mother. In the California Supreme Court, the one justice—the court's only woman justice—who dissented from the contract law basis of awarding custody to the Calverts ruled in their favor on the basis of the "best interests of the child."

18. Chasnoff, *Choosing Children.*

19. This expense is in addition to initial registration, the cost of classes to learn how to do self-insemination, medical examination, and shipping, if required.

20. See my discussion of the case of Mary K. and Jhordan C. in Laqueur, "The Facts of Fatherhood" 215, passim. See *Jhordan C. v. Mary K.* for the results of the litigation.

21. There is a similar New York state case, also decided against the nonbiological mother, that I might have discussed here (572 NE 2d 27, NY 1991). I have focused on Michelle and Nancy because the National Center for Lesbian Rights kindly sent me the briefs and other documents relevant to the case (No. 642975-5 CA 1st App. Dist. Filed 20 March 1991).

22. The feelings are so fervent on this subject that the major San Francisco gay and lesbian newspaper, edited by an adopted child, did not in the early 1990s allow advertisements for anonymous sperm donation. The Sperm Bank of California, where I did my fieldwork, is banned despite the fact that its director is a well-known lesbian activist, that it has pioneered the most complete questionnaire of any sperm bank for eliciting the history of the donor, and that it allows release of the donor's name once the child is eighteen.

23. Qtd. in Newbourgh; Philips; "Community Forum"; Cobb.

WORKS CITED

Aeschylus. *The Eumenides.* Trans. Richard Lattimore. Vol. 3. Greene and Lattimore, *Greek* 1–41.

———. *Seven against Thebes.* Trans. David Green. Greene and Lattimore, *Complete* 263–302.

Atwood, Margaret. *The Handmaid's Tale.* New York: Ballantine, 1987.

Blackstone, William. *Commentaries on the Laws of England.* Chicago: University of Chicago Press, 1979.

Chesler, Phyllis. "What Is a Mother?" *Ms.* May 1998: 36–40.

Choosing Children. Dir. Debra Chasnoff. Frameline. 1984.

Cobb, Nancy. "Who Is My Donor Dad?" *Boston Globe* 10 Sept. 1992.

"Community Forum." *Coming Up!* Nov. 1987.

Corea, Gena, ed. *Man-made Women: How New Reproductive Technologies Affect Women.* Bloomington: Indiana University Press, 1987.

———. *The Mother Machine: Reproductive Technologies from Artificial Insemination to Artificial Wombs.* New York: Harper, 1985.

Fildes, Valerie A. *Breasts, Bottles, and Babies: A History of Infant Feeding.* Edinburgh: Edinburgh University Press, 1986.

———. *Wet Nursing: A History from Antiquity to the Present.* New York: Blackwell, 1988.

Fleming, Ann Taylor. "Our Fascination with Baby M." *New York Times Magazine* 29 Mar. 1987: 87.

Foundation on Economic Trends. *Surrogacy Information Packet.* Washington, D.C., 1990.

Freud, Sigmund. *Der Mann Moses und die monotheistische Religion. Gessammelte Werke.* London: Imago, 1950.

———. *The Standard Edition of the Complete Psychological Works of Sigmund Freud.* Ed. and Trans. James Strachey. 24 vols. London: Hogarth and the Institute of Psycho-Analysis, 1953–1974.

Gest, Ted. "Finally, a Ruling—'M' Is for Melissa." *U.S. News and World Report* 13 Apr. 1987: 60–61.

Grene, David, and Richmond Lattimore, eds. *The Complete Greek Tragedies.* Vol. 3. Chicago: University of Chicago Press, 1959.

———. *The Greek Tragedies.* 3 vols. Chicago: University of Chicago Press, 1960.

Human Fertilisation and Embryology Act. Sec. 27. 1990.

In re Baby M. 109 NJ 396. 1988.

Jay, Nancy B. *Throughout Your Generations Forever: Sacrifice, Religion, and Paternity.* Chicago: University of Chicago Press, 1992.

Jhordan C. v. Mary K. 179 CA 3rd 386, 224 CR 530. CA Super. Ct. 1986.

The Jockey Club. *Rules and Requirements of the American Stud Book.* Lexington, KY: The Jockey Club, 1998.

Johnson, Anna. Letter to Crispina Calvert. 23 July 1990. 234 3rd 1557. CA Ct. App. Oct. 1991: 1663–64.

Johnson v. Calvert. 19 CA Rptr. 2d (CA 1993); cert. Denied, 114 Super. Ct. 206. 1993.

Kramer, Heinrich, and James Sprenger. *Malleus Maleficarum.* Trans. Montague Summers. New York: Dover, 1971.

Laqueur, Thomas W. *Making Sex: Body and Gender from the Greeks to Freud.* Cambridge: Harvard University Press, 1991.

———. "The Facts of Fatherhood." *Conflicts in Feminism.* Ed. Marianne Hirsch and Evelyn Fox Keller. New York: Routledge, 1990.

Los Angeles Times. 23 Oct. 1990: 1.

Macpherson, William. *A Treatise on the Law Relating to Children.* London: 1841–1842.

Michael H. v. Gerald D. 491 US 110. 1989.

Newbourgh, Celeste. "Time to Come out of the Adoption Closet." *Coming Up!* Oct. 1987.

O'Hanlan, Katherine A. "In the Family Way: Insemination 101." *Advocate* June 1995: 40–43.

Petchesky, Rosalind Pollack. "Fetal Images: The Power of Visual Culture in the Politics of Reproduction." *Feminist Studies* 13.2 (1987): 263–92.

Philips, Randa. "The Need to Know Does Not Go Away." *Coming Up!* Oct. 1987.

Phillips, Roderick. *Putting Asunder: A History of Divorce in Western Society.* Cambridge: Cambridge University Press, 1988.

Poynter, F. N. L. "Hunter, Spallanzani and the History of Artificial Insemination." *Medicine, Science, and Culture.* Ed. Lloyd G. Stevenson and Robert P. Multhauf. Baltimore: Johns Hopkins University Press, 1968. 99–113.

Rassam, Yasmine. "'Mother,' 'Parent,' and 'Bias.'" *Indiana Law Review* 69.4 (1994): 1165–92.

Roberts, Elizabeth. "Making Babies (in Public)." Honors thesis. University of California, Berkeley, 1993.

Rose, Mark. "Mothers and Authors: *Johnson v. Calvert* and the New Children of Our Imaginations." *Critical Inquiry* 22.4 (1996): 613–33.

Scheper-Hughes, Nancy. *Death without Weeping: The Violence of Everyday Life in Brazil.* Los Angeles: University of California Press, 1992.

Schiebinger, Londa. "Why Mammals Are Called Mammals: Gender Politics in Eighteenth-Century Natural History." *American Historical Review* 98.2 (1993): 382–412.

Schneider, David M. *American Kinship: A Cultural Account.* 2nd ed. Chicago: University of Chicago Press, 1980.

Seligson, Susan V. "Seeds of Doubt: A Successful Donor-Insemination Service Catering to Heterosexual Single Women and Lesbians Raises Some Difficult Questions." *Atlantic Monthly* Mar. 1995: 28–32.

Sophocles. *Antigone.* Vol. 1. Grene and Lattimore 177–228.

Stacey, Judith. *Brave New Families: Stories of Domestic Upheaval in Late-Twentieth-Century America.* New York: Basic, 1990.

Chapter 4

Mediating Intimacy

Black Surrogate Mothers and the Law

DEBORAH GRAYSON

In January 1990, Mark Calvert and Crispina Calvert, a middle-class couple of white and Filipino ancestry, hired Anna Johnson, a working-class woman of African-American and European descent, to serve as their gestational surrogate. In their arrangement, the Calverts were to pay Johnson $10,000 plus medical fees not covered by insurance. They also agreed to purchase a $200,000 life insurance policy for Johnson, who at the time had a four-year-old daughter, and pledged to provide her with emotional support. For her part, Johnson consented to allow herself to be implanted with the zygote formed from Mark Calvert's sperm and Crispina Calvert's egg. Pursuant to the terms of the contract, Johnson agreed to carry the resulting fetus to term and, upon its birth, to relinquish the baby and "all parental rights" to the Calverts (Anna 372). During the time of the contract and before the child was born, relations between Johnson and the Calverts began to break down. By August 1990, when she was eight months pregnant, Johnson announced that she would file suit against the Calverts. In her lawsuit, Johnson sought to terminate her contract and to be declared the baby's legal parent. This lawsuit marked the first time that a surrogate mother without a genetic link to the child she had carried fought for custody of that child (see Allen, "Black"). In September 1990, Anna Johnson gave birth to a baby boy. The next month, Judge Richard Parslow ruled that she had no rights whatsoever to the child she had delivered. Comparing Johnson's role in the birth of baby Christopher to those of a foster mother and a wet nurse, Parslow stated that the surrogate contract that Johnson had signed was enforceable, terminated Johnson's temporary visitation rights, and awarded full custody of baby Christopher to the Calverts.[1] Both the California Court of Appeals and the California Supreme Court upheld Judge Parslow's ruling, arguing that Johnson had neither a legal claim nor maternal rights to the infant. In October 1993, the supreme court refused to hear the case, a move that assured the Calverts full custody of the baby.[2]

Can a woman be the mother of a child with whom she has no genetic connection—as was the case for Johnson? Or does the genetic material of the egg and the sperm donated to create the child determine who its natural parent or parents are? When does a woman become a mother—while she is pregnant or after she has delivered a baby? What of the bodily experience of pregnancy? Does a woman's participation in pregnancy—her carrying the fetus in her uterus—have any bearing on assessing who is the "true" or "natural" mother? In light of the choices made available by new reproductive technologies, can we sensibly argue, as was done in *Anna J. v. Mark C.*, that genes alone should be the determining factor in defining parental rights and relationships, or that custody disputes should be decided solely on the basis of the parental intent of the persons who supplied the genetic material? Who and what is a mother? Can a child, as Justice Joyce Kennard asked during the California Supreme Court hearing of *Johnson v. Calvert*, have two biological mothers (Johnson 506–18)?

Since the 1970s, medical technologies have changed the reproductive body and our relationship to it, in particular, altering the process of reproductive decision making. With assisted reproductive technology, whose acronym, ironically, is ART, it is possible for a child to have at least two biological mothers (see Polikoff). Through the use of ART, biological motherhood has been separated into competing components of genetics and gestation, a separation that has given rise to disputes over motherhood and its meanings. As a growing number of couples elect to hire gestational mothers to have their children, more and more people are finding themselves involved in legal battles over what used to be considered the definitive "fact" of maternal identity (see, for example, Stumpf; Pollitt; Cahill). In *Johnson v. Calvert*, the parties disputed this question. Both sides wanted the courts to decide whether the "natural" mother of the baby was Anna Johnson, the woman who carried the child in her womb and gave birth to it, or Crispina Calvert, the woman who, though unable to give birth, intended for the child to be born, supplied the ova, and made the necessary arrangements for the child to be (re)produced.

In this essay, I argue that what happens in *Johnson v. Calvert* is symptomatic of a general crisis in American culture over what constitutes a family. The first section addresses the ways in which the law tries to regulate familial property and the norms of what makes a family and explores the incoherence of the logic of courts and the law in making these determinations. Surrogacy extends the boundaries of intimacy and of traditional notions of familial kinship patterns by dispersing what was once thought of as a unified entity—mother—and making it into something without a definitive aspect or dimension. No longer belonging simply to the realm of private acts and decision making of couples, the procreative process has also become a collaborative process that takes place in the public spaces of the laboratory and the clinic. Within these public spaces, assisted reproductive techniques such as artificial insemination, in vitro fertilization, embryo transfer, and surrogacy allow a multitude of individuals to participate in a couple's attempts to conceive. For many couples, procreation now includes the participation of additional parties such as health care professionals, surrogates, donors, and, increasingly, the state. Now, not only is birth a process mediated by the intervention of physicians but conception has become a more complex, drastically mediated process as well. The "private act of love, intimacy, and secrecy" of creating a child, as Sarah Franklin argues, has become a "public act, a commercial transaction, and a professionally managed procedure" (336).[3] Nevertheless, despite the growing public and collaborative process of procreation, the courts in *Johnson v. Calvert* and other such cases have attempted to maintain the priority of the meta-notion of a private, genetically-based family.

The second section addresses the complicated and never fully articulated relations among gender, economics, and race and the ways they get expressed in the family form. I delineate the euphemized quality of the discourse of reproduction that enables the family form to take such discursive priority that race, gender, and class hierarchies are ignored. Although these hierarchies are central to reproduction, assisted or otherwise, their stories are not being told because the family is perceived as an interlocked unit—an intimate, guarded entity that serves as a stand-in for the issues that do not get worked out. Facilitating the lack of resolution of matters of family in *Johnson v. Calvert* is the iconicity of Anna Johnson's pregnant black body as a signifier for a set of sublimated meanings about family and race. Johnson's body is at once too much body—one laden with multiple meanings—and too little body—one reduced to scant meaning. She enters the public discourse, as Valerie Hartouni notes, as a "densely scripted figure" that is "occupying and occupied by the category 'black woman'" (75). Indeed, during and after the various trials, Johnson was depicted as everything from a welfare queen and con artist to an extortionist. Her body is, then, both a site of explanation and a body that creates, in a new way, a problem of meaning.

Predictably, Johnson's body is the only body that is explicitly raced in what is presented as merely a story of two mothers. The racial identity of Crispina Calvert, a Filipina American and the other mother of baby Christopher, is never acknowledged. In the eyes of the court and in the public debate surrounding the case, Crispina becomes white.[4] Mark Calvert, the father and a white man, is not negatively defined by race. The signs of race, specifically the signs of black race, operate as an often silent but nevertheless powerful narrative motive within the trial and in its surrounding publicity. Race, particularly black race, is pre-scripted in this case by existing narratives in current and historical memory in the United States that define mothers and motherhood as bearers of social, cultural, and racial identity (see Doyle 10–34). Motherhood, in *Johnson v. Calvert,* is a tightly policed border where racial, class, and sexual hierarchies are defined and maintained in the name of familial affiliation.

Finally, in the third section, I suggest ways to move beyond the limited definitions of who and what is a mother. At issue is the question of whether the national public can imagine a public family. What does family stand for in American culture? More specifically, how does surrogacy raise doubts about tacit knowledge of race and familial kinship? I argue that more diverse definitions of mother and, by extension, of father and of family are both possible and necessary to accommodate the different methods used to reproduce and introduce babies into families. Drawing on Patricia Hill Collins's concept of "shifting centers" in her analysis of motherhood and reproduction ("Shifting" 59), I argue that practices of assisted conception such as surrogacy require that we find ways to acknowledge rather than diminish or ignore the participation of all parents in these processes even if the effect is to destabilize previously held notions of the family.

I.

"'I am not a slave. *Semper Fi.*'"
—Anna Johnson, letter to Geraldo Rivera[5]

There are many things that are neither new nor historically unique in *Johnson v. Calvert*. For centuries, a fundamental concern of black women has been the struggle over reproduction. As Darlene Clark Hine has argued, the "productive and reproductive capacities" of black women have been central to determining which women can be gendered through motherhood (915). Johnson's decision to enter into the surrogate agreement and her subsequent struggle to win custody of the child she bore as a result of this arrangement bring to the forefront once again the issues Hine raises in her historical analysis. The contemporary situation of Anna Johnson represents a centuries-old struggle in which black women attempt to gain personal autonomy in the face of hegemonic social degradation. For a black woman to enter into a surrogate contract reanimates issues that, at least in some ways, have been juridically resolved.

The disputes over surrogacy and maternity in *Johnson v. Calvert* make clear some of the negative consequences of dividing biological motherhood into competing components of gestational and genetic motherhood, outcomes that, as critics have noted, include the degradation of pregnancy and the exacerbation of class differences and racial inequality.[6] In addition, however, disputes over surrogacy and maternity render more visible already existing fractures in cultural constructions of pregnancy as a disembodied experience.[7] With the development of techniques in medical imaging that make fetal life visible, the growth of areas of medical specialty such as neonatology and the increasing arguments for fetal and father's rights mean that the experience of pregnancy is slowly being divested of its physical and emotional significance (see Bordo; Newman; Duden; Hartouni). In surrogate arrangements, for instance, pregnancy is presented as a form of alienated labor where women's reproductive capacities are viewed as "services" that can be separated from their material persons. Women who agree to be gestational mothers are expected to transform their bodies, or rather their body parts, into empty vessels distinct from their physiological and emotional selves. This notion of woman as fetal container is a growing phenomenon in current cultural discourse on pregnancy.

Johnson decided she could not simply be a carrier or a container for the Calverts. Citing California's Uniform Parentage Act, Johnson and her attorneys claimed that a legal precedent had been set for her to establish her right to be a mother to the child she carried even though she was not genetically related to the child.[8] Section 7003.1 of the act states that " 'the parent and child relationship may be established . . . between a child and the natural mother . . . by proof of her having given birth to the child' " (qtd. in Anna 377). While providing drastically different reasons for their findings, the rulings given by the trial court, the appellate court, and the California Supreme Court majority denied Johnson's claim that she was the natural mother of the baby.

Rejecting Johnson's interpretation of the Uniform Parentage Act, Judge Parslow, the presiding judge in the trial court, held that the statute does not say that a woman who gives birth to a child is its natural mother. According to him, the act merely states that, in addition to blood testing, one way to establish a parent-child relationship is by giving birth. Characterizing Johnson as a foster mother and a wet nurse rather than as a natural mother, Judge Parslow unequivocally stated that he was not going to find that the infant had two mothers—a situation he describes as "ripe for crazy-making" ("California" 37). Instead, noting that blood tests of the Calverts demonstrated that there was a 99.999 percent probability that the Calverts were Christopher's parents and that Johnson offered no evidence that the blood tests were inaccurate, Parslow held that Mark Calvert and Crispina Calvert were the natural parents of baby Christopher because blood tests proved they were his genetic parents. For Judge Parslow, genetic maternity was the definitive form of motherhood. Or, to put it another way, for the judge, total ownership of the fetus depended on the condition of genetic ancestorship.

Like the trial court, the California Court of Appeals held that baby Christopher could have only one natural mother and that the basis for determining this natural parentage should be genetics.[9] In their interpretation of the Uniform Parentage Act, the appellate court ruled that genes were incontestable evidence of parentage. Providing a more detailed analysis of the Uniform Parentage Act in their ruling, the appellate court argued that its specialized provision authorizing biological evidence such as blood as proof of parentage allowed them to conclude that the genetic relationship was conclusively more persuasive than the gestational relationship. As Randy Frances Kandel demonstrates, when viewed in this way, disputes arising from surrogate arrangements will always inevitably favor the genetic mother as natural parent over the gestational mother. By focusing solely on biological markers such as blood to determine parentage, Kandel explains, the courts suggest that it is possible that "natural parenthood" can be "reduced to a single simple biological principle" (176).

Kandel, as well as others, point to the kind of reasoning both the trial court and the appellate court used in their rulings on *Johnson v. Calvert* as examples of genetic essentialism, a mode that Dorothy Nelkin and M. Susan Lindee describe as "a way to talk about the boundaries of personhood, the nature of immortality, and the sacred meaning of life" (41). According to Nelkin and Lindee, genetic essentialism "promises to resolve uncomfortable ambiguities and uncertainties" brought about by existing boundaries of class, race, gender, and, I would add, family (43). Increasingly, the courts are using biological concepts to settle custody disputes involving infants born to gestational mothers, controversies over adoptions, and situations where babies have been switched at birth.[10] Whereas previously the "best interests of the child" theory was used in child custody suits, genetic evidence is now more often favored by the courts.[11]

In the trial court, Judge Parslow referred to Anna Johnson as a "genetic hereditary stranger" to baby Christopher, a descriptive category that has since been used in at least one other custody case (Nelkin 2121). According to Parslow,

"Who we are and what we are and identity problems particularly with young children and teenagers are extremely important. We know that there is a combination of genetic factors. We know more and more about traits now, how you walk, talk and everything else, all sorts of things that develop out of your genes. . . . They have even upped the intelligence ratio of genetics up to 70 percent now." (qtd. in Dolgin 685)

Similarly, the appellate court argued that:

There is not a single organic system of the human body not influenced by an individual's underlying genetic makeup. Genes determine the way physiological components of the human body, such as the heart, liver, or blood vessels operate. Also . . . it is now thought that genes influence tastes preferences, personality styles, manners of speech and mannerisms. (Anna 380)

Issues related to the significance of biological predisposition that are being contested within scientific communities are being presented in the courts as if they were accepted fact. But, as Marilyn Strathern notes, the "simple idea that one person [passes] on a characteristic to another, like a piece of property" has been changed by a "sense of the complex way in which elements combine" as our "primitive knowledge of the inheritance of characteristics is being displaced by knowledge about genetic mapping" and other scientific manipulations ("Displacing" 356; see also *Reproducing*).[12] Relying on genetics as the only basis for determining parental status rather than as one component in a larger social, cultural, and legal context is problematic because, despite the appeal of using scientific evidence to resolve complex legal, cultural, and social issues related to reproduction and family, questions remain about the facticity of this evidence. Scientific communities and the public are still debating whether genetic knowledge constitutes knowledge at all.

Shifting from the biological reasoning of the two lower courts, the California Supreme Court argued that the Uniform Parentage Act did not indicate a preference for blood test results over giving birth as evidence of natural parenthood. For this court, in instances where genetic consanguinity and childbirth do not coincide in the body of one woman, the woman who intended to procreate the child and to raise it as her own is the natural mother.[13] In the California Supreme Court hearing of the case, it would seem, a distinction was made between the "ruling 'head' and the laboring 'body'" (Doyle 21). In this court's estimate, the Calverts' decision to have a child takes precedence over the work of Johnson's laboring body, since the intended parents' initial decision to have a child was the reason that the child was brought into being. The California Supreme Court believed that Johnson's entry into a surrogate agreement was not equivalent to exercising her right to make procreative choices. Instead, according to this court, Anna Johnson was agreeing to provide a service to Mark Calvert and Crispina Calvert, the intended parents, and she should have had no expectation that she would be able to raise as her own the child she carried.[14]

In a strongly worded dissent, Justice Kennard, the lone woman on the bench, disagreed with the majority that the woman who intends to have a child and contributes the ovum should automatically be considered its natural mother, and she found fault with the majority's reliance on the rule of intent to resolve *Johnson v. Calvert*. In her dissenting opinion, Kennard states that in its justification for its intent test, the majority equated children or the right to children with intellectual property.[15] For Justice Kennard, both the genetic and gestational mothers have substantial claims to legal motherhood. And yet, as she points out, California law bestows the "rights and responsibilities of parenthood to only one 'natural mother'" with no provision for what to do when a situation indicates that a child has more than one (Johnson 507).

The "originator of the concept," or rule of intent, argument, a position Justice Kennard describes as comfortingly familiar to the courts in instances where they are called to justify the law's protection of intellectual property, is, in her view, wrong for determining parenthood and parental rights because it suggests that children and the right to children can be viewed as property comparable to a book, a software program, or any other invention. In addition, she argues, using the rule of intent to "break the tie" between the genetic and gestational mother of the child also implies that "property transactions governed by contracts . . . ought presumptively to be enforced and, when one party seeks to escape performance, the court may order specific performance" (Johnson 514).[16]

In addition to the objections Justice Kennard outlines in her dissenting arguments, the rule of intent raises several other issues that need to be considered. What about Johnson's parental intentions as gestational mother? One could argue that, like the Calverts, Johnson also intended to procreate, demonstrating this intention when she allowed herself to be implanted with the Calverts' zygote, carried it to full development at some risk to herself, and then changed her mind about relinquishing the baby once she had delivered it.[17] The courts, as I have mentioned, believed that Johnson should have had no expectation that she would be able to keep and raise the child. After all that Johnson had invested in the pregnancy, the courts decided somehow that her desire to be a mother to the child was "unnatural." And, while the Calverts obviously wanted Christopher and fought long and hard to keep him, there have been instances when genetic parents have reneged on contractual agreements with gestational mothers and refused to take the intended child once it had been delivered.[18] Finally, the rule of intent, as legal scholar Anita Allen persuasively argues, is often inconsistently applied because it assumes an equality among individuals that does not yet exist. Gestational mothers who renege on their contracts—poor mothers, lesbian mothers, black mothers, and other mothers thought to be functioning outside middle-class, father-centered families—often find themselves without support or legal recourse in child custody disputes. Courts consistently rule against these groups of women, favoring instead configurations of the family that fit the nuclear family model of white, middle- to upper-middle-class heterosexual couples.[19]

At least for now, fetuses do develop inside and pass through the bodies of women.[20] While the fetus that the gestational surrogate carries is not genetically related to her, it is also not wholly other to her. Instead, the bodies of both are interconnected in complex and contradictory ways that the court rulings rendered in *Johnson v. Calvert* do not begin to address.[21] For the most part, the courts have only superficially begun to speak to the issues that reproductive technologies raise for how families are constructed and defined in American culture.

The rulings of courts deal not only with the material world but also with the socialization of citizens and the development and maintenance of traditions. In their decisions, the courts have tended to "promote traditional views about marriage, procreation, and family relationships" that may dissuade individuals from entering into situations where traditional views will not be upheld (Karst 628). Surrogacy and other assisted reproductive practices call into question much of what we, as a society, have come to believe about personal identity, intimate relationships, and the beginnings of life. The larger problem for the legal system in *Johnson v. Calvert* and similar cases, then, is how to maintain traditional two-parent, heterosexual families in the face of the ways ART is changing this privileged family. Arguments about genetic relation or rules of intent in custody battles serve essentially as a means to contain the proliferation of meanings that medical technology makes and its ways of constantly altering knowledge about intimate relations.

II.

"'our blackest nightmare'"
—Mark Calvert, as quoted in a weekly tabloid[22]

Race served as both a pretext and a subtext in the debate surrounding *Johnson v. Calvert*. The court decisions, the media coverage, and the public's response to the case were all predictably informed by race, despite arguments by parties involved that race "played no discernible role." Following the laws of racial designation and naming set in place in American culture long before such things as surrogacy were possible, the fact that having a black ancestor, let alone a black mother, makes one black is reason enough to assume that race informed the courts' and the public's perceptions in the outcome of the case. To find that Johnson could be a legal and natural mother to Christopher would have meant that the court would have had to make Christopher black. Surrogacy, like race, forms yet another example of the ongoing crisis of representation where "legal definitions contradict physical signs and social codes," a crisis that is only heightened by the two (Saks 40).

Ironically, in the discourse surrounding the case, Johnson is the only person who is described as having a problem with race. Attorneys for the Calverts painted Johnson as so captivated by whiteness that she wanted to have a white baby. Johnson, of course, does have a problem with race but it is not a problem of fetishizing whiteness. Instead, Johnson's issue with race has to do with the fact that as a black woman she is defined by and is thought to embody race. But what does race mean here? Hartouni describes Johnson as "enter[ing] the public discourse an already densely scripted figure whose deviance, whatever its particular form, was etched in flesh." Indeed, the portrayal of Johnson in the courts and in the media as a fraudulent welfare mother, con artist, and extortionist play on beliefs long held by the public that black women are "less fit mothers, less caring mothers, and less hurt by separation from their children" than non-black women.[23]

In representing Johnson as a welfare cheat, the media and the Calverts' attorneys employed a form of shorthand not only for her blackness but also for the kind of person she was, and particularly for the kind of mother she would be. The unsubstantiated charges that Johnson had defrauded the government by receiving welfare payments she was not entitled to made it easier for some to make the point that she was untrustworthy, dishonest, and, therefore, an unfit parent. By identifying Johnson as a welfare recipient, a point that was repeatedly mentioned in press coverage throughout the trial, no one had to make explicit the racial grounds for objecting to Johnson's fitness as a mother, in any sense, to Christopher. The welfare mother, as Wahneema Lubiano notes, "can be seen as exemplifying the pathology of the category 'black woman'" (324). The representation of black woman and welfare mother as the same, and as equally pathological, operates in *Johnson v. Calvert* as a narrative means to shape public opinion regarding the intersection of race, motherhood, and surrogacy. Race, in *Johnson v. Calvert*, signifies not only blackness but blackness as difference and deviance.

The strongest images of black women circulating in American culture center on black women's motherhood. Figures such as the mammy, the matriarch, and the welfare mother or welfare queen, all of which have their roots in nineteenth-century cultural discourse, continue to dominate current discussions of black women's motherhood (see Ikemoto; Collins, *Black;* Omolade; White). As a legal and economic construct, slavery conferred a "breeder" status on black women and their reproductive capacities.[24] By reversing English law, which determined an individual's legal status through the father, a peculiar system of racial specification and naming was designed under slavery that forced children to follow the condition of their mothers. Establishing the child's legal status through the mother allowed slave owners to classify their biracial offspring as blacks and as slaves. In this way, slave owners ensured that their slave labor force would be increased and that there would be no legal consequences for them regarding the biracial children that they fathered.

Promoting a breeder status for black women also served to sever their biological motherhood from their social and cultural functions as mothers. Black women were expected to "perform the physical tasks of motherhood" as nannies or wet nurses, for example, but they were not entrusted with the "moral duty" of providing children, their own or anyone else's, with "proper values" (Ikemoto 483). How ironic, then, that Judge Parslow, in the first hearing of the case, referred to Anna Johnson as a "wet nurse." In addition to having demeaning racial undertones, by employing the available language of servitude, the phrase also works to resituate Johnson in her place as laboring black body. In so doing, the history of conflating African-American women's reproductive labor with their labor as workers is recalled.[25]

Current practices and perspectives on black women, fertility, and reproduction parallel earlier events in black women's reproductive history. Advances in reproductive technology, including new methods of birth
control, are frequently used by the state as a means of legal and social
control of black and, often, poor women's reproduction.[26] While all
women are increasingly subject to regulatory incursions with respect to reproductive technologies, black women, other women of color, and poor
women are disproportionately affected by this type of intervention in the
form of hospital and prison detention, forced sterilization (both temporary
and permanent), and court-ordered medical procedures such as cesarean
sections.[27] Gestational surrogacy invites the singling out of black women
for exploitation not only because a disproportionate number of black
women are poor and thus more likely to turn to leasing their wombs as a
means of income but also because it is incorrectly assumed that black wom-
en's skin color can be read as a sign of their lack of visual relation to the
children they would bear for the white couples who seek to hire them.[28]
Black women have long been asked to raise white children without having
any parental rights to them. Now, it would seem, they can be asked to birth
white children and have no claim to them.[29] This is to say that from the
point of view of white reproductive contract law, black women's surrogacy
is the most alienated of labor. In contemporary national discourse on family and family values, and with its focus on welfare queens and their "over-
reproduction," black female maternal labor is granted no value except, of
course, that given to it contractually in surrogate arrangements. In situations where black women's maternity occurs for themselves and not for the
benefit of others, it is deemed socially harmful.[30]

III.

"It looks just like us."
—Crispina Calvert[31]

The issues contested in *Johnson v. Calvert* highlight the increasingly public struggle over assisted reproduction and its effects on the family. Both
the public and the courts continue to grapple with whether surrogacy
should be legal, who should be able to gain from the process, and what its
ramifications are for the construction of the family. Other questions that
the case raises include determining what is fair and right for children in
considerations of child custody disputes. What are the ultimate consequences for a culture that views its children as property—as "things" that
people can barter, sell, or have "rights" to? What does it mean for the
court to decide what is in the "best interests of the child," especially
when the child's interests appear to serve as a cover for the ideological and
political interests of individuals and institutions seeking to model specific

behaviors and relationships? In California, the National Conference of Commissioners on Uniform State Laws did attempt to respond to these questions by proposing the Uniform Status of Children of Assisted Conception Act, a piece of legislation that addressed most of the issues involved in *Johnson v. Calvert.* The legislation was never enacted. The failure to do so, I suspect, has a lot to do with the reluctance of the legislature to go on record as taking a position on an issue that is considered to be such a complex moral problem. As bioethicist James Nelson points out, politicians "'run the other way'" when legislation on surrogacy comes before them because, on the one hand, to make surrogate contracts unenforceable would be to limit the options of infertile couples attempting to have children, and, on the other hand, to support surrogacy would mean that they might be seen as facilitating the "'exploitation of women [and the cheapening] of the family'" (qtd. in Kasindorf 13).[32]

What was once described as the biologically rooted, racially closed, heterosexual, middle-class family has been disrupted by the new knowledge that ART has made available. With assisted conception, as Strathern notes, there increasingly "exists a field of procreation whose relationship to one another and to the product of conception is contained in the act of conception itself and not in the family as such" ("Displacing" 352). In a discussion of the manner in which reproductive technologies are "displacing knowledge" about familial kinship, Strathern argues that "making visible the detachment of the procreative act from the way the family produces the child adds new possibilities to the conceptualization of intimacy in relationships" ("Displacing" 353). In so doing, these technologies displace our sense of what we have come to know about health, life, and death. Still, the legal system has been slow to address how this expansion in knowledge, and the resulting proliferation of meanings, put people in the position of having to make novel choices—to make different kinds of decisions based on this transformed information (see "Displacing" 347).

The belief the courts held in *Johnson v. Calvert* that a child may have only one mother is inconsistent both with the new facts of life that technology has made possible and with some of the courts' own models for reconfiguring the family in light of this technology. Models of family that are different from the nuclear family model were already available for the courts to choose from. These existing models would have allowed the courts to acknowledge the parental rights of the Calverts and Johnson without diminishing the role of either. Courts have acknowledged the division of procreative mothering from social mothering in decisions on adoption and stepparenting, for example. In both instances maternal status is extended to at least one other woman. In addition, in cases of egg donorship the courts have held that the woman who gestates and gives birth to a child formed from the egg of another woman and who intends to raise that child as her own is considered the natural mother of that child.

Various cultural models of mothering and communal parenting were also available for the courts in developing their response to the case. Patricia Hill Collins presents one such model in her account of the more inclusive, more collaborative parenting effort that takes place within African-American communities. For Collins, the concept "motherwork" delineates a category that "soften[s] the dichotomies in feminist theorizing about motherhood that posit rigid distinctions between private and public, family and work, the individual and the collective" ("Shifting" 59). This notion of motherwork draws upon traditions in African-American communities where multiple models of mothering relationships exist. While European-American models of mothers and mothering are often limited to blood relationships, within many African-American communities mothering is conceptualized as a form of cultural work that incorporates the mothering relationships of nonblood relations as well (James 44; see also Collins, "Shifting," *Black;* Polatnick). In addition to bloodmothers, mothering roles such as "othermother" or "community othermother" may be assumed by those who, in addition to blood relatives, take on the responsibilities of kin in black communities (James 44, 47; see also Stack). In this configuration of family, Johnson's claim to motherhood would not have been viewed as unreasonable or unnatural. This particular model of motherhood allows for both Anna Johnson and Crispina Calvert to be viewed as mothers to Christopher without diminishing the role of either.

In *Johnson v. Calvert,* the California Supreme Court did acknowledge that there was undisputed evidence that both Crispina Calvert and Anna Johnson could be mothers to baby Christopher. But the court then went to great lengths to describe why both women could not be considered mothers to the child and why Crispina Calvert, rather than Anna Johnson, must be considered the natural mother. In the communal, capacious model of motherhood that Collins describes, the weight of biological or genetic ties and their significance for defining familial relationships is shifted. Within African-American communities, "those who tend, care for, [or] carry [children] are by definition those with authentic claims to be named owner of the things or people whose growth they nurture" (Petchesky 398).

Because black women's gender has never included privacy, they are always forced to act as if the distinction between public and private is irrelevant in their day-to-day lives.[33] As Collins's example demonstrates, black women's maternity and kinship have already found multiple definitions. Instead of making biology or blood ties the definitive form of motherhood (or of fatherhood, for that matter), biology or blood ties represent simply one way of establishing familial relationships and bonds. Heredity is not given privileged status in this configuration of family. However, the alternative models of maternity and fictive kin that black women have constructed have been pathologized. Instead of being viewed as a useful example of the ways extended family can work, communal parenting in black communities is viewed by the national public as aberrant behavior, charged with being a kind of neglect, or with enabling and promoting family forms that are not father-centered.

IV.

The emergence of assisted reproductive technology that both is "conflated with" and "displaces . . . nature" has disrupted naturalizing assumptions made about the categories of "mother," of "family," and of "nature" itself (Franklin 334). As new conception narratives have arisen, the previously protected realm of categories such as these is, through technology, suddenly made visible and available for (re)interpretation and (re)inscription. The Calverts along with other parties involved in the case—the judges, the lawyers, the press, and the public—presuppose that the Calverts should get to keep baby Christopher because he is "like" them and that their desire for "likeness" in their child is a "natural" desire (see P. J. Williams 226). "Likeness" for the courts, for the Calverts, and, no doubt, for other contracting couples serves as a visual metaphor for kinship and the right to ownership of children.

Crispina Calvert's comment that baby Christopher looks just like her and her husband, a comment that was repeated like a mantra in the press, is both a statement about belonging and a statement about exclusion. It is about belonging because it represents a claim by Crispina Calvert that Christopher, because of how he looks, is a part of both her and her husband. Christopher's "likeness" to the Calverts is believed to demonstrate his "blood," his genetic and racial heritage, and therefore to reflect his link to the Calverts. Crispina Calvert's comment is also about exclusion because Christopher's likeness serves not only as a (meta)physical and conceptual link indicating rights to his parentage but also because "likeness" operates as a sign for blood—for the closed racialized membership of family and race (see B. F. Williams; Haraway 213–66).

In addition to skin color, which has not always been a reliable sign of racial demarcation, a practice of using blood as a pseudoscientific explanation for race has existed for centuries in the United States (see Jordan). Political and social movements and now medical technologies have complicated and redefined theories of blood and its value for determining identity. The continuing legacy of miscegenation laws which used, among other things, the trope of blood, specifically the "one drop rule," to maintain distinctions and separations among groups of people places a high value on white skin—white "blood"—because those who can have it are strictly limited and monitored. As is also evident in the history of designating blood as racial and therefore a familial marker, the boundaries of these rules shift and can be contradictory based on the needs and desires of the ruling class.

To say that Johnson could be a mother to the baby she bore would be to indicate a willingness on the part of the courts and the public to relinquish or, at minimum, to blur racial-familial boundaries. As Laura Doyle argues, in a "race-bounded economy the mother is a marker of boundaries, a generator of liminality; in giving birth, mothers reproduce both children and, through the lives of their children, the life of the racial divide" (27). The notion of reproducing children "like" oneself, then, reproduces the specific rights and privileges of particular cultural groups. These rights and privileges are connected to systems of value. White mothers and white children are considered valuable in a marketplace where white skin is valued; for black mothers and black children, the converse is true. Much of the legal and popular discussion of *Johnson v. Calvert* draws on the historical devaluation of black women as mothers. Black women are frequently blamed for the effects of poverty on their children. They also serve as scapegoats in public policy for legal decisions related to issues of family, custody, and reproduction. The image of black women as drug-using, child-abusing welfare recipients who live to breed at taxpayer expense is illustrative of this phenomenon.

In *Johnson v. Calvert,* we find a shift in the definitions and valuing of maternity, bodily integrity, and family. The courts are willing to reaffirm the primacy of a closed, privatized, and homogenous family and all of its attendant qualities even if this means that they make inconsistent and contradictory decisions. Like it or not, reproductive technologies have destabilized this notion of family. The fact that what constitutes a family is now variable poses a problem for the efforts of the courts to limit and hierarchically arrange bounded, private families. Even in their contortions to maintain this closed version of family, the courts have in their rulings helped to open the door to different forms of family.

The tension in *Johnson v. Calvert* between what constitutes a family versus what constitutes a mother is linked by questions of race. Although innovative reproductive and other medical and scientific technologies make this link more tenuous, race is the one remnant from the past that remains animated. We are left with a highly entrenched racialized image of the family. As a culture we continue to trip over the notion of reproduction as a racial act, and this is demonstrated by the dogged reliance—in the courts and popular media—on pseudonatural categories of the body. People rely on these categories to maintain intelligibility in the face of dramatic social and cultural change. In particular, persons in the law and the medical sciences must rename and reconfigure the so-called nuclear family in order to shore up the face of reproductive technology. In the rhetoric they adopt, the bodies of women who are poor, who are of color, or who reproduce outside of this family form are marked as "degenerate" or merely "not proper." Their devalued social positions are thus solidified, while everything else around them changes. New reproductive technologies have simply made it possible to develop a new vocabulary that can continue to utilize these familiar representations.

NOTES

I wish to thank Lauren Berlant, Charles E. Moore, and Susan M. Squier for their incisive comments on earlier drafts of this essay. I would also like to thank my colleagues at the Georgia Institute of Technology, who provided valuable suggestions. A shorter version of this essay was presented at the Biotechnology, Culture, and the Body conference held at the Center for Twentieth Century Studies, University of Wisconsin-Milwaukee, 24–26 April 1997.

1. The Calverts named the baby Christopher. Johnson had given him the name Matthew.

2. For various accounts of the events in the case as they were reported in the media, see Sachs; Tifft; Armstrong; Chu, Matsumoto, and Benet; and Kasindorf.

3. See also part 1 of a four-part series on infertility published in the *New York Times* (Gabriel). This article on high-tech pregnancies and the fertility market focuses on clinics and hospitals with specialties in in vitro fertilization (IVF). This branch of medicine is reported to be a sector of a "virtually free-market branch of medicine" that is a "$350 million-a-year business" (A1). The article describes mostly affluent couples paying $25,000 or more for procedures, usually IVF, to assist them in conception. Few insurance companies cover IVF, making most of the financial burden fall on the couples. Prices for the procedure described in the article include a $2,000–$3,000 fee for egg donors for women who are unable to produce eggs and to a median cost of $7,800 for one procedure of IVF that lasts about the length of a menstrual cycle. Since most couples are not successful on the first try, many couples end up trying three to four more times before giving up.

4. Crispina Calvert's "honorary white status," no doubt, can be attributed to the racist stereotype of Asian Americans as members of a "model minority."

5. Qtd. in Kasindorf 31.

6. George Annas notes, for instance, that the women who bear the children in surrogate arrangements are often "lower-middle-class and lower-class" women (27). Statistics from the Congressional Office of Technology Assessment survey, *Infertility: Medical and Social Choices,* demonstrate that surrogate mothers tend to be less educated and less financially secure than those who hire them. Only a small percentage of the women waiting to be hired as surrogates have attended college and a large percentage of these women earn less than $30,000 annually (United States).

7. Emily Martin makes clear in her analysis how frequently women see themselves as separate from their bodies. Among the women she interviewed, Martin describes a "fair amount of fragmentation and alienation in women's general conceptions of body and self" of which they "do not seem aware" (89).

8. The California version of the Uniform Parentage Act was introduced in 1975. The purpose of the act was to eliminate legal distinctions between legitimate and illegitimate children. It came about as a result of rulings by the United States Supreme Court that mandated equality between legitimate and illegitimate children. The Uniform Parentage Act "bases parent and child rights on the existence of a parent and child relationship rather than on the marital status of the parents" (Johnson 497). Although the act predates the situations that assisted reproductive

technology (ART) brings about, Johnson cited the act in an attempt to establish her parental rights.

9. Legal scholar Randy Frances Kandel argues that in the first two rulings on *Johnson v. Calvert* by the trial court and the court of appeals, the judges "put the cart before the horse." She asserts that in both decisions the courts attempted to resolve the prior issue of whether Crispina Calvert or Anna Johnson or both women could be the natural mothers by first settling the second issue of whether it is in the best interests of the child for both mothers (assuming the child had two mothers) to have custody rights (174).

10. In "After Daubert," Nelkin questions the reliability of court testimony that utilizes genetic evidence. She argues that testimony in the area of genetics should be more closely examined because of its growing appeal in court cases and its potential impact on legal decision making.

11. In fact, as Kandel, Nelkin, and Rochelle Cooper Dreyfus point out, the courts could have used the "best interests of the child" theory to settle the custody dispute in *Johnson v. Calvert*. This ruling would have been in line with the courts' attempts to maintain the traditional nuclear family that Crispina Calvert and Mark Calvert seemed to be able to provide. The courts could have also determined that Johnson had waived all parental rights to the child when she signed the surrogate contract. The question, as Randy Kandel so aptly puts it, is, "Why, then, did the courts feel compelled to resolve the 'natural' parent issue using the [Uniform Parentage] act?" (178). See also Dreyfus and Nelkin.

12. In tension with the emphasis on genes as indicators of parenthood is the language common in many surrogate contracts that stipulates that gestational mothers must refrain from smoking, drinking, or engaging in any other activities that might endanger the fetus. Gestational mothers are also frequently required to agree to follow all doctors orders, including those that force them to submit to invasive procedures or to curtail their normal physical activities. All of these strictures seem to suggest an awareness of how the environment of the birth mother's body is interconnected with the fetus.

13. For a more detailed analysis of the rule of intent, see Shultz. For a discussion of the rule of intent as it pertains specifically to *Johnson v. Calvert*, see Johnson.

14. On this issue, Elizabeth V. Spelman's discussion of people of color and white women as mere body comes to mind. According to Spelman, those individuals defined by these categories are typically closely associated with the body and basic bodily functions—"sex, reproduction, appetite, secretions, and excretions"— and as "given over to attending the bodily functions of others (feeding, washing, cleaning, doing the 'dirty work')" (127). In Johnson's case, the work of her body includes the "dirty work" of gestation and birthing.

For discussions of reproductive freedoms, specifically as they refer to the right to procreate, see Gostin; Neff; and Rutherford.

15. For an excellent analysis of the representation of procreation as analogous to authorship, see Rose.

16. Legal scholar Anita Allen suggests that one way to respond to this issue is to view surrogate arrangements as unenforceable personal commitments ("Privacy").

17. Johnson had had a history of problem pregnancies—two miscarriages and two stillbirths before and after her daughter was born—a fact she failed to reveal to the Calverts when she entered into the agreement with them. Furthermore, some have argued that because the fetus and the gestational mother are unrelated, the gestational mother is at higher risk for severe complications during pregnancy such as ectopic pregnancy, preeclampsia, and diabetes. See Kandel 189.

18. In one situation, a gestational mother gave birth to twins, a boy and a girl.

The family, however, was only interested in the girl and left the boy behind. The gestational mother sued and won the right to retain custody of both children. She later ended up on drugs and lost both children. The children were then placed in foster care. In another example, a child was born HIV positive. Upon learning of the child's HIV status, the contracting parents refused to accept the baby. It is clear that in this situation a thorough preconception medical and psychological screening did not take place. See Charo; Kasindorf.

19. In one recent example, through the use of artificial insemination a lesbian couple became the parents of two children. After the dissolution of their relationship and, ultimately, a custody battle for the children, a court ruled that both women should be denied parental and visitation rights. See "Lesbians Denied"; Boxall. In another instance, Mary Frank Ward, a mother of three, went to court to attempt to get additional child support from her ex-husband for their youngest child. The ex-husband, a convicted felon who had battered and eventually murdered his first wife, sued for custody of the child and won. The court argued that because Mary Ward and her oldest daughter were both lesbians and had live-in lovers, her home was a bad influence for her youngest child. A Florida court of appeals upheld the decision by the lower court. Ward, who ultimately gave up her fight for custody of her youngest child, recently died of a heart attack. See Navarro; Scheer; "Lesbian Who."

20. While scientists are currently able to construct artificial wombs, they have only managed it for animals, not humans. Scientists in Japan have developed a technique called extrauterine fetal incubation (EUFI). Using goat fetuses, the scientists have "threaded catheters through the large vessels in the umbilical cord and supplied the fetuses with oxygenated blood while suspending them in incubators that contain artificial amniotic fluid heated to body temperature" (Klass 117). The goat fetuses were able to survive in this environment for three weeks although team physicians had difficulty with circulatory failure in the experiments as well as experiencing other technological problems. While scientists are quoted as saying that the "ideal situation for the immature fetus is growth within the normal environment of the maternal organisms" (117), they continue to pursue the technology for constructing artificial wombs for humans. Arthur Caplan, director of the Center for Bioethics at the University of Pennsylvania, predicts that "Sixty years down the line . . . the total artificial womb will be here," arguing that this procedure is "technologically inevitable" (119).

21. In the context of a discussion about surrogacy, to make the link between gender and the body, as Carol Bigwood suggests, need not lead to the determination of the category "woman" or "mother" as a fixed or closed biological identity. Instead, the female body in this instance is a body that is "open, sensate, procreative"—a body not forced into a pseudomale body(lessness) (Petchesky; Bigwood).

22. Mark Kasindorf notes in his article that "a tabloid weekly outraged Mark Calvert by quoting him as calling Johnson 'our blackest nightmare'" (31).

23. Even the so-called ideal black mother figure, the mammy, a selfless nurturer of white children (under the supervision of the white mistress), has been portrayed as "careless and unable to care properly for her own children" (Roberts 313).

24. Angela Davis notes that since slave women were classified as "breeders" rather than mothers, "their infant children could be sold away from them like calves or cows" (7).

25. Anita Allen provides a careful analysis of why slavery and surrogacy are not the same thing ("Surrogacy").

26. Norplant, for instance, has been used as a criminal penalty against women of childbearing age who have been convicted of child or drug abuse. In addition, legislation has been proposed in Louisiana and Kentucky, to name just two states,

that would offer financial incentives to women on welfare who "voluntarily" agree to use Norplant. Ironically, while access to publicly funded abortions is limited in most states, all but two states, California and Massachusetts, fund Norplant through Medicaid (see, for instance, Krauss; Field; Nelson, Buggy, and Weil; Kolder et al.; Rhoden).

27. Attorney Deborah Krauss notes that in instances where the pregnant woman is a "member of a racial minority or disadvantaged economic group," physicians are more likely to obtain court-ordered obstetrical interventions (531). Eighty percent of the patients who were forced to undergo court-ordered cesarean sections were members of minority groups. Teaching hospitals play a critical role in this situation. Every documented request for a court-ordered intervention involved women who were patients at teaching hospitals or who received public assistance (531).

28. Critics of the Calverts have argued that this is the reason why they chose Johnson to be their surrogate. To be fair, reports on the events of how Johnson and the Calverts came together have Johnson approaching the Calverts and offering to be their surrogate. Reading skin color as a sign of genetic claim, we should all know, is not indicative of anything. The activity of passing should have taught us this. Passing is successful because so many people continue to rely on skin color as a visual sign of race. In many instances, when attempting to make a visual identification of blackness using this scheme, people would be wrong in their assumptions.

29. The now (in)famous *Baby M.* case demonstrated to the mostly white couples who seek gestational surrogates that if they choose healthy white women who are unrelated to them to be their surrogate, they run the risk that the surrogate contract will be voided. Unlike Johnson, in the *Baby M.* case, Mary Beth Whitehead was the genetic and gestational mother of Baby M. Undergoing what is now referred to as "traditional surrogacy," Whitehead was artificially inseminated with Stern's sperm. Whitehead was denied custody, but she was granted visitation rights based on her genetic tie with the child. Still, the possibility remains that a surrogate mother under these circumstances would be granted full custody.

30. Ironically, while black women have an infertility rate that is one and a half times higher than white women, they are the least likely to benefit from ART or other infertility "treatments." Cost tends to be a prohibitive factor in many instances. Rather than being supported in their desire to reproduce, black women's attempts at reproduction are most often perceived as dangerous, something that should be controlled. But, as I will demonstrate in the next section, another reason why black women make up a small percentage of the women who utilize ART could be their reliance on alternative models of mothering, in which genetic relation is not considered imperative to establishing kinship ties.

31. Crispina Calvert to Mark Kasindorf (Kasindorf 11).

32. In his article, Kasindorf also reveals that the United States Congress has ignored antisurrogacy legislation placed before it.

33. On this point, see Hortense Spillers, "Mama's Baby, Papa's Maybe: An American Grammar Book." Clearly, this old adage as it is described in Spillers's title has been reversed in *Johnson v. Calvert*, which seems to be a case of "Papa's Baby, Mama's Maybe."

WORKS CITED

Allen, Anita. "The Black Surrogate Mother." *Harvard Blackletter Journal* 8 (1991): 17–31.

————. "Privacy, Surrogacy, and the Baby M Case." *The Georgetown Law Journal* 76 (1988): 1759–92.

————. "Surrogacy, Slavery, and the Ownership of Life." *Harvard Journal of Law and Public Policy* 13 (winter 1990): 139–49.

Anna J. v. Mark C. et al. 286 CA Rptr. CA Ct. App. 4 Dist. 1991.

Annas, George. "Fairy Tales Surrogate Mothers Tell." *Law, Medicine, and Health Care* 16 (spring 1988): 27–33.

Armstrong, Scott. "California Surrogacy Case Raises New Questions about Parenthood: Mother Seeks Custody, but Has No Genetic Link to the Child." *Christian Science Monitor* 25 (Sept. 1990): 1+.

Bigwood, Carol. "Renaturalizing the Body (with the Help of Merleau-Ponty)." *Hypatia* 6.3 (1991): 54–73.

Bordo, Susan. *Unbearable Weight: Feminism, Western Culture, and the Body.* Berkley: University of California Press, 1993.

Boxall, Bettina. "Laws Mean Lesbian Custody Battles Often Are One-Sided; Under Rigid Definition of Parenthood, Partner Who Didn't Bear Child Usually Has Little Recourse." *Los Angeles Times* (27 Jan. 1997): A1+.

Cahill, Lisa Sowle. "The Ethics of Surrogate Motherhood: Biology, Freedom, and Moral Obligation." *Law, Medicine, and Health Care* 16 (spring 1988): 65–71.

"California Judge Speaks on Issue of Surrogacy." *The National Law Journal* 13 (Nov. 1990): 36.

Charo, R. Alta. "Legislative Approaches to Surrogate Motherhood." *Law, Medicine, and Health Care* 16 (spring 1988): 96–112.

Chu, Dan, Nancy Matsumoto, and Lorenzo Benet. "A Judge Ends a Wrenching Surrogacy Dispute, Ruling That Three Parents for One Baby Is One Too Many." *People Weekly* (5 Nov. 1990): 143–44.

Collins, Patricia Hill. *Black Feminist Thought: Knowledge, Consciousness, and the Politics of Empowerment.* Boston: Unwin, 1990.

————. "Shifting the Center: Race, Class, and Feminist Theorizing about Motherhood." *Representations of Motherhood.* Ed. Donna Bassin, Margaret Honey, and Meryle Mahrer Kaplan. New Haven: Yale University Press, 1994. 56–74.

Davis, Angela. *Women, Race, and Class.* New York: Vintage, 1983.

Dolgin, Janet L. "Just a Gene: Judicial Assumptions about Parenthood." *UCLA Law Review* 40 (1993): 637–94.

Doyle, Laura. *Bordering on the Body: The Racial Matrix of Modern Fiction and Culture.* New York: Oxford University Press, 1994.

Dreyfus, Rochelle Cooper, and Dorothy Nelkin. "The Jurisprudence of Genetics." *Vanderbilt Law Review* 45 (March 1992): 313–48.

Duden, Barbara. *Disembodying Women: Perspectives on Pregnancy and the Unborn.* Cambridge: Harvard University Press, 1993.

Field, Martha A. "Controlling the Woman to Protect the Fetus." *Law, Medicine, and Health Care* 17 (summer 1989): 115–29.

Franklin, Sarah. "Postmodern Procreation: A Cultural Account of Assisted Reproduction." Ginsburg and Rapp 323–45.

Gabriel, Trip. "High-Tech Pregnancies Test Hope's Limit." *New York Times* (7 Jan. 1996), local ed.: A1+.

Ginsburg, Faye D., and Rayna Rapp, eds. *Conceiving the New World Order: The Global Politics of Reproduction.* Berkeley: University of California Press, 1995.

Gostin, Larry. "A Civil Liberties Analysis of Surrogacy Arrangements." *Law, Medicine, and Health Care* 16 (spring 1988): 7–17.

Haraway, Donna J. *Modest_Witness@Second_Millenium.FemaleMan©_Meets_OncoMouse™: Feminism and Technoscience.* New York: Routledge, 1997.

Hartouni, Valerie. "Fetal Exposures: Abortion Politics and the Optics of Allusion." *Camera Obscura: A Journal of Feminism and Film Theory* 29 (1992): 130–49.

Hine, Darlene Clark. "Rape and the Inner Lives of Black Women in the Middle West." *Signs: Journal of Women in Culture and Society* 14 (1989): 912–20.

Ikemoto, Lisa. "The Code of Perfect Pregnancy: At the Intersection of the Ideology of Motherhood, the Practice of Defaulting to Science, and the Interventionist Mindset of the Law." *Critical Race Theory: The Cutting Edge.* Ed. Richard Delgado. Philadelphia: Temple University Press, 1995. 478–97.

James, Stanlie M. "Mothering: A Possible Black Feminist Link to Social Transformation?" *Theorizing Black Feminisms: The Visionary Pragmatism of Black Women.* Ed. Stanlie M. James and Abena P. A. Busia. New York: Routledge, 1993. 44–54.

Johnson v. Calvert. 19 CA Rptr. 2d (CA 1993); cert. Denied, 114 Super. Ct. 206. 1993.

Jordan, Winthrop. *White over Black: American Attitudes toward the Negro, 1550–1812.* Chapel Hill: University of North Carolina Press, 1968.

Kandel, Randy Frances. "Which Came First: The Mother or the Egg? A Kinship Solution to Gestational Surrogacy." *Rutgers Law Review* 47 (fall 1994): 165–239.

Karst, Kenneth L. "The Freedom of Intimate Association." *Yale Law Journal* 89 (1980): 624–92.

Kasindorf, Mark. "And Baby Makes Four: *Johnson v. Calvert* Illustrates Just about Everything That Can Go Wrong in Surrogate Births." *Los Angeles Times Magazine* (20 Jan. 1991): 10+.

Klass, Perri. "The Artificial Womb Is Born." *New York Times Magazine* (29 Sept. 1996): 117–19.

Kolder, Veronika E. B., et al. "Court-Ordered Obstetrical Interventions." *New England Journal of Medicine* 316 (19) (7 May 1987): 1192–96.

Krauss, Deborah. "Regulating Women's Bodies: The Adverse Effect of Fetal Rights Theory on Childbirth Decisions and Women of Color." *Harvard Civil Rights–Civil Liberties Law Review* 26 (1991): 523–48.

"Lesbian Who Sought Custody Dies." *New York Times* (23 Jan. 1997), local ed.: A19.

"Lesbians Denied Custody after Break-Up." *New York Times* (24 March 1991): A22.

Lubiano, Wahneema. "Black Ladies, Welfare Queens, and State Minstrels: Ideological Wars by Narrative Means." *Race-ing Justice, En-gendering Power: Essays on Anita Hill, Clarence Thomas, and the Construction of Social Reality.* Ed. and intro. Toni Morrison. New York: Pantheon, 1992. 323–63.

Martin, Emily. *The Woman in the Body: A Cultural Analysis of Reproduction.* Boston: Beacon, 1992.

Navarro, Mireyz. "Appeals Court Rebuffs Lesbian in Custody Bid: Child Will Stay with Father Who Killed." *New York Times* (31 Aug. 1996), local ed.: A7.

Neff, Cristyne. "Woman, Womb, and Bodily Integrity." *Yale Journal of Law and Feminism* 3 (1991): 327–53.

Nelkin, Dorothy. "After Daubert: The Relevance and Reliability of Genetic Information." *Cardozo Law Review* 15 (1994): 2119–28.

Nelkin, Dorothy, and M. Susan Lindee. *The DNA Mystique: The Gene as Cultural Icon.* New York: Freeman, 1995.

Nelson, Lawrence J., Brian Buggy, and Carol Weil. "Forced Medical Treatment of Pregnant Women: 'Compelling Each to Live as Seems Good to the Rest.'" *The Hastings Law Journal* 37 (May 1986): 703–63.

Newman, Karen. *Fetal Positions: Individualism, Science, Visuality.* Stanford: Stanford University Press, 1996.

Omolade, Barbara. *The Rising Song of African-American Women.* New York: Routledge, 1994.

Petchesky, Rosalind Pollack. "The Body as Property: A Feminist Revision." Ginsburg and Rapp 387–406.

Polatnick, M. Rivka. "Diversity in Women's Liberation Ideology: How a Black and a White Group of the 1960s Viewed Motherhood." *Signs: Journal of Women in Culture and Society* 26 (1996): 679–706.

Polikoff, Nancy D. "This Child Does Have Two Mothers: Redefining Parenthood to Meet the Needs of Children in Lesbian Mother and Other Non-Traditional Families." *Georgia Law Review* 78 (1990): 468–73.

Pollitt, Katha. "When Is a Mother Not a Mother?" *The Nation* (31 Dec. 1990): 825+.

Rhoden, Nancy K. "The Judge in the Delivery Room: The Emergence of Court-Ordered Cesareans." *California Law Review* 74 (1986): 1951–2030.

Roberts, Dorothy E. "The Value of Black Mothers' Work." *Critical Race Feminism: A Reader.* Ed. Adrien Katherine Wing. New York: New York University Press, 1997. 312–16.

Rose, Mark. "Mothers and Authors: *Johnson v. Calvert* and the New Children of Our Imagination." *Critical Inquiry* 22 (summer 1996): 614–33.

Rutherford, Charlotte, Esq. "Reproductive Freedom and African-American Women." *Yale Journal of Law and Feminism* 4 (1992): 255–84.

Sachs, Andrea. "And Baby Makes Four: A New Custody Battle Intensifies the Debate over Surrogacy." *Time* (27 Aug. 1990): 53.

Saks, Eva. "Representing Miscegenation Law." *Raritan* 8 (fall 1988): 39–69.

Scheer, Robert. "Warped View of What's Fit as Family Life." *Los Angeles Times* (10 Dec. 1996): B7.

Shultz, Majorie Maguire. "Reproductive Technology and Intent-Based Parenthood: An Opportunity for Gender Neutrality." *Wisconsin Law Review* 297 (1990): 298–398.

Spelman, Elizabeth V. *Inessential Woman: Problems of Exclusion in Feminist Thought.* Boston: Beacon, 1988.

Spillers, Hortense. "Mama's Baby, Papa's Maybe: An American Grammar Book." *Diacritics* (summer 1987): 65–81.

Stack, Carol B. *All Our Kin: Strategies for Survival in a Black Community.* New York: Harper, 1974.

Strathern, Marilyn. "Displacing Knowledge: Technology and the Consequences for Kinship." Ginsburg and Rapp 346–63.

———. *Reproducing the Future: Essays on Anthropology, Kinship, and the New Reproductive Technologies.* Manchester: Manchester University Press, 1992.

Stumpf, Andrea. "Redefining Mother: A Legal Matrix for New Reproductive Technologies." *The Yale Law Review* 96 (Nov. 1986): 187–208.

Tifft, Susan. "It's All in the (Parental) Genes: A California Court Rules that Bearing a Child Is Not Motherhood." *Time* (5 Nov. 1990): 77.

United States Congressional Office of Technology Assessment. *Infertility: Medical and Social Choices, OTA-BA-358.* Washington, D.C.: GPO, 1988.

White, Deborah Gray. *Aren't I a Woman: Female Slaves in the Plantation South.* New York: Norton, 1985.

Williams, Brackette F. "Classification Systems Revisited: Kinship, Caste, Race, and Nationality as the Flow of Blood and the Spread of Rights." *Naturalizing Power: Essays in Feminist Cultural Analysis.* Ed. Sylvia Yanagisako and Carol Delaney. New York: Routledge, 1995. 201–36.

Williams, Patricia J. *The Alchemy of Race and Rights.* Cambridge: Harvard University Press, 1991.

Chapter 5

Body Boundaries, Fiction of the Female Self

An Ethnographic Perspective on Power, Feminism, and the Reproductive Technologies

GILLIAN M. GOSLINGA-ROY

> Our bodies; ourselves: bodies are maps of power and identity.
> —Donna J. Haraway, *Simians, Cyborgs and Women*

INTRODUCTION

A common theme in the feminist literature on the reproductive technologies has been that their advent has broken apart reproduction into its genetic, biological, and social aspects.[1] Certainly, with the realization of gestational surrogacy in the mid-1980s, the splintering of what had been historically a "unified" and "natural" reproduction within a woman's body appears complete. Gestational surrogates, implanted with the embryos of their couples, fulfill the strictly biological or, more accurately, the physiological aspect of reproduction, while their couples fulfill both the social and genetic aspects. The professional language of assisted reproduction upholds these divisions: surrogates are referred to as "carriers" or "womb donors," pointing to their instrumentality in reproduction, while "intended" or "recipient" couples are the genetic (that is, "real") parents.

In this essay, I complicate this narrative. While the splintering of reproduction into its social, genetic, and physiological aspects may be analytically true of the reproductive technologies, my fieldwork of a gestational surrogacy arrangement suggests that at the level of embodied practice these separations are not ontologically stable; rather, they require ongoing discursive administrations to remain separate. At an obvious level, this is because genetic and biological aspects of reproduction are also social categories, and all social categories are ultimately embodied processes, deeply implicated in power and history. But less apparent is the way in which the physiological, the genetic, and the social function as naturalized abstractions when discursively deployed. As abstractions, they do not adequately account for the complexities of real-time biographical experience. And yet they often pose as realist descriptions of social processes, even while "freezing time out of the picture" (Braidotti). My analysis of a gestational surrogacy arrangement speaks to this slippage between real-time embodied practice, what I call biographical embodiment, and the often panopticized popular and scholarly discourses on the reproductive technologies, what I call, in contrast, biological embodiment, for reasons that will become clear in this essay. I also deliberately eschew gender as a sole analytical framework, for the issues the reproductive technologies raise far surpass the gender/sexual ghetto into which they are usually thrust.

The research that grounds this essay took place between March 1995 and August 1996 as part of a project in visual ethnography.[2] During this period, I intimately followed on videotape a couple and the gestational surrogate with whom they contracted to carry their genetic child. The arrangement was open, which means the surrogate and the couple were known to each other,[3] and brokered through the contracting couple's fertility clinic of ten years. My methodology included participant observation in all the environments the women and their families frequented (their homes, the homes of relatives and friends, the fertility clinic, surrogacy center, obstetricians' offices, and the hospital at the birth), as well as many informal and formal interviews with the participants and the professionals taking part in the arrangement (the fertility doctor, obstetricians, and surrogacy center psychiatrists and therapists). My approach was inspired by the life history method (Langness and Frank; Frank, "Embodiment"; The Personal Narratives Group) and integrated a phenomenological approach with a feminist commitment to collaboration with one's informants (see Mbilinyi). Interviews and significant events (for example, doctor's visits, the baby shower, and the birth and the good-byes the following day) were videotaped in a cinema verité style and edited into an ethnographic film, *The Child the Stork Brought Home.*

WHY GENDER IS NOT ENOUGH

Let me illustrate this important point with an explosive discussion I had with Julie Thayer (all names are pseudonyms), the gestational surrogate I followed. This conversation took place after one of her monthly surrogate support group meetings, roughly halfway through the pregnancy. As was our habit, Julie called me after the meeting.

"I know you are going to think I'm racist," she blurted out, "but I've just got to tell you this, at the risk of shocking you badly. There was this woman at the meeting today—she's Hispanic and she's pregnant with the embryos of a *black* couple! I couldn't believe it! You're always asking me how I feel about my body and carrying Pamela's baby—I'm okay with that, but a black baby! I couldn't do that."

"Why not?" I asked.

"Come on, Gillian, can you picture little, blonde, blue-eyed me giving birth to a black child?"

Julie was right in thinking I would be shocked. She knew that my partner was a man of color; she had met him during a shoot for which he was the cinematographer. We had several intense (and tense) telephone conversations. Later, Julie explained that she did not think of herself as racist, but the idea of carrying a black child had not once occurred to her until that point:

"It feels foreign to me, different," she elaborated. I could carry a Japanese baby or a Chinese baby because they are white to me. Society sees them as white. But a black child is more difficult. I'm already surrounded by controversy: I married a man thirty-two years older than me; I work in a late-term, problem-pregnancy abortion facility; and I'm a surrogate. To give birth to a black child would add one more controversial aspect to my life and I'm not ready to be on the front page of the *National Enquirer*."

Clearly, Julie's objections to carrying a black child skirted what had been a strong and visceral racist reaction. I was struck by how carrying someone else's child had provoked no gut reaction in her whatsoever (as it did in the feminist in me), but the thought of carrying a black child made her experience the surrogacy as an intimate and at once public violation of her bodily and moral boundaries, worthy of front page *National Enquirer* attention. Julie's reaction, predictably, took the shape of race in America: blackness as the contaminating Other; her fear of miscegenation had been a knee-jerk reaction. On the one hand, the feeling she described, appropriately, was one of foreignness. On the other hand, at a visceral level, her white couple's child (or the child of any couple coded as white) was not foreign, but rather it was familiar, the "same" as she. Whiteness I had to conclude was the invisible glue that held together her narrative of gestational surrogacy.

In this narrative, Julie went to great lengths to clarify her genetic unrelatedness to this child as its gestational mother. Whenever accused of "baby selling," which was fairly often, Julie would confidently retort, "How can I sell a baby which is not mine in the first place? I'm not genetically related to this child. It is not mine to sell!" She invoked biomedical wisdom to confirm that not even her blood came in contact with the fetus; nor were any fluids exchanged when the child was conceived: five choice embryos had been transferred into her womb with a sterilized pipette. But, while genetic unrelatedness and assisted reproduction practices helped Julie conceptually separate herself from the child inside her body and thus safeguard her sexual, bodily, and personal integrity, at an embodied level, a crossing of her boundaries had occurred, only this crossing was not felt because it happened within a "white" world. If one factored in blackness, genetic distance, allegedly organic and immutable, was ipso facto replaced with a suffocating but equally organic closeness; the boundary crossing suddenly became visible. Clearly, this child was not really ontologically distinct from Julie, as her invocation of genetics implied.

Had I not had these conversations with Julie I might have missed how literally and viscerally power maps bodies and identities. I might have also confidently continued to see the reproductive technologies as primarily an issue of gender rather than, more accurately, as a complex and shifting knot of power and embodiment into which race, class, sexuality, and ethnicity are always inextricably and tightly interwoven. Bodies *are* maps of power and identity as Donna J. Haraway has suggested (180), but the legends can be misleading.

Precisely because it transgresses many embodied cultural mappings, gestational surrogacy involves its participants in an unpredictable and emotional journey, where the legends often are ambiguous. The practice does not neatly fit into existing or even new conceptualizations of social or kin relations, or of a person, a body, or a pregnancy (Strathern, "Displacing"). A gestational surrogacy arrangement is at once a public contract between persons (a market relation) and what has been culturally coded as a private and intimate experience between persons (a family relation) (Ragoné). A gestational surrogate is a non-mother mother, a liminal creature growing and birthing a child that is not hers. Similarly, a genetic mother is also a non-mother mother, one who does not grow the child in her own body, and yet she is, culturally speaking, the child's "real" mother. Institutionally, obstetrician offices and hospitals are not accustomed to the intricacies of a two-mother pregnancy; nor are insurance companies or the law equipped to deal with such pregnancies. These breaches are potentially volatile. Hardened categories of old are loosened; in this loosening, relationships of power become evident, perhaps even reversible, as for example when lesbians and gay men become "natural" parents through the reproductive technologies (Hayden; Lewin) or white women, through accidents in in vitro laboratories, give birth to black babies (Petchesky, "Body" 402).

Over the course of the surrogacy I followed, the most visible breaches were around class and gender. As Pamela and Paul Martin began to exercise their class privilege with respect to the surrogate as the genetic and contractual, and therefore only, parents of the child, they were resisted by the surrogate, who brought to the arrangement an explicit personal agenda of her own. I begin with this agenda because it challenges commonly held views that surrogates, most of whom are from the lower middle classes (Ragoné 91), are lured by the economics of these arrangements[4] and/or have fallen prey to the compelling ideology of women as altruists, unwittingly becoming "breeders" and "rented wombs" in the service of male biomedical power (Corea, *Mother, Man-made;* Rowland; Ragoné). While assisted reproduction positions surrogates, and women in general, as means to various ends, most notably more biomedical research (See

Franklin, "Deconstructing," "Postmodern"; Steinberg), to assume that surrogates are ipso facto "breeders" or "rented wombs" is to obscure both how these women might be turned into breeders and how they may resist these categorizations.

A SURROGATE'S PERSPECTIVE: NEGOTIATING BODY BOUNDARIES

Julie Thayer, a young woman of twenty-five and mother of a ten-month-old son, had no delusions about her role in the gestational surrogacy arrangement. She would freely joke about being a "cow," an "incubator," or an "oven." Her bellybutton toward the end of the pregnancy was a "pop-up timer." She fully appreciated that her role in this arrangement was to gestate, to be a "breeder," a "vessel." She had chosen gestational surrogacy instead of egg donation or traditional surrogacy because this way she would "just be nesting this baby." She would not be obligated to the child by a genetic relationship and this, she figured, would make it easier for her to give the child away, and also easier for her family, who might otherwise feel she would be giving away a piece of them, too.

But Julie also saw that "nesting this baby" was an irony of her situation, of being a mother who is not a mother. She did not consider herself simply a womb for her couple's desire. She would still have obligations to the child in her, the commitments, as she once put it, of a "grandmother," or those of distant nurture. Her intention in becoming a surrogate was explicit: she wanted to help make a family, not just a baby. To this end, she had carefully chosen a surrogacy center which arranged open surrogacies, where surrogates and couples can get to know one another. She had been specific in her application about the couple with whom she would make this family: both partners had to want a relationship with her; they could not be "Hollywood" types because she did not want "to have to deal with egos" (which in Los Angeles is always a consideration); and they had to want to raise this child themselves, without recourse to nannies or mother's helpers. These conditions would fulfill her agenda of being a part of a certain kind of family.

The idea of surrendering the privacy of her womb for this goal was empowering for Julie, even though, in practice, it often required difficult physical and emotional sacrifices on her part, as well as on the part of her family. Her buttocks, for example, were like a pinpricked cushion the first two months of the arrangement because of the daily regimen of hormone injections, which her husband helped administer. She took time away from her husband and young son for numerous appointments with doctors, lawyers, and counselors. Her husband had to contend with the insurance paperwork. They had to abstain from sexual relations. She would give birth and no doubt endure engorged breasts once more as she had with her son.

The fee, the standard $12,000 plus expenses, was insufficient to compensate for her time, professionally speaking, but would ensure that her family would not be burdened financially by her choice to become a surrogate.[5] As Julie quipped, "It would be stupid to do it *for* the money and stupid to do it for *no* money." When I asked after a shoot in her home midway through the pregnancy why she was not growing resentful about these impositions on her body, her person, her family, and her time—they seemed overwhelming to me—she replied that these "sacrifices"—and this was the term she used—would not make sense to most people, but they were, in the frame of her life, necessary and good.

Twice, Julie explained to me, her life had been spared by uncanny, bodily intuitions. She impulsively walked off the fated Lockerbie, Scotland, airplane which blew up midair, killing six of her friends and classmates; she had, ten days or so before this tragedy, impulsively walked off a train, with two friends in tow, which then crashed on the outskirts of London. In her early twenties, she found companionship, love, and fulfillment with a man thirty-two years her senior, defying her family and gossip with their marriage. The marriage was very good, a blessing considering her troubled childhood and adolescence growing up in an alcoholic family. Her partner in the first year of their relationship had exploratory surgery for cancer. Luckily, the alarm proved false. Julie was supposed to be infertile from a chronic ovarian cyst condition and a botched teenage abortion, but, shortly after her partner's cancer scare, she became pregnant with their son even though they were using contraception. These were gifts she had received by way of her body, she told me, gifts which recalled for me Marcel Mauss's famous analysis: they carried with them a moral obligation to reciprocate in kind. Becoming a surrogate was an opportunity to reciprocate these blessings also by way of her body through tangible sacrifices. The physical and emotional demands were in this sense necessary and good.

The choice of surrogacy also grew out of Julie's life experience. Surrogacy was something she had wanted to do because of her work as a counselor-nurse in a late-term, problem-pregnancy abortion clinic. There, Julie counsels women through difficult abortions and grieving, since these are "very wanted pregnancies." One of the counseling procedures is particularly emotionally trying for Julie:

"We do what are called intact dilation and evacuation which means the babies come out whole and intact. [They are euthanized in the womb.] We then wash the baby and dress it and bring it back to the parents. It's like a closure for them, but it's very hard. It's very hard to have to take this baby out of their arms and turn around and walk out of the room. One of the things I told Pamela [the genetic mother] that I'm very much looking forward to is handing her that baby at the birth. This is a baby she gets to take home."

To give a woman a baby "she could take home," then, would be Julie's ultimate reward, not as a surrogate but as a person: this was a gift of life, a present commensurate with the blessings she had received in her own life.[6] The sacrifices required of Julie as a surrogate would give flesh to her gratitude; they embodied her desire to reciprocate.

But the surrogacy offered many other, more mundane rewards as well: full breasts on her small person, the thrill of being a "scientific experiment," and, significantly, the enjoyment of a pregnancy unweighted by gendered expectations of motherhood. Julie emphasized these merits throughout the arrangement: she was free to enjoy her pregnant body without the worries of preparing for the arrival of a child, the financial expense, and the anticipated stress of caring for a newborn. These expectations fell on Pamela, the child's real mother. Sharing motherhood with another woman was, in this respect, liberating and empowering. Julie thus retained full sovereignty over her self, even as her body was turned over to biomedical processes and the desires of her couple.

RE-ENVISIONING BODY TERRITORIALITIES

To fully appreciate how Julie viewed surrogacy, I find it necessary to re-envision the individualist body territoriality that much of feminism and biomedical discourse share.[7] The idea that a body is an organism that "ends at the skin" (Haraway 168) and which an "I" inhabits as an owner would her property is inadequate to convey the complexity of Julie's self/body relationship. As Rosalind Pollack Petchesky argues, these are notions "imprisoned in the conventional (bourgeois and Lockean) notion of property that [involve] exclusivity, isolation, objectification, and self-interest. From this vantage point," Petchesky rightly concludes, "owning the body necessarily means reducing it to a commodity" ("Body" 396). In other words, the body that ends at the skin is a body whose agency exists within the confines of commodification; if it is not already a commodity, then it is always a commodity-in-potentia. It is an individualized and privatized body, ahistorical and nonbiographical. As feminists of color and the Third World have pointed out, second wave feminists—the majority of whom are white and of the middle class—have also produced a privatized and individualized female self, owner of a body defined by its unique capacity to reproduce (Mohanty, Russo, and Torres). In this construction, "'alienating the unique physiological, emotional, and creative capacity of [the female] body'" necessarily means to alienate oneself as a woman (Pateman qtd. in Petchesky, "Body" 395).

These assumptions dangerously narrow feminist discussions of the reproductive technologies around the paradigmatic abstractions of choice and consent, polarizing possibilities for agency. Women are either perceived as the victims of the reproductive technologies because of the "fictive device of the contract and voluntary consent" (Pateman; see also

Corea, *Mother, Man-made;* Rowland; Steinberg) or they are argued to enjoy greater reproductive freedom and authority over their bodies since they can now control, as women, the market value of their reproductive processes (Shalev). Surrogates suffer a particularly egregious fate in these debates: as the service side, contractual partners of couples they are almost always portrayed as victims, in Gena Corea's powerful phrase, as "breeders."

From my point of view, what is missing in these analyses is an appreciation of real-time, embodied practice, of biographical embodiment. Embodiment, as Haraway argues, is "not about a fixed location in a reified body, female or otherwise, but about nodes in fields, inflections in orientations, and responsibility for difference in material-semiotic fields of meaning" (195). The conflation of woman with her biology and the narrow view of reproduction as exclusively and privately located inside a female body-that-ends-at-the-skin enables the abstract language of ownership that frames these debates about surrogacy. What again and again is rendered invisible in the process are the historical, discursive, and biographical variations women experience between and within their reproductive histories. More important, these conceptions obscure how power is exercised in and through these histories, as in, for example, the naturalization of whiteness as an epistemology of embodiment for the surrogate and the couple I followed. Paradoxically, then, the body-that-ends-at-the-skin, so biomedically and culturally real, is a disembodied body. This Lockean, liberal conceptualization forgets biographical time.

Petchesky proposes a feminist revision of the body as property which allows an ownership of the body that does not abstract out the singularity of time. This revision is not unlike the anthropological concept of *usufruct,* where considerate usage, and not private ownership, guarantees rights of access and decides obligations. Following recent black feminist theory,[8] Petchesky argues for "an ethic of women's bodily integrity that is communal and extended rather than individualized and privatized," one that does account for discursive pressures ("Body" 399). She writes,

> When the "objects of property" speak they remind us the language of self-creation, self-propriety, and freedom is always a story told and retold, and to reject that language wholesale is to leave those without property nothing at all to own. . . . Instead of discarding the rhetoric of property, persons and bodies, we need to enlarge its frame of reference, to broaden who and what counts as owners and the moral and communal spaces in which they define their selves. As my examination of Leveller tracts and slave narratives suggests, "self-propriety" may refer to a concept of property that is inclusive rather than exclusive and a model of the body that is extensive rather than insular. It may become a rallying point through which individuals establish their connection to a larger group and community. (400)

In many ways Julie's body/self concept exemplifies what Petchesky defines as "self-propriety." As a surrogate, Julie saw her body as an inclusive and extensive communal and moral space which she deliberately extended to encompass the desires of more than herself and her immediate kin. She consciously expanded this space to encompass the desires of her couple, as a means of reciprocating the riches that had broadened her sense of self. This meant at the practical level that she would deliberately share the pregnancy with her couple in such a way as to make them feel like parents during the gestation, as they would have if the mother had carried the child herself. Julie involved the Martins, for example, in decisions about obstetricians' and doctors' visits; she frequently shared stories with Pamela about what the baby was "doing jumping around her uterus for that day"; Julie gladly agreed to play at home a tape the Martins made for their child so that the baby inside her would become familiar with their voices; Julie invited the Martins to be present in the birth room with her and even arranged with her doctor to have the father, Paul Martin, help ease the baby out of her, considering this would be "a unique experience for a father."

Julie invited the Martins into her "domestic network" (Stack): she extended to them the courtesies and obligations she would a close relative or a friend. Julie's greatest challenge, she explained halfway through the pregnancy, was "bringing the ideas of four persons together." This was especially difficult because, as she put it, "what is happening is happening inside *my* body." As both a mother and wife, Julie clarified for me, she had learned to share her body/self with her son and husband. Now, as a surrogate, she had chosen to further expand the boundaries of her body/self to include the Martins and their child. This involved careful negotiations, as she needed to ensure her needs were met as well. Julie's agency thus encompassed far more than the instrumentality of her pregnant body.

At the level of embodied experience, the surrogacy for Julie foregrounded the social dimension of reproduction as well as the personal and biographical. Although this shared dimension was in tension with the singularity of her body, Julie's sense of self was not identified piecemeal with her body or womb, at least not as long as the child inside her was coded as white. Within whiteness, the boundaries of her body were remarkably fluid and negotiable. She claimed for herself the right to decide where and how these boundaries should be drawn. Empowerment and disempowerment were not a matter of how she chose to dispose of her body as her property, but rather they had everything to do with her ability to preserve the "sanctity of [her] own personal boundaries" even as she opened up the collective of her body to include the Martins' desires (Petchesky, "Body" 400).

In summary, Julie knew she was a "cow" but imagined it possible, through a deliberate and thoughtful sharing of the pregnancy with her couple, not to be treated like a cow. She was unprepared for the way in which her couple, the world at large, and I, too, at first, would insist on framing her experience and aspirations through the narrow and predetermined lens of domination, in which there can only be one natural woman—the reproductive woman; one natural mother—the genetic mother; one natural family—the nuclear, private family; and one way to be a surrogate—a "cow." These rigidly unitary and essentialist representations of human lives, what I call biological embodiment because the symbols are materialized in practice, naturalize power in such a way as to predetermine possibilities for action, and, thus, strangle the existential richness and complexity of being embodied at specific intersections of historical biographies and social geographies.

THE PERSPECTIVE OF THE INFERTILE: NATURALIZING CLASS[9]

Biological embodiment, in contrast, framed the narrative of the Martins' practice of surrogacy and it disempowered the autonomy Julie brought to the arrangement by removing her desires from the realm of what really counted: the fulfillment of this couple's biological destinies as a mother and a father. From where the Martins stood in the arrangement, the surrogate was ultimately a final (and fortunately successful) means to an end.[10] Pamela Martin expressed it thus early on in the pregnancy: "It really doesn't matter how this baby gets here. In the end, I'm going to be its mother."

As Gay Becker and others such as Sarah Franklin have noted ("Postmodern," "Deconstructing"), the discourse of infertility casts infertility as a condition (disease) of the body and fertility as a condition of the self. In this construction, biomedical technological interventions on bodies in matters of reproduction happen with the full consent of desiring infertile selves. The Martins, diagnosed with unexplained infertility in their mid-thirties, spent the better part of twelve years and huge sums of money trying various fertility technologies. Pamela describes eight in vitro tries, at over $8,000 each, and countless other hormonal therapies, which caused deep bouts of hormonally induced depression and eventually the removal of a dysfunctional thyroid. Paul underwent a single therapy, a corrective course of vitamin A for low sperm count. The asymmetry in their fertility treatment should not come as a surprise, since biological embodiment locates infertility inside women's reproductive bodies. Adoption was not a welcome option because Paul, an only child, was anxious to perpetuate his genes, and because Pamela felt she would not be able to bear another loss (after three early-term miscarriages) should the adopted mother decide to take back her child, as adoption laws in California allow in the first year.

This personal history illustrates the gendered use of the reproductive technologies when considered from the perspective of the infertile—having located reproduction inside a woman's body, interventions must logically be on this body; the history also reveals the strong undercurrent of individualism in the biomedical definition of infertility (Becker). This undercurrent combines with the discourse of rights proper to the making of the bourgeois self in Euro-American culture (Foucault, *History*) to make infertility a humiliating personal affront to one's rightful biological destiny. As Becker notes in her study of the infertile population, emphasizing the "continuity of the self rather than cultural continuity [means] the individual has [now] responsibility for attempting that continuity and permanence" (401). The Martins' aggressive pursuit of a technological solution to their infertility exemplifies this responsibility. For them, achieving fertility was a matter of personal willpower and determination, fed by the hope of a technological cure (and use of substantial financial means), and essential to their sense of completion and success as individuals. Their pursuit of fertility was also conventionally gendered along class lines:

> *Paul:* I certainly didn't stop to feel sorry for myself when it turned out we had difficulties conceiving naturally; like every other seemingly insurmountable obstacle that I have confronted in my life it just—I just assumed that I, in this case we, would find a way to achieve what we wanted to achieve. I don't know the word "defeat." I don't know the words "give up." They are not in my vocabulary, either in business or in my personal life.

> *Pamela:* Of course, I wanted to do this for my husband and my marriage. . . . I think I would have been just as happy adopting. I mean, I just wanted a baby. It didn't have to be mine. Paul is an only child, as am I, but his parents have died and it was much more important for him to have a child genetically related to him to continue his bloodline.

Thus, on the one hand, infertility mattered to Paul because it was an obstacle which stood in the way of "his Darwinian impulse to reproduce his genes" (Paul Martin, same interview). In an ironic twist of patriarchy, it was not important to him that his baby would be a girl; his genetic information would live on in her just as it would have in a boy, and this was what was significant because Paul wanted "his genes represented in the human genome pool" (same interview; also see Franklin, "Postmodern" 331). Infertility mattered to Pamela, on the other hand, because she wanted to reproduce herself as a mother, a social and gendered role. She wanted to reproduce her capacity to nurture, an aptitude she saw herself as manifesting naturally in her maternal disposition toward her friends and cats: "I just wanted a baby. It didn't have to be mine [genetically]." Fertility for Pamela, unlike for her husband, was a matter of the heart; and infertility pitted her heart against an uncooperative, mysterious body.

When Paul proposed the idea of gestational surrogacy after they had exhausted all other venues, Pamela explains that she resisted the idea first on the ground that: "I really wanted to be the one pregnant. I wanted to nurture *my* child *in utero*, to be the one in control of *my* pregnancy." Her reasons were layered. Unlike a traditional surrogate, a gestational surrogate would have no biological, hence legal, claims on her baby. This was good, but, in the absence of such a tie, how could Pamela be sure that a gestational surrogate would do right by her child? More important, Pamela would not be able to bond "naturally" with her child *in utero*.

What changed Pamela's mind were conversations she had with Cheryl Saban, a two-time mother through gestational surrogacy and author of *Miracle Child: Genetic Mother, Surrogate Womb*. Saban explained to Pamela that gestational surrogacy could be wonderful if one had a surrogate with whom one was close. Bonding with the baby would occur post-birth, not to worry; during the pregnancy, if Pamela was close to her surrogate, she would be able without too much trouble to get a sense of "what [the surrogate] was eating, if she was jogging in the first trimester," or generally doing anything that would harm the baby. Trusting another woman with her child would then not be so difficult. "This is why," Pamela said, "it was extremely important for me to have a relationship with the surrogate and I was extremely lucky to get Julie, who also wanted a relationship with the mother—me—and is very forthcoming with information about her diet and habits, and everything she does."

What is implicit in Pamela's concerns is that she, as the child's "real" mother, remained its only custodian, even though her child was in another woman's body. Trusting her surrogate, Julie, did not mean trusting the person, Julie, as Julie might have hoped. It meant, rather, trusting that Julie would do right by her child, exactly as Pamela would have done right by the child if she had been the one carrying it. The subtitle of Saban's book captures well this projection: it unself-consciously reads *Genetic Mother, Surrogate Womb*. What changed Pamela's mind, then, was appreciating that in an open gestational surrogacy arrangement, the surrogate would only be a womb in relation to Pamela and thus would not jeopardize Pamela's standing as the child's true mother, both genetic and nurturing.

Pamela considered herself lucky to have been matched with Julie. Not only was Julie forthcoming about her diet and lifestyle habits but she was also, Pamela would make sure to bring up often, not the "maternal type." Pamela had drawn this conclusion from conversations she had had with Julie in their initial meetings. As they discussed the possibility of being matched, Pamela explained, Julie had repeatedly stressed that she had not bonded with her own son *in utero*, and that the "maternal thing" had not kicked in until after he was born. But when I asked Julie about her lack of maternal instincts, it became clear that Pamela had at least partly misunderstood her. Anticipating that women in Pamela's position would be concerned about bonding, Julie had sought to reassure Pamela that she no longer "believed in that romantic love women are supposed to have with their babies *in utero*" because she had not been able to bond with her son. Julie told me, "I knew that if anything went wrong with him or the pregnancy, and I would have to terminate, I'd be devastated. So I didn't allow myself to bond with him until I knew everything was going to be okay with him, and that was pretty much *after* he was born." This element of choice—or agency—Pamela had not at all grasped. While the exposure to grieving families in the late-term abortion clinic had prompted Julie to deconstruct in her pregnancy rigidly biologized accounts of motherhood, nothing in Pamela's life experience had challenged her strong belief in biologically induced "romantic love" between mothers and unborn infants. In Pamela's mind, Julie could not be a natural maternal type because she had not automatically bonded with her son, and this, from Pamela's perspective, was indeed a reassuring thing.

Pamela thus framed Julie as a doubly safe womb, one which she could have access to on command and one which would not inappropriately bond with her child. Because of this, Pamela would never be able to appreciate the complexity of Julie's agency in the surrogacy. In this blind spot, Pamela was fully supported not only by the recent reconceptualization of reproduction through a genetic template (which further strips the biographical dimension from reproduction) but also by visual technologies such as ultrasound; by the surrogacy center which encouraged Pamela to assume her responsibilities as the real mother by working closely with her "professional" surrogate; by the hundreds of radiant baby snapshots that adorned the fertility clinic hallways and screamed immaculate conception; and by subtle institutional changes, such as the New Arrivals sign which replaced the Maternity Ward sign of old over the door of the hospital's nursery during our videotaping. Everything around Pamela said how her baby arrived and who carried this baby did not in the least matter: what mattered was the fulfillment of her natural desire to be a mother. Even Julie's willingness to work closely with her confirmed that this was Pamela's private journey: she could survey Julie, as she would have surveyed her own body if she had been the one pregnant. This unself-conscious privileging of desire through surveillance is decidedly a bourgeois technology (see Foucault, *Discipline*).

Subjecting Julie to surveillance, however, turned out to be not as easy as those initial conversations promised. There were many instances when Pamela felt acutely the powerlessness of her position as a non-mother mother, for example, when Julie's weight gain in the first four months was slow, causing her to fear for the health of the baby, or when Julie decided at the Alpha Fetal Protein test that she would switch obstetricians, because she did not like the high-risk doctor Pamela had asked her to see. "It's my baby, I'm the mother; but it is her pregnancy, so I'm the one that has to be flexible," Pamela told me, torn between her obligations as a mother and her humanist obligations to Julie, whom she could see was going through many physical ordeals in lieu of her.

Pamela's anxieties and guilt were exacerbated by her uncritical dependence on a biologized notion of motherhood which precluded the possibility of a genuine sharing of the pregnancy, a co-mothering; what Julie was proposing. Pamela's inability to expand the boundaries of her privatized, individualized, and gendered body pitted her against Julie, sometimes explicitly, sometimes implicitly. I now turn to three episodes in the arrangement which illustrate this poignant conflict of interest: the child's conception, ultrasound checkups, and birth.

THE EMBRYO TRANSFER: CONTESTED CONCEPTIONS

Pamela's acceptance of gestational surrogacy hinged, as we have seen, on her ability to retain in language (as the "genetic mother") and in practice (through surveillance) the semblance of her status as the only mother of this child. Visual technologies such as the microscope and ultrasound imagery helped facilitate Pamela's role while further marginalizing Julie's active agency in the pregnancy. Contrasting each woman's account of the embryo transfer vividly captures this marginalization. The microscope and the reproductive laboratory—as the new procreative site—implied a virtual reproduction which masked the embodied experiences of both women, but especially Julie's.

Whenever Pamela talked about the embryo transfer, she would tell the story of Leftie. First, she would explain, her eggs and her husband's sperm were mixed in a petri dish in the laboratory. Then, her doctor and fertility specialist, Dr. Peters, had let her see through a microscope, right before their transfer into Julie's uterus, the five choice embryos he had picked. One of these embryos wiggled as Pamela peered through the lens: "Gee, I wonder if this one in the left corner will get us pregnant!" she remembers exclaiming. The vitality of this embryo—its wiggle, ironically, was an optical illusion according to the reproductive laboratory director whom I later interviewed—seemed to exhibit a will to live commensurate with Pamela's desire for a child and the force that would get her and Paul pregnant. When "they" did get pregnant at that first transfer, it was Leftie who got them pregnant with the help of their "excellent embryologist."

In Pamela's biologized narrative of the conception of her child, Julie never featured as an agent. There was no mention of the prayer the two women shared over Julie's belly after the transfer, or of Julie's efforts before and after the transfer to ensure its success. Julie had not only dangerously increased her hormone dosage in preparation for the transfer, before catching a mistake in the fertility's clinic prescription, she also had spent three anxiety-ridden days in bed after the transfer, with her legs propped up, to make sure the embryos would not flush out of her uterus, after noticing some abnormal bleeding.

Statistically, an embryo transfer has one in four chances of success. Julie got pregnant at the first transfer. She, unlike Pamela, attributed the success of their first transfer to the "vivid goodness of the day," in which their joint prayer was tantamount:

"It was such a special day. Everybody was so nice, the nurses, everybody. When Dr. Peters transferred the embryos into me, I joked that this was more thrilling than when my husband knocked me up with Kyle, and he blushed. That was fun. Afterwards Pamela and I were left alone and she prepared the sandwich they had brought for me, and I thought that was very sweet of her. I don't know why but a feeling came over me [Julie began to cry here]—I am not religious or anything and Pamela and I never did discuss religion, but the feeling came over me to say a prayer. Since I don't pray very often, I thought if I did it would be a little sacrilegious or something, so I asked her if she would and she did, asking God if it was in His heart to make it possible for her to get pregnant. Everything was so wonderful, so special that day, that I would have found it hard to believe if I hadn't gotten pregnant."

This specialness—an existential, embodied quality—confirmed for Julie the righteousness of her actions. Since her wish for the embryo transfer to succeed was so strong, vivid, and tangibly expressed in the actions of the day, and since she knew her motives to be sincere, goodness had to follow in the shape of the fulfillment of Pamela's and her joint desire, "maybe not at the first transfer, but it had to happen, eventually." It was this joining of desires that Julie felt made the transfer successful, along with the concrete actions both Pamela and she had taken to make the transfer happen. In Julie's narrative of the conception, the biogenetic events did not feature at all, rather the biographical gestures of both women did.

Dr. Peters mentioned to me in a later interview that patients often bring to their embryo transfers "fertility" objects (such as African or Asian statuettes or pendants and rings that siblings or friends wore when they got pregnant) or they perform various kinds of ritualized behavior which they believe will influence the outcome of the transfer, including prayer. The biogenetic representation of reproduction is therefore rarely experienced in biogenetic terms alone, and this was true of Pamela. Her public account may have been purely biogenetic, but her private account, it turned out, was not. When I much later discussed with Pamela how Julie had understood their "special day," Pamela recoiled: "But I spent *years* praying for this child. My church group spent *five* years praying for me." Pamela was horrified that Julie thought that she had had anything intimate to do with the "miracle of her child." This miracle had sprung from Pamela's heart and her heart alone.[11] So Pamela, too, brought with her a biography to the transfer, but it was one which was individualized and privatized and excluded Julie much in the same way the biogenetic account had. There just was no room for Julie as an agent in the making of "Pamela's" child.

According to David Schneider, conception is symbolically coded in American culture as the "spiritual union [between husband and wife]. It is a union of the flesh, that is a personal union, and out of that union a new person is formed. The word for such a spiritual union is love" (49). Clearly, Julie and Pamela's union at the transfer was not sexual, but in the weeks immediately following the transfer, when the two women were in contact daily as hormone tests monitored the viability of the pregnancy, love did seem to have been sparked between the two women. Pamela would describe Julie as "like a sister." She would hug Julie at every visit and call her "sweetie." Pamela's gratitude took the shape of a ruby and diamond heart pendant, which she gave Julie on the occasion of the baby's first heartbeat, another deeply personal moment. (Significantly, Paul was not present at the transfer or the first heartbeat.) But this "sisterly love" was for Pamela largely circumstantial, and it was always mediated by the image of Leftie, the wiggling embryo and long-awaited fruit of her and her husband's desire and labor. Conception had taken place in the reproductive laboratory and in their hearts before the transfer. For Julie, however, conception had crystallized in their joint prayer over her belly in the immediacy of that "special day." The feeling of love Julie experienced and Pamela's references to her as "a sister" did seem to Julie to be the kin-like closeness she had hoped for in pairing up with Pamela. Julie could not see that she had already been displaced by Leftie, and that this kin-like closeness would not morally obligate Pamela or Paul to her in any way. Also, Julie did not understand that the gifts Pamela gave her and the money the Martins were paying her at every successful stage of the pregnancy—the embryo transfer, first heartbeat, confirmation of a viable fetus at three months, and the birth—were meant to discharge all the obligation they felt toward her for carrying their child.

ULTRASOUND IMAGES AS THE FETUS'S BODY

Feminists have been concerned with how fetal images give fetuses a new individuated ontological status apart from their mothers' bodies, but the women who use the technologies often find fetal images appealing, especially middle-class women for whom management is an important trope of motherhood (Petchesky, "Fetal"). As the ones responsible for the health and well-being of their families, fetal images give many women a greater sense of control over their pregnancies, thereby reducing anxiety. They also give women a pleasurable new form of visual bonding, which they can share with family and friends by circulating sonogram pictures and stories (see Taylor, this volume). The fetus has thus acquired a double-edged materiality through fetal imagery, one that at once separates women from their fetuses and yet naturalizes this separation by appealing to class and gendered expectations.

Gestational surrogacy profits from this double-edged materiality: on the one hand, sonograms allowed Pamela to bond with "her" fetus even though she could not feel it inside her body; on the other hand, sonograms encouraged Julie to objectify the fetus outside her body. This was especially true in the beginning of the pregnancy, when Julie could not feel the baby move but sonograms revealed a healthy heartbeat (which appeared as a pumping black sac), and soon after this they showed a tiny creature Pamela named Flippers, because of the shape of its developing arms. Not surprising, these ultrasound photographs and others came to populate Pamela's pregnancy scrapbook. She also always carried in her organizer the latest picture, carefully annotated with the pregnancy week. When at week sixteen the baby girl sported a "little pug nose just like Pamela's" and "long legs like her Daddy's" and looked like a "human with arms and legs," the pictures came to stand in literally for the body of the child. This marked a turning point in Julie and Pamela's relationship. The closeness they had experienced after the transfer began to dissipate as Pamela, now the confirmed mother of a healthy baby girl, began to assert her new confirmed identity as a mother-to-be. The ultrasound images, spit out in a long string of photographs at each doctor's visit, were the only private contact she could have with her baby, unmediated by Julie's body or presence.

While Julie acknowledged that watching the ultrasound images made her feel, particularly early on, that she was an observer to this process (like me, the filmmaker-anthropologist, she once pointed out), they did not altogether obliterate her embodied and erotic attachment to the child within her or the pleasure of being pregnant, that "neat feeling when the baby kicks you" or "that pregnancy glow, which just makes you feel good." Julie did experience the baby growing inside her as a separate being, but not as a being separate from her body. As the pregnancy progressed, she began to enjoy the baby's developing personality, finding it amusing, for example, that the baby would often kick her son or husband whenever they pressed against her. Julie's erotic attachment was to the process of individuation her body was going through as a woman with child, the extraordinariness of being two in one body. The ultrasound images were just one aspect of a complex relationship that was, in terms of embodiment, more social than biological.

Pamela did not share this ability to distinguish between subtleties of attachment. Inside Julie's body, Pamela's child was alone and among strangers, "away from her parents who want to love her and hold her," as she once put it. She longed for the day her baby would come "home." As the pregnancy progressed, Julie began to notice with chagrin that Pamela was less and less likely to feel the child through her belly, awkwardly touching her only when Julie pressed her, and more and more likely to insist that ultrasounds be done at every doctor's visit.

These ultrasound pictures, along with books on pregnancy and mother-hood and discussions Pamela had with her girlfriends who were mothers (many through gestational surrogacy like her), came to sum up Pamela's referent world. Julie grew annoyed with Pamela's unwillingness to trust her professional and personal knowledge of pregnancy—after all, Julie worked in reproductive medicine and was a mother—as well as her immediate em-bodied sense of the child's well-being and progress. She disapproved of Pamela's dependence on ultrasounds, being aware that frequent use of this technology had not been proven safe for fetuses. A nagging tension emerged between the two women as Julie stubbornly continued to attempt to share the pregnancy, and Pamela continued to attempt to retain control of the pregnancy, safeguarding whatever privacy she could hold onto. As Julie resisted the surveillance and her growing marginalization, Pamela's frustration grew. "Next time we do this," Pamela shared with me at around week thirty-four, "we'll get a country bumpkin!"

Julie's *person* was interfering with Pamela's efforts to be the best mother she knew how under the circumstances: the child's only mother. Julie's refusal to perform as "just a womb"—her desire, for example, to attend the baby shower (to see Pamela showered with gifts, Julie explained) or to know the Martins' home address in order to send baby gifts (as would any friend celebrating the arrival of a child)—presented too great a challenge to Pamela's unitary and private understanding of motherhood. Biologized surrogacy positioned the surrogate as a means to Pamela's end of becom-ing a mother, a vessel whose only function was, in theory, to nourish her baby. But, in practice, the biographical and social dimensions asserted themselves with disturbing impunity on the person of Julie. Pamela refused Julie's requests to attend the baby shower on the grounds this was the "only day during the pregnancy she could be a mother"; Pamela refused to give Julie her home address, for fear that her surrogate might show up at her door after the baby was born.

THE BIRTH: FULFILLMENT OF THE CONTRACT

In her ethnography of surrogate parenthood, Helena Ragoné suggests that the post-birth "feelings of loss" commonly experienced by surrogates pri-marily have to do with the "loss of specialness" their surrogate status con-fers on them (80–81). Surrogates, she argues, are typically working-class women who, by receiving remuneration for what are essentially devalued homemaking skills in American culture, elevate their gender status by be-coming professional homemakers. As professional homemakers, they re-ceive the attentions of wealthier, middle- to upper-class couples and are celebrated at their surrogacy centers as commendable altruists. Once the contract is fulfilled at the birth and the arrangements terminate, these attentions stop, hence the overwhelming feelings of disappointment and loss.

Problematic assumptions underlie this view: for example, that working-class women devalue homemaking skills and uncritically aspire to the world of their middle- to upper-class couples. Apart from this, my research suggests these feelings of loss have their origin in a more complex working of class and culture. Ragoné's analysis fails to grant legitimacy to the desires and expectations of surrogates or to their embodied experience of the surrogacy, including the experience of giving birth, which for many women is an intense physiological, personal, and emotional act, not to mention dangerous.

From the perspective of the discourse of biological embodiment, a gestational surrogacy arrangement ends at the birth with the handing over of the child, and thus the completion of the intended nuclear family as contracted. In this intended family, the surrogate has no "official" role.[12] In keeping with this construction, the Martins whisked off their newborn daughter to their private hospital suite after witnessing the birth, leaving Julie to recover alone, the whole evening, in her maternity room. This dismissal—complicated by the fact that Julie recognized the Martins' need for "time alone with the child"—shocked Julie to her core. She noted with dismay that the maternity room Pamela had arranged for her was the furthest possible from their suite in the New Arrivals ward. All at once, Julie knew that the close relationship she had wanted to have with Pamela and believed Pamela had wanted to have with her was immaterial, or at best, within the total control of the Martins, who now made it clear that they wanted little to do with her. Paul left the hospital without saying good-bye; Pamela's gratitude, when Julie came by their room the next day to say her good-byes, narrowed Julie's contribution to their child's life to a physiological function. To Pamela's repeated "You baked her"; "Look at what a good *job* you did: she's beautiful"; "You gave her all your nourishment," Julie insisted, through her tears, "Look at the good job *we* did." Her own aspirations of this moment—the sense of having shared a pregnancy, of being a part of the making of a family, of having invested not only of her body but also of her self—were subsumed within Pamela's completely utilitarian assessment of having done the job well.

At the monthly surrogate support group, where Julie shared her feelings of disappointment two weeks after the birth, the pressure to deny her aspirations continued. When she expressed her anger at having been dismissed so "businesslike," one of the counselors responded:

"What I am understanding more and more is that even if you really like the couple you're working with, and hopefully you do, actually every couple falls off the pedestal at some point. There's sort of this romance stage. No couple is perfect all the time. And frankly no surrogate is special all the time, too. As nice as they are sometimes, they're not what you expected. That often doesn't mean that you don't wish you'd had the baby."

And another counselor reminded Julie, "She [the baby] is still here and she's here because of you. That was a great thing that you did."

In this rhetoric, Julie's expectations of the arrangement were put right back onto her shoulders, leaving intact the class privilege of her couple to decide unilaterally, and without consideration of Julie's explicitly stated desires and needs, the fate of their relationship. The moral reward of a "job well done," of being the one to understand her couple's imperfections and the sympathies of her "colleagues"[13]—these things should be reward enough. But these rewards were privatized and individualized, and what Julie had expressly wanted was to share the making of this family: to "be included," as she put it in our final interview. No, she did not want to "be invited to Sunday barbecues," but Julie did want to remain, through Christmas and birthday cards, photograph exchanges, and an occasional phone call or visit, at the periphery of the Martins' domestic network, the network she had labored to grow. The language of professional achievement the surrogacy center promoted erased this dimension of Julie's experience, naturalizing instead the right of the Martins to keep her out.

I would venture that the feelings of loss many surrogates experience after the birth do not have to do with a loss of the specialness surrogacy confers on them, as Ragoné suggests (although for some surrogates this may play a part), but rather these feelings result from the rude loss of power many surrogates experience when their shared and communal embodied experience of the surrogacy/pregnancy—including their social attachment to the child they bear—is forced to closure by both centers and parents in a way over which these women have little control. Furthermore, for surrogates a surrogate pregnancy does not end with the birth. Engorged breasts and other physiological reminders drag on for weeks or even months afterward. In Julie's case, dramatically, her bladder collapsed within days of the birth, requiring painful surgery that had to wait two months for her uterus to heal from the labor.

CONCLUSION

The complexity of Julie's agency was in the end fully eclipsed by the authority of her couple's biologized and privatized narrative. Appropriately, Pamela's baby scrapbook of the pregnancy, replete with sonogram pictures of every developmental phase of her daughter, would end with a birth certificate which identified Pamela as the birth mother of the child. Her daughter's first photo album would open with a picture of the wall clock in the birthing room, giving the exact time of birth, and it would be followed by an intimate picture of Pamela with her daughter at her breast, in their hospital suite, looking up lovingly at her husband sitting next to them. In this representation of the baby's birth, Julie had completely vanished.

The scrapbook, the décor of fertility clinics, the ultrasound images, and the practices and language of the assisted reproductive technologies which render contributors in the making of a family socially anonymous (Strathern, "Displacing") were all material representations which naturalized the desires of the Martins while erasing the particular desires of their surrogate, Julie. They were, and are, an exercise of domination, in the Foucauldian sense, because they privilege one narrative at the expense of others. For surrogates, whose involvement in the procreative and reproductive process can never be purely instrumental, this erasure can be profoundly wounding, emotionally and physiologically, and often impossible to forget because it may be indelibly marked on their bodies. The corrective surgery Julie had for her collapsed bladder has left a scar inside her vagina that will forever remind her that she was a surrogate for the Martins. Other, less dramatic markings pepper her body, such as stretch marks. The Martins, in contrast, have no indelible trace of Julie, not in the child she bore for them (there is no physical resemblance since the child is not genetically related to Julie) and not in the memorabilia they put together to celebrate the arrival of their long-awaited child. A short seven months after the birth, Pamela told me she hardly ever thinks of Julie now that they are a family; although, she added, "I will always be grateful for what she did for me."

All said and done, Julie did end up being a breeder for the Martins. But she did not become a breeder because of carrying and giving birth to a child that was not her own. Nor did she become a breeder because she was paid to carry this child. In the final count, Julie was a breeder because the Martins' personal biographies, along with the discourse and practices of the assisted reproductive technologies, are steeped in a potent cocktail of biogenetic and class ideologies and customs, a 350-year-old Euro-American brew called biopower (by Foucault, *History*) and biological embodiment (by myself), which made it virtually impossible for Paul and Pamela to begin to comprehend Julie's expectations and desires, or to see her as anything but a womb. Had something, somewhere, sometime during the surrogacy cracked through the naturalized veneer of these ideologies and practices for the Martins, this essay would be telling a happier story.[14]

The recipe for this brew includes any number of rigidly unitary discourses and practices, such as the natural supremacy of the privatized nuclear family; the use of money to discharge moral obligations; the biologized notion of motherhood suggesting that women make babies inside their bodies (Strathern, *Gender*); the naturalization of a splintered reproduction into genetic, biological, and social aspects (as though these domains exist outside biographical time and personal histories); the pathologization of infertility; and the inability of a privatized and individualized discourse to honor real-time biography, ambiguity, multiplicity, and change . . . the list goes on.

We urgently need to problematize oversimplistic, abstracting, and unitary representations of embodiment, power, and technology. In real-time experience, domination is an always happening hegemonic process, negotiated in the daily rhythms of biographies and the visceral guts of bodies, and not in the abstract spaces of culture or society or patriarchy or discourse or technology or biology or the body. These hegemonies are partial and often reversible, as, for example, when Julie, because of her experiences at the late-term abortion clinic and her experimentation with pregnancy, was able to de-biologize motherhood for herself and articulate the social gestures which went into bonding with a child *in utero*. Even her initial racist gut reaction was open to some reversibility: Julie surprised me right before she gave birth when she offered to be a surrogate for me and my partner—a man of color—should my "infertility" prove to be enduring.

"A cyborg body is not innocent," writes Donna Haraway, "it was not born in a garden; it does not seek unitary identity and so generate antagonistic dualisms without end (or until the world ends); it takes irony for granted. Cyborg imagery means refusing an anti-science metaphysics, a demonology of technology, and so means embracing the skillful task of reconstructing the boundaries of daily life, in partial connection with others, in communication with all of our parts" (180–81). Julie was attempting just this scenario in her choice to become a surrogate. She expanded the boundaries of her body to include the aspirations and hopes of another couple. Julie embraced the task of reconstructing the boundaries of her daily life and the daily life of her couple, in partial connections with them, in communication with all of her parts. If Julie failed in her vision, it was only because Pamela and Paul Martin refused to consider her and the work she did as a part of their collective body.

NOTES

I wish to thank Gelya Frank, my mentor and friend, whose work on embodiment and life histories has inspired both my research methodology and analysis. I am grateful to Dion Farquhar, Donna J. Haraway, Amy Stamm, Paul E. Brodwin, and especially Lisa Rofel, for their generous and insightful comments on earlier drafts of this essay. The anonymous readers of *Feminist Studies* provided invaluable suggestions on a later draft. I am, happily, in everyone's debt.

1. See Corea, *Mother, Man-made;* Farquhar; Franklin, "Postmodern"; Ragoné; Rowland; Snowdon; Stanworth; Strathern, "Displacing," *After,* and *Reproducing.*

2. The project was my master's thesis, completed at the Center for Visual Anthropology at the University of Southern California, Los Angeles. I produced a visual ethnography and a written thesis entitled "Epistemologies of Embodiment in the New Reproductive Technologies: A Gestational Surrogacy Case Study."

3. Open surrogacy arrangements are in contrast to closed surrogacy arrangements, where couples and surrogates have little to no contact with each other until the day of the birth.

4. A traditional surrogate who donates her own ovum to the pregnancy receives a fee of $15,000, plus medical and other out-of-pocket expenses (maternity wardrobe, childcare should it be an issue, et cetera). Gestational surrogates, because they do not donate their own gametes, receive anywhere between $10,000 and $12,000, plus medical expenses, et cetera. If one considers that pregnancy is a twenty-four-hour condition, these fees translate to below minimum wage.

5. Julie, a counselor-nurse, earns well above minimum wage.

6. Broker agencies shamelessly capitalize on this desire. Ragoné notes in her study of surrogate motherhood that one of the agencies doubled their responses from potential surrogates when they changed their advertisement from "Help an Infertile Couple" to "Give the Gift of Life" (32). Hillary Hanafin, a prominent psychologist in the field and head psychologist at one of the most successful broker agencies in the country, has found that most surrogates have experienced losses in their lives, such as abortions or having to put children up for adoption (personal communication with the author).

7. Dion Farquhar offers an insightful reading of the shared ideologies between liberal and feminist discourses on the reproductive technologies (*Other*).

8. Petchesky draws from Patricia Williams and Patricia Hill Collins.

9. I borrow the phrase "naturalizing power" from the title of Yanagisako and Delaney's thought-provoking edited collection of essays, in which the contributors explore how capitalism and patriarchy have been naturalized as culture and biology.

10. The surrogacy was financially costly also, about $65,000 including the surrogacy center fee of $17,000, Julie's fees and expenses, hospital expenses, psychotherapists fees for Julie, the fertility clinic's fees and medical costs, and some obstetrician costs (most of these were covered by Julie's insurance, with the Martins paying the deductibles and differences).

11. Helena Ragoné subtitled her ethnography of surrogate motherhood *Conception in the Heart,* speaking to a common belief among intended mothers that their babies were first conceived in their hearts (126).

12. In traditional surrogate arrangements, where the surrogate mother is both the birth and genetic mother, this fiction of the nuclear, privatized family is more difficult to sustain than in gestational surrogacy because physical resemblances emerge over time between the surrogate and the child, serving as constant reminders of the origins of that child (see Ragoné 121).

13. An analysis of the way in which surrogacy centers frame the experience of surrogacy is revealing of the inherent contradictions between the representation of the experience in purely instrumental terms and the embodied experience of surrogacy. The centers work hard to keep the surrogates' contributions to their pregnancies purely biological or biogenetic. Surrogates are strongly encouraged to attend the monthly surrogate meetings and are assigned personal psychologists to help them cope with the social ambiguity of their positions as surrogates and problems which may arise with their couples. But the meetings and counseling relationships also provide the agencies with an opportunity to ideologically frame the surrogates' experiences in a way which helps guarantee that these women will hand over the babies without resistance at the birth (see also Ragoné 38–50). The psychologists, not incidentally, are paid for by the contracting couples, and the couples themselves are not required to do any form of therapy.

14. In truth, the video I ended up producing did threaten to shatter the natu-

ralized veneer of the Martins' narrative by allowing Julie her voice and representing her agency over the course of the arrangement. The Martins' reaction has been in turn to threaten me with a lawsuit for fraud and libel should I distribute the film. The ethical and political issues of this development are discussed elsewhere (Goslinga-Roy).

WORKS CITED

Becker, Gay. "Metaphors in Disrupted Lives: Infertility and Cultural Constructions of Continuity." *Medical Anthropology Quarterly* 8 (1994): 383–410.
Braidotti, Rosi. *Nomadic Subjects: Embodiment and Sexual Difference in Feminist Theory.* New York: Routledge, 1994.
Collins, Patricia Hill. *Black Feminist Thought: Knowledge, Consciousness, and the Politics of Empowerment.* Boston: Unwin, 1990.
Corea, Gena. *Man-made Women: How Reproductive Technologies Affect Women.* Bloomington: Indiana University Press, 1987.
——. *The Mother Machine: Reproductive Technologies from Artificial Insemination to Artificial Wombs.* New York: Harper, 1985.
Crowe, Christine. "Whose Mind over Whose Matter? Women, *In Vitro* Fertilisation and the Development of Scientific Knowledge." McNeil, Varcoe, and Yearley 27–57.
Eisenstein, Zillah. *The Female Body and the Law.* Berkeley: University of California Press, 1988.
Farquhar, Dion. *The Other Machine: Discourse and Reproductive Technologies.* New York: Routledge, 1996.
Foucault, Michel. *Discipline and Punish: The Birth of the Prison.* Trans. Alan Sheridan. New York: Vintage, 1977.
——. *The History of Sexuality: An Introduction.* Trans. Robert Hurley. New York: Vintage, 1980.
Frank, Gelya. "Becoming the Other: Empathy and Biographical Interpretation." *Biography* 8.3 (1985): 189–210.
——. "On Embodiment: A Case Study of Congenital Limb Deficiency in American Culture." *Culture, Medicine, and Psychiatry* 10 (1986): 189–219.
Franklin, Sarah. "Deconstructing 'Desperateness': The Social Construction of Infertility in Popular Representations of New Reproductive Technologies." McNeil, Varcoe, and Yearley 200–229.
——. "Postmodern Procreation: A Cultural Account of Assisted Reproduction." Ginsburg and Rapp 323–45.
Gallagher, Catherine, and Thomas Laqueur, eds. *The Making of the Modern Body: Sexuality and Society in the Nineteenth Century.* Berkeley: University of California Press, 1987.
Ginsburg, Faye D. *Contested Lives: The Abortion Debate in an American Community.* Berkeley: University of California Press, 1989.
Ginsburg, Faye D., and Rayna Rapp, eds. *Conceiving the New World Order: The Global Politics of Reproduction.* Berkeley: University of California Press, 1995.
Goslinga-Roy, Gillian. "The Voyeur and the Agent: On the Politics and Ethics of Videotaping a Gestational Surrogacy Arrangement." Annual Meeting of the American Anthropological Association. Chicago. 1999.
Gostin, Larry, ed. *Surrogate Motherhood: Politics and Privacy.* Bloomington: Indiana University Press, 1990.

Hanafin, Hillary. "Surrogate Mothers: An Exploratory Study." Diss. California School of Professional Psychology, Los Angeles, 1984.

Haraway, Donna J. *Simians, Cyborgs, and Women: The Reinvention of Nature.* New York: Routledge, 1991.

Hayden, Corinne. "Gender, Genetics, and Generation: Reformulating Biology in Lesbian Kinship." *Cultural Anthropology* 10 (1995): 41–63.

hooks, bell. *Black Looks: Race and Representation.* Boston: South End, 1992.

Langness, Lewis L., and Gelya Frank. *Lives: An Anthropological Approach to Biography.* Novato: Sharp, 1981.

Lewin, Ellen. *Lesbian Mothers: Accounts of Gender in American Culture.* Ithaca: Cornell University Press, 1993.

Mandelbaum, David G. "The Study of Life History: Gandhi." *Current Anthropology* 14 (1973): 177–206.

Martin, Emily. *The Woman in the Body: A Cultural Analysis of Reproduction.* Boston: Beacon, 1987.

Mauss, Marcel. *The Gift: Forms and Functions of Exchange in Archaic Societies.* New York: Norton, 1967.

Mbilinyi, Marjorie. "'I'd Have Been a Man': Politics and the Labor Process in Producing Personal Narratives." The Personal Narratives Group 204–27.

McNeil, Maureen, Ian Varcoe, and Steven Yearley, eds. *The New Reproductive Technologies.* London: Macmillan, 1990.

Mohanty, Chandra, Ann Russo, and Lourdes Torres, eds. *Third World Women and the Politics of Feminism.* Bloomington: Indiana University Press, 1991.

Pateman, Carole. *The Sexual Contract.* Stanford: Stanford University Press, 1988.

The Personal Narratives Group, eds. *Interpreting Women's Lives: Feminist Theory and Personal Narratives.* Bloomington: Indiana University Press, 1989.

Petchesky, Rosalind Pollack. "The Body as Property: A Feminist Re-vision." Ginsburg and Rapp 387–406.

———. "Fetal Images: The Power of Visual Culture in the Politics of Reproduction." *Feminist Studies* 13 (1987): 263–93.

Ragoné, Helena. *Surrogate Motherhood: Conception in the Heart.* Boulder: Westview, 1994.

Rowland, Robyn. *Living Laboratories: Women and Reproductive Technologies.* Bloomington: Indiana University Press, 1992.

Saban, Cheryl. *Miracle Child: Genetic Mother, Surrogate Womb.* Far Hills, NJ: New Horizon, 1992.

Schneider, David. *American Kinship: A Cultural Account.* Chicago: University of Chicago Press, 1968.

Shalev, Carmel. *Birth Power: The Case for Surrogacy.* New Haven: Yale University Press, 1989.

Snowdon, C. "What Makes a Mother?: Interviews with Women Involved in Egg Donation and Surrogacy." *Birth Issues in Perinatal Care* 21 (1994): 77–84.

Stack, Carol B. *All Our Kin: Strategies for Survival in a Black Community.* New York: Harper, 1974.

Stanworth, Michelle, ed. *Reproductive Technologies: Gender, Motherhood, and Medicine.* Minneapolis: University of Minnesota Press, 1987.

Steinberg, Deborah. "The Depersonalisation of Women through the Administration of '*In Vitro* Fertilisation.'" McNeil, Varcoe, and Yearley 74–122.

Strathern, Marilyn. *After Nature: English Kinship in the Late Twentieth Century.* Cambridge: Cambridge University Press, 1992.

———. "Displacing Knowledge: Technology and the Consequence for Kinship." Ginsburg and Rapp 346–63.

———. *The Gender of the Gift: Problems with Women and Problems with Society in Melanesia.* Berkeley: University of California Press, 1988.

———. *Reproducing the Future: Essays on Anthropology, Kinship, and the New Reproductive Technologies.* New York: Routledge, 1992.

Williams, Patricia. *The Alchemy of Race and Rights.* Cambridge: Harvard University Press, 1991.

Yanagisako, Sylvia, and Carol Delaney, eds. *Naturalizing Power: Essays in Feminist Cultural Analysis.* New York: Routledge, 1995.

Chapter 6

An All-Consuming Experience

Obstetrical Ultrasound and the
Commodification of Pregnancy

JANELLE S. TAYLOR

The transformation of any society should be revealed by the
changing relations of persons to objects within it.
—Jean Comaroff and John L. Comaroff,
"Goodly Beasts and Beastly Goods"

In the past twenty-odd years, a number of different prenatal diagnostic
technologies have, with startling rapidity, gone from highly experimental
to virtually routine medical procedures in the United States. Ultrasound,
alphafetoprotein (AFP) blood screening, and amniocentesis, which may
with varying degrees of accuracy detect a range of abnormalities and other
health problems in the fetus, have become regular features of prenatal
care for the majority of American women who enjoy access to any form of
health care.[1] In the context of ongoing conflicts over abortion, the routinization of prenatal diagnostic technologies has aroused concern from many
quarters that embryos and fetuses are being reduced to the status of commodities. Feminists have voiced the further concern that women are, in the
process, being reduced to the status of unskilled reproductive workers, who
produce these valued commodities through their alienated labor.

In these responses to prenatal diagnostic technology, we may discern a
number of related cultural anxieties about perceived threats to boundaries
between persons and things—the intrusion of technology into the body;
the incorporation of biological reproduction (of people) into the structures of industrial production (of things); and the specter of treating
human beings as if they were commodities. These discussions are framed
by two key assumptions: (1) that reproduction is best understood by analogy to production, and (2) that we can and should clearly distinguish persons from commodities.

In this essay, I draw upon an ethnographic study of obstetrical ultrasound (probably the most widely used of all the prenatal diagnostic
technologies) to question these assumptions.[2] The case of obstetrical

ultrasound shows that: (1) reproduction has increasingly come to be constructed as a matter of consumption; and (2) in the process the fetus is constructed more and more as a commodity at the same time and through the same means that it is also constructed more and more as a person. I shall first sketch out the argument by analogy to production, which is a compelling and influential strand within the feminist critiques of reproductive medicine, and I will show how this line of analysis explains the advent of obstetrical ultrasound. I shall then discuss a number of other aspects of ultrasound that do not easily fit into this framework, and which I believe are better understood in terms of consumption. I suggest that there are at least four different (if interrelated and overlapping) ways in which reproduction has come to be construed in terms of consumption: (1) consuming on behalf of the fetus, (2) consumption as pregnancy, (3) consuming pregnancy as a commodified experience, and (4) consuming the fetus.

REPRODUCTION AND THE ANALOGY TO PRODUCTION

The same span of time that has witnessed this proliferation of prenatal diagnostic tests and other new reproductive technologies has also seen the rise of compelling and influential feminist critiques of reproductive medicine. One important position staked out within these arguments turns on the analogy between reproduction and production, specifically identifying ways that the analogy to industrial factory production operates within medical discourse and shapes medical practice. Scholars such as Emily Martin (*Woman*, "Egg"), Ann Oakley (*Captured*, "History"), Dorothy C. Wertz and Richard W. Wertz (*Lying-In*), and Barbara Katz Rothman (*Tentative*, *Recreating*), among others, have argued that as pregnancy and childbirth have come under the aegis of the male-dominated medical profession, these natural processes have come to be represented and understood in terms of analogies drawn from the world of industrial production. Doctors have come to be positioned as "managers" relative to reproduction; fetuses appear as valuable "products"; and women are like reproductive "workers." These metaphors operate within medical discourse, and they structure the medical treatment of reproduction—with negative consequences for women. In this view, the routinization of ultrasound and other prenatal diagnostic technologies seems to support this process on every front: bolstering the power of doctors as managers, enhancing the value of fetuses as products, and further alienating women as reproductive laborers.

Doctors as Managers

Let us first consider the claim that doctors are like managers. In *The Woman in the Body*, Emily Martin shows how medical textbooks use imagery which likens labor to factory production:

Medical imagery juxtaposes two pictures: the uterus as a machine that produces the baby and the woman as laborer who produces the baby. Perhaps at times the two come together in a consistent form as the woman-laborer whose uterus-machine produces the baby. What role is the doctor given? I think it is clear that he is predominantly seen as the supervisor or foreman of the labor process. (63)[3]

Martin argues that such texts, by teaching doctors to view themselves as managers, encourage caesarian sections and other surgical and technological interventions—which are often alienating, dangerous, and unnecessary—whenever pregnancy or labor appear to deviate from a fixed, abstract schedule of production.[4] From this perspective, the rapid routinization of prenatal diagnostic technologies emerges as the latest step in a long historical process by which doctors have established their power and authority over pregnant women.

And indeed, ultrasound technology as used in prenatal care does seem to have this effect. Obstetricians have eagerly embraced ultrasound as a relatively inexpensive, non-invasive, and presumed safe means of obtaining a great deal of information not otherwise available about the position and appearance and activity of the fetus. The availability of such information does not, however, explain why the data is considered valuable or useful. Ann Oakley has argued convincingly that a large part of ultrasound's appeal for doctors is that it allows them technologically to bypass pregnant women as a source of knowledge about pregnancy. For example, instead of asking a woman for the date of her last menstrual period (and having to rely upon her word), a doctor may seek an ultrasound estimation of "gestational age."[5] One might also note that the generation of so much data also vastly expands the "need" for experts, such as doctors, to manage it.[6] In this regard, it does seem plausible to interpret the routinization of ultrasound as a strategic move by doctors to solidify their position as "managers" of reproduction construed as analogous to industrial production—especially in light of recent studies suggesting that the routine use of this technology does not improve pregnancy outcomes (Ewigman et al.).

Fetuses as Products

A second comparison through which the feminist argument by analogy to production proceeds is the claim that fetuses are like products. In this light, prenatal diagnostic testing represents a way in which doctor/managers try to make sure that this product, whose production they oversee, is of consistently high quality. Barbara Katz Rothman writes that:

genetic counseling, screening and testing of fetuses—[serve] the function of "quality control" on the assembly line of the products of conception, separating out those products we wish to develop from those we wish to discontinue. Once we see the products of conception as just that, as products, we begin to treat them as we do any other product, subject to similar scrutiny and standards. (*Recreating* 21)

Again, obstetrical ultrasound does seem to fit readily into this argument. An ultrasound examination can provide information about a variety of major and minor problems in the development and health of the fetus. However, with ultrasound as with other prenatal diagnostic tests, medicine has at present no treatment to offer for the majority of problems that can be detected, other than the option of elective abortion; and indeed, one of the primary justifications given in the medical literature for offering ultrasound screening to all women on a routine basis is the expectation that fetuses exhibiting anomalies will be aborted. One article advocating routine ultrasound screening of all pregnancies concludes that, "Long-term gains would include identification of a major anomaly and termination of pregnancy, thus avoiding the birth of a child with an anomaly who is likely to survive but with a poor quality of life" (Gabbe 72). Certainly, the point of view from which a diagnosis of fetal anomaly followed by abortion could be described as a "gain" is not that of a woman who has had to undergo this experience![7] This passage does reflect, instead, a managerial view of the fetus as a product that must be subjected to "quality control."

As this view of the fetus as a product takes hold among pregnant women, and in society at large, it is argued that respect for persons necessarily diminishes. In Rothman's words, the routinization of prenatal diagnostic testing is symptomatic of "the expansion of a way of thinking that treats people as objects, as commodities" (*Recreating* 19). This assumes, of course, that commodification is directly corrosive of personhood—to the extent that we think of fetuses as commodities, we fail to regard children fully as persons. I shall return to this point later to show how an ethnographic approach to obstetrical ultrasound may challenge this view.

Women as Workers

For the moment, however, let us first turn to the third crucial step in the feminist argument by analogy to production—the claim that women are like workers. When doctors act as managers and fetuses appear as products, then women come to be regarded as unskilled workers, alienated from their reproductive labor. "As babies and children become products, mothers become producers, pregnant women the unskilled workers on a reproductive assembly line. . . . What are the causes of prematurity, fetal defects, damaged newborns—flawed products? Bad mothers, of course—inept workers" (Rothman, *Recreating* 21). In a similar vein, Martin argues that with new reproductive technologies, "the possibility exists that the woman, the 'laborer' will increasingly drop out of sight as doctor-managers focus on 'producing' perfect 'products'" (*Woman* 145).

As deployed in medical practice, then, the analogy between reproduction and production appears clearly to work to women's detriment. How-ever, the same analogy may also be used analytically to ground feminist demands for change and calls to action—on the logic that "women have control over the means of reproduction (at least for the present . . .) in the form of their own bodies" (Martin, *Woman* 143). Or, in Carol A. Stabile's words, "only women can carry out the work that is pregnancy" (94).

FROM PRODUCTION TO CONSUMPTION

But is pregnancy really all work and no play, so to speak? An ethnographic study of obstetrical ultrasound leads me to think not.

The argument by analogy to production represents an important strand within the broader feminist critiques of reproductive medicine. This posi-tion does hold considerable merit, and, as I have tried to show, it can go some distance toward accounting for the routinization of ultrasound within obstetrics. This analysis, however, takes doctors perhaps a bit too much at their word by assuming that obstetrical ultrasound is about just what they say it is about—prenatal diagnosis and medical management. As with any technology, its use in social practice does not correspond exactly to any one set of intentions.[8]

Many aspects of obstetrical ultrasound as it is actually used have little or no strictly "medical" function, but they are nonetheless significant in terms of how women experience ultrasound examinations, pregnancy, and the fetus. In the United States, the medical task of obtaining certain views and measurements of the fetus and placenta is combined with a number of other practices—a pregnant woman is usually allowed to bring a com-panion (often her husband or boyfriend) into the examining room; and the sonographer typically shows them the screen, points out features of the fetus, offers the pregnant woman the option of finding out fetal sex if it can be visualized, and gives her an image to take home. While it is true that the examination is often fraught with anxiety over the possibility of a "positive" diagnosis of fetal anomaly or death, it is also true that many women look forward to and enjoy ultrasound, and even seek it out.[9]

What are we to make of women's acceptance, or even embrace, of ultra-sound? In terms of the analogy to production, we might ask: if prenatal diagnostic technologies help reduce women to the position of unskilled laborers in the reproductive process and expand the power exercised over them by doctor-managers, then why do more women not actively resist? This, of course, is a version of the old question: Why don't the workers revolt?

I would like to suggest that perhaps part of the reason is because the same transformations that have positioned women as workers relative to reproduction have also offered up to them the pleasures of reproduction construed as consumption. The seeds of this approach already lie within the feminist critique by analogy to production. For if women are the un-skilled workers on the reproductive assembly line, and doctors are the fore-men supervising their work, and the fetus is the valuable commodity being produced, then it seems only natural to ask: Who are the consumers? What do they consume and how? In the discussion to follow, I shall suggest sev-eral ways in which reproduction has come to be construed in terms of consumption.

AN ALL-CONSUMING EXPERIENCE

Consuming on Behalf of the Fetus

It began to dawn on me that ultrasound has a lot to do with consumption, only after many months of listening to women tell me so. For a period of just under one year, I conducted ethnographic research in the context of an ob/gyn ultrasound clinic based in a Chicago hospital, in the course of which I interviewed over one hundred women and, with their permission, was present as an observer during their ultrasound examinations. I was interested in seeing how the use of ultrasound in medicine reflected and was shaped by meanings attached to ultrasound in the broader culture. Among the questions that I would ask women was whether they were inter-ested in finding out the sex of the fetus if it could be visualized—and if so, why. Time and time again, women told me that they were looking forward to the ultrasound examination because they were hoping to learn the sex, so that they could start buying things for the baby.

"I'm just curious, and I want to shop."

"I hope it's a girl, I'll get dresses."

"I'd rather know, to see what type of clothes to buy."

This kind of response seemed so banal that I kept looking for deeper or more interesting answers. It was only after some time that I began to recognize that there were more serious and absorbing issues at stake in the relationship between ultrasound technology and women's desire to pur-chase pink or blue baby clothes—namely, the question of how reproduc-tion has come to be constructed as a matter of consumption.

The purchase during pregnancy of mass-produced consumer goods in-tended for the anticipated child is, on one level, simply a rather unremark-able extension to the period before birth of the phenomenon of parents buying stuff for their kids. In the early part of the twentieth century, at-tempts to target children's needs as a potential marketing opportunity seemed to threaten an intrusion of the profane world of business into the sacred space of the home, as Daniel Thomas Cook demonstrates in his historical study of the children's apparel industry. This potential conflict was resolved, according to Cook, "by recasting the expression of mother-hood as consumer practice" ("Mother" 519). By now, eighty years later,

Americans take for granted that consuming on behalf of one's children is an important parental responsibility. And the material trappings of middle-class childhood in contemporary America are legion—including not only baby clothes but toys, books, toiletries, strollers, furniture, car seats, breast pumps, baby monitors, and much else.[10]

Many women resist engaging in consumption of baby goods during pregnancy out of a sense of caution that something might go wrong, preferring to wait until just before the birth and then to buy only what they consider basic necessities. In the context of American society, where ubiquitous advertisements promise sex, happiness, and success through the purchase of goods, the notion that consumption during pregnancy could be seen as unwise or even dangerous ("tempting fate") might seem to be a curious superstition, an incongruous relic of a premodern system of beliefs. This sort of ambivalence toward consumption on behalf of the fetus is perhaps better understood, however, as a tacit recognition of the extent to which consumption of commodities functions to construct identity in contemporary American society. Buying things for the fetus on some level amounts to recognizing it as an individual consumer, a baby, a person. A woman might not wish to grant such recognition, or she might wish to embrace and proclaim it, depending upon a number of factors, including how far along she is in her pregnancy.[11]

Most women who carry their pregnancies to term usually do end up engaging in some consumption of goods "on behalf of the fetus," if only passively, thanks to the American custom of the baby shower. The baby shower—which has evolved from a ritual passing on of knowledge among women to become more an example of what Leigh Eric Schmidt calls "consumer rites" in American culture—conventionally takes place in the last month or two of pregnancy. One item that family and friends often like to buy as a shower gift is clothing for the baby, and convention dictates that girls should wear pink clothes and boys should wear blue (although we might note that this convention is relatively recent, dating back only to the 1940s [see Paoletti and Kregloh]). In this context, given that women undergoing prenatal diagnostic tests in this country are commonly offered the option of finding out fetal sex, it is easy to see that women's attitudes toward, and experience of, ultrasound and other prenatal diagnostic technologies might be shaped by cultural imperatives of consumption.

Consumption as Pregnancy

Not only do women engage in consumption during pregnancy, however, but consumption to a significant degree constitutes the experience of pregnancy, especially in its early stages. Arjun Appadurai suggests that: "consumption creates time and does not simply respond to it. . . . Where repetition in consumption seems to be determined by natural or universal seasonalities of passage, always consider the reverse causal chain, in which consumption seasonalities might determine the style and significance of 'natural' passages" (*Modernity* 70). Building upon this insight, I would like to consider here the ways in which transformations in consumption in fact serve to create the "time" of early pregnancy and determine the significance of the "natural" passage into pregnancy. Specifically, consumption constitutes the chief avenue of control for women who are just beginning to conceptualize themselves as mothers, and also it is the primary performative arena where they may exemplify their mothering skills for others.[12]

As Faye D. Ginsburg points out in *Contested Lives,* her ethnographic study of the American abortion debate, women's relationship to reproduction has been transformed in recent decades, such that even conservative defenders of traditional values now frame motherhood not as women's biological destiny but as the result of a conscious decision to embrace their reproductive potential. For a great many women of all political persuasions, the decision to embrace their reproductive potential, and to enter a state of being possibly pregnant, follows an extended period of engaging in sexual activity while using birth control. This transition is often both marked and effected by a transformation in the meaning and practice of consumption. Long before it becomes possible to feel the fetus moving or to see the belly bulging, often before pregnancy is confirmed or even attempted, the transition to hoped-for motherhood may be experienced as a shift to a new, more highly disciplined regime of consumption—it is the movement from being an individual consumer to a mother-as-consumer.

This change concerns, in the first place, consumption in its most literal meaning—eating and drinking. Indeed, for many women, a change in patterns of consuming food, drink, and drugs both precedes pregnancy and in some sense "causes" it. The daily ritual of swallowing a birth control pill, for example, gives way to the swallowing of vitamins high in folic acid, foods full of nutrients, and waters labeled and sold as free of contaminants. Many women temporarily forswear (or try to) some of the daily pleasures of consumption, such as giving up alcohol, tobacco, caffeine, and other substances thought to be harmful, as well as avoiding over-the-counter drugs.[13] For many, the knowledge that one is pregnant comes as soon as several days after a missed period, in the form of a mass-produced, disposable, over-the-counter device a woman must go buy in her local drugstore: the home pregnancy test. And one watches for early symptoms of pregnancy to manifest themselves as disturbances in normal patterns of consumption of food: nausea and cravings.

It is not only patterns of consumption that change when it begins to be on behalf of the fetus but so too does the meaning of consumption. Suddenly, consumption is invested with new levels of moral significance—consumption is cast as an act of maternal love and an expression of a woman's strength of character and powers of self-discipline, even as consumption is also seen to literally create the fetal body. As a popular book on diet during pregnancy advises women, "Not only are you what you eat, but your baby is, too" (Eisenberg, Murkoff, and Hathaway 16). Do you love your future child enough to give up coffee? Are you dedicated enough to resist sweets? How diligent are you in reading the labels of everything you eat and drink? A woman's first maternal duty, it would seem, is to act as an intelligent and effective filter between the fetus within her and the world outside, letting in the good and keeping out the bad; and the primary locus of this responsibility is her choice of what substances to consume.[14]

To transform one's patterns of consumption in this way is difficult, of course, demanding constant vigilance and self-denial. It is perhaps not surprising, therefore, that many women whose experience of early pregnancy is defined in good measure by the burdens of newly significant consumption might welcome the prospect of an ultrasound examination as providing tangible evidence that (as so many women put it) "there really is a baby in there"—a baby whose existence gives meaning to the sacrifice of all those passed up (or thrown up!) desserts, cigarettes, cups of coffee, and glasses of wine.[15]

Indeed, careful consumption of food even appears to promise a means by which the individual woman (and her fetus) may transcend the effects of environmental damage—harm wrought, in large part, as a by-product of existing societal and global patterns of consumption. One popular book on diet during pregnancy advises women that, "Even with so much in our environment out of our control . . . today, having a healthy baby is, most of the time, more up to us than up to chance. . . . Eating . . . is an area in which there are enormous possibilities for control" (Eisenberg, Murkoff, and Hathaway 12). Consumption of food, in other words, is presented as the way to deal with the problems that consumption of goods causes. Women engaged in reproducing the next generation are thus encouraged to regard their scope of effective action as limited to the realm of consumer choices, which do not challenge broader social and economic structures— and which if anything serve only to reproduce them. We may see here one example of how "reproduction, in its biological and social senses, is inextricably bound up with the production of culture" (Ginsburg and Rapp 2). We may also see how consuming on behalf of the fetus reprises long-standing cultural associations between consumption and gender. The idea that consumption generally and consumption of food in particular are avenues through which women may exercise "control" has, of course, a lengthy history in American society; the conscientious pregnant woman of today is treading a path already well-worn by women shoppers and dieters through the years.[16]

Consuming Pregnancy as a Commodified Experience

The sense that consumption carries heightened significance during pregnancy is not, however, limited to the physical consumption of food and other substances. The experience of pregnancy has come to be a more or less standardized product, available in a range of varieties to suit individual consumer preferences. This, I suggest, is a third way in which reproduction is linked to consumption: consuming pregnancy as a commodified experience.

Thanks in large part to the impact of feminism and the women's health movement in this country, women who enjoy good medical coverage may now choose from among several options: what kind of care to seek during pregnancy and from whom, what kind of education or preparation, and where and how to plan on giving birth. Robbie E. Davis-Floyd points out that, "At present in our society, the culturally recognized spectrum of possible beliefs about pregnancy and birth is encompassed by two basic opposing models, or paradigms, which are available to pregnant women for the perception and interpretation of their pregnancy and birth experiences— the technocratic and wholistic models" (155). While most women still pursue care from doctors and give birth in hospitals, growing numbers choose to seek prenatal care from nurse-midwives, and a small but significant minority give birth at home under the care of lay midwives.

As with the choice of the provider and setting for prenatal care and birth, educational classes for pregnant women may also range widely along the ideological spectrum, from those which hew closer to the "technocratic" model of birth to those which reflect more "wholistic" forms. In addition to childbirth education classes, women may attend other classes and groups, on topics ranging from exercise during pregnancy to breastfeeding to infant massage. Pregnant or trying-to-become-pregnant women have also been discovered as a large and lucrative market for educational or "self-help" books offering information and advice on pregnancy, conception, childbirth, and infant care—again, written from positions between the technocratic and wholistic ideological poles. A recent glance at the pregnancy section of a branch of the national bookstore chain Borders, for example, turned up no less than 114 titles.

The image of *Spiritual Midwifery* (Gaskin) nestled up to *Which Tests for My Unborn Baby?* (De Crespigny and Dredge) on the bookstore shelf hints at the measure to which reproduction has come to be construed in terms of consumption. Women who pursue a wholistic approach to pregnancy and childbirth might be regarded, in terms of the feminist argument by analogy to production, as workers in revolt; I suggest, however, that we must (also) recognize them as consumers in action. The ideological opposition between wholistic and technocratic models of pregnancy and childbirth plays out against the backdrop of a consumer culture and a class structure that remain fundamentally unchallenged. When a woman plans and contracts for a particular kind of prenatal care, for a specific type of birth, she is (to the extent that such a thing is possible) engaging to purchase a distinct pregnancy experience. Women who choose a "high-tech" approach and those who choose a "natural" approach may find each other's decisions incomprehensible, unenlightened, or silly, but the differences that separate them are ultimately less significant than the commonalities—namely, that a woman demonstrates her powers and her talents as a consumer, and engages in the construction of her identity, by the manner in which she consumes her pregnancy and birth.[17]

In this context, as part of the consumption of pregnancy as a commodified experience, women's demand for ultrasound is perhaps more comprehensible. Women from different positions along the ideological spectrum have come to regard ultrasound as an important standard feature of the larger package deal, if not always for the same reasons. As a glitzy high-tech medical procedure, ultrasound may seem to be evidence that one is receiving the best that modern medical science has to offer. But given that usually no problems are detected, the ultrasound examination also seems to promise an emotionally gratifying moment of "reassurance" and "bonding" that many have come to characterize as a not-to-be-missed part of the experience of pregnancy.[18]

Consuming the Fetus

I have suggested a number of ways in which I believe reproduction has come to be constructed not only in terms of production but also as a matter of consumption. Women engage in consumption of baby clothes and other mass-produced goods on behalf of the fetus during pregnancy; consumption (especially of food) to a considerable extent constitutes the experience of pregnancy, especially in its early stages; and women approach prenatal medical care and education as the consumption of a commodified pregnancy experience. At this point, I would like to return to the most obvious, but also the most unsettling, implication of this argument—namely, the suggestion that women "consume" their fetuses. What might it mean to say that the fetus is a commodity? In what sense does one consume such a commodity?

Much of the theoretical literature on commodities turns upon the distinction between use value and exchange value—commodities are items that are produced not in order to be used directly but in order to be exchanged for money (which may then be used to purchase other commodities) in a capitalist market. Appadurai, in an important departure from this traditional definition, proposes that we understand the commodity not as something that belongs to a certain class of objects (those produced for exchange) but as something which occupies, even temporarily, the "commodity situation—the situation in which its exchangeability is its socially relevant feature" (*Social* 13). In the case of the fetus, however, both of these approaches seem somewhat inadequate.

First of all, in the case of the fetus, exchange cannot be neatly separated from or opposed to consumption. Ultrasound (along with other prenatal diagnostic technologies) objectifies the fetus in ways that not only make possible certain exchanges but also make the fetus available for the pregnant woman to possess and enjoy in new ways. An ultrasound examination produces visual imagery and medically authorized information about the fetus—its size, sex, appearance, heart rate, position, activities, and so forth—which endow it with an objectified existence. In this form, the fetus may be "consumed" by a pregnant woman, who may take considerable pleasure in being able to not only feel its presence within her body but also see it, name it, show it around to others, and construct for it the rudiments of a personal and social identity (a point to which we shall return in a moment). At the same time, the fetus as externalized in this form may in certain contexts be said to have an "exchange value." For example, a fetus displaying some particularly rare and interesting anomaly may have substantial exchange value within the obstetrician's professional world, leading to publications, research presentations, and other professional rewards. The pregnant woman carrying this same fetus would be compelled first to either accept or reject the equivalence being made between the fetus within her and the fetus as externalized through ultrasound diagnosis and imagery, and then to weigh the value of this fetus against the possible futures (or the possible future fetuses) for which it might be exchanged. And, in cases where no problems are detected, the fetus may be said to have other sorts of exchange values in the context of a pregnant woman's social world. The image, and additional bits of information gleaned from the ultrasound examination, are exchanged with others and valued as tangible evidence of the presence of the fetus as a new person to be incorporated into networks of family and kin.

Clearly, just as exchange is entangled with consumption in the case of the fetus, so too is commoditization inextricably bound up with personification. The visual image of the fetus on the screen, the take-home Polaroid snapshot, the diagnosis, the medically authorized knowledge that it is a boy or a girl, and the narrative descriptions the sonographer provides in the course of the ultrasound examination all contribute to the process by which a pregnant woman and the people around her construct for her fetus a social identity. Lisa Meryn Mitchell, drawing upon ethnographic fieldwork conducted in an ultrasound clinic in Montreal, has detailed the ways in which "sonographers' accounts of the fetal image for parents describe the fetus as a social being, with a specific social identity, and possessing intention, consciousness, emotion and communicative ability" (184). My research confirms this, as I regularly observed how visible physical features or movements of the fetus were translated into terms that served to create for it a personality and an identity. To take just one example, consider the following dialogue which took place during an ultrasound examination among Sondra (the sonographer), Rosa (a twenty-year-old African-

American woman, seven months pregnant with her first child), and Rosa's friend (the younger sister of the baby's father):

Sondra: Here is the head. . . .

Rosa: That's the head? Look at that! Fills up the whole screen! That's big! Looks like his father. If you argue a lot, and fight, he'll come out looking just like his father. Look at that head!

Friend: That's just a superstition. (To Sondra) Can you tell which way he's facing?

Sondra: Baby's facing up, see, here's his profile.

Rosa: Oh, look at that! Baby's got a huge forehead! Baby's got an *ugly* head!

Sondra: Baby's gonna hear you say that, you'll get it when he gets out!

Rosa: Look at that big old forehead, looks just like his father—oooh!

Friend: What's wrong with him?!

Rosa: He's no good.

Friend: She's talking about my brother right in front of me. . . . I'm gonna have an *ugly* nephew!

One need not take such comments, which seemed clearly to be good-humored teasing, literally to nonetheless see this conversation as illustrating how, even before birth (and even in jest!), the information and visible traces of the fetus that the obstetrical ultrasound offers have been incorporated into the social and cultural processes through which it is constructed as a son, a nephew, an "ugly" child, a child who can hear and may respond to things his mother says, and a boy who is "just like his father."

Through obstetrical ultrasound as it is practiced in the United States, then, the fetus is commoditized and personified, "produced" as an object for exchange and for consumption. To adopt the terminology Igor Kopytoff proposes, ultrasound commoditizes the fetus (that is, makes it exchangeable) in the same movement that it also singularizes it (that is, endows it with a singular social identity).

MIXING METAPHORS: THE HUNGRY CONSUMER AND THE PREGNANT WOMAN

To consider the fetus in terms of questions about commodification and consumption might seem an obvious enough move to make. Consumption practices are involved in the experience of pregnancy in a number of ways, as we have explored above. Furthermore, popular debates about abortion and new reproductive technologies, like theoretical discussions of consumption and commodification, revolve around the question of how best to understand and adjudicate the relationship between "persons" and "things."[19] However, applying notions of consumption and commodification to the topic of reproduction turns out to be somewhat tricky, as we have seen. I would like to suggest here that this difficulty emerges, at least in part, out of the way in which implicit metaphors operate in social theory (no less than in medical practice!). The body metaphors understood in theories of consumption, I suggest, mix badly with the metaphorical associations of procreation.[20]

The concept of consumption, now taken within social theory to refer to all kinds of uses of goods and services, still retains heavy traces of its older association with eating (Williams). Indeed, *Merriam-Webster's Collegiate Dictionary*, tenth edition, defines "consumer" as: "one that consumes: as **a**: one that utilizes economic goods **b**: an organism requiring complex organic compounds for food which it obtains by preying on other organisms or by eating particles of organic matter" (249). This association between consumption (that is, purchase and use) of commodities and consumption (that is, ingestion) of food has, of course, deep historical roots. Sidney Mintz has traced an important part of this history in his account of the rise of sugar, arguing that the development of a taste for sucrose among the British poor was a crucial first step in their historical transformation into modern industrial workers and consumers (180; see also Sahlins). As Marshall Sahlins points out, this linkage has also left its mark upon the theoretical literature on consumption. Modern economic theory bears the traces of this connection, insofar as it presumes a view of the individual as a creature driven by needs which are construed as bodily hungers.

> In the world's richest societies, the subjective experience of lack increases in proportion to the objective output of wealth. Encompassed in an international division of labor, individual needs were seemingly inexhaustible. Felt, moreover, as physiological pangs, as deprivations like hunger and thirst, these needs seem to come from within, as dispositions of the body. The bourgeois economy made a fetish of human needs in the sense that needs, which are always social in character and origin and in that way objective, had to be assumed as subjective experiences of pain. ("Sadness" 401)

Mintz and Sahlins are, of course, two among many anthropologists whose work has enriched discussions of consumption with a critical focus on the social and cultural dimensions of goods and commodities, their exchange and use, and the needs they are taken to satisfy.[21] A large body of literature has also grown up around the topic of gender and consumption, highlighting especially the role of women as primary consumers of commodities for the household.[22] There is also a vibrant, emerging literature within anthropology that underscores specifically the intersections between changing regimes of consumption and processes of commodification of bodies and persons, primarily in non-Western contexts.[23] Too often, however, consumption continues to be defined exclusively as the purchase and use of (mass-produced) material things. For example, Daniel Miller's recent ethnographic study of shopping in North London draws fascinating connections between consumption and kinship, particularly the mother-child relationship, but it nonetheless proceeds from an understanding of consumption as having to do only with material things (*Theory*); indeed, in Miller's recent review article on the anthropology of consumption and commodities, "almost all of the cited literature . . . consists of studies in anthropology and material culture" ("Consumption"). The anthropological literature on consumption thus still tends to preserve largely intact the image of a hungry subject inhabiting a world of needed and desired objects external to oneself.

Indeed, hunger is inscribed as the paradigm for consumption already within Karl Marx's definition of the "very queer thing" that is the commodity: "A commodity is, in the first place, an object outside us, a thing that by its properties satisfies human wants of some sort or another. The nature of such wants, whether, for instance, they spring from the stomach or from fancy, makes no difference" (199). Wants that "spring from the stomach" stand here for real physical need (to the extent that any such unmediated needs might be said to exist), while other sorts of desires, wants that spring "from fancy," are refigured as a sort of mental or spiritual hunger. The consumer, then, is a metaphysically hungry individual, a person whose desires are oriented toward some "object outside," whether food or another material object symbolically standing in its place.[24]

All of this makes it difficult to conceptualize reproduction in terms of consumption. The hungry consumer of social theory inhabits a body which, if not necessarily male (for after all, women too eat and feel hunger), is at least not easily imagined as a specifically pregnant body. The fetus is, moreover, emphatically not "an object outside us," and although I would argue that it may nonetheless satisfy "human wants of some sort or another," it is troublesome to imagine these wants as if they "spring from the stomach." How can the bearing of children be likened to the ingestion of food? The suggestion seems to invoke that most frightening of all monsters, the mother who eats her children. But if the awkwardness of looking at reproduction in terms of consumption serves to "wake up sleeping metaphors" (Martin, "Egg" 501), then the attempt may bring to light unexamined aspects of the theoretical apparatus with which we approach the topic of consumption. In particular, the exercise raises questions about how the implicit metaphor of the hungry consumer may hinder our critical understanding of ongoing transformations—transitions which are at once technological, social, cultural, and political—in the meaning and practice of reproduction.

CONCLUSIONS

As I have tried to show, obstetrical ultrasound plays a part in constructing the fetus more and more as a commodity at the same time and through the same means that it is also constructed more and more as a person. Furthermore, pregnant women are positioned as "consumers" through the same processes that have positioned them as reproductive "laborers." The feminist argument by analogy to production cannot easily account for the ways that obstetrical ultrasound technology has come to figure in the experience of pregnancy in contemporary U.S. society. I have argued that we must therefore consider reproduction in light of questions of commodi-fication and consumption, even if doing so requires that we struggle against gendered body metaphors implicit within these theoretical con-structs. The point, in doing so, is not to replace one analogy with another, much less to argue that consumption is liberating or free, but to view things from another angle—to try to see ultrasound, and the transformations in reproduction within which the routinization of ultrasound is situated, from the point of view of the women engaged in the all-consuming experience that pregnancy has become.

NOTES

Versions of this essay were presented at the 1997 annual meetings of the Society for Social Studies of Science (4S) and the American Anthropological Association (AAA), as well as at the conference Biotechnology, Culture, and the Body held at the Center for Twentieth Century Studies. Thanks to members of those audiences, especially Paul E. Brodwin, José Van Dijck, Caroline Seymour-Jorn, and Mary Maho-wald, for helpful and instructive comments and questions. Appreciation also to Daphne Berdahl, Jean Comaroff, Linda Layne, Lynn Morgan, and Michael Rosen-thal for comments on earlier drafts. The research on which this essay is based was supported in part by a Jacob K. Javits Graduate Fellowship from the U.S. Depart-ment of Education. I am particularly grateful to the women and the medical profes-sionals whose generosity and cooperation made this research possible. A revised version of this chapter is forthcoming in *Feminist Studies* under the title "Of Sono-grams and Baby Prams: Prenatal Diagnosis, Pregnancy, and Consumption."

1. Ultrasound is a technology that employs high-frequency sound waves to cre-ate visual images of internal bodily structures. Although used in many branches of medicine, it has been particularly important in the field of obstetrics, where it is utilized to visualize the ovaries, cervix, fetus, placenta, umbilical cord, and so forth. AFP blood screening tests for levels of alphafetoprotein in maternal blood; abnor-mal levels of this hormone may indicate that the fetus suffers neural tube defects (such as spina bifida or anencephaly) or Down's syndrome. The AFP blood test alone is not definitive, however, and in case of a "positive" result further testing is recommended. Amniocentesis is a procedure in which a doctor inserts a needle through a pregnant woman's abdomen to extract a small amount of the fluid con-tained in the amniotic sac surrounding the fetus; this fluid is then analyzed for the possible presence of a (steadily increasing) number of chromosomal and genetic disorders. Other less commonly used prenatal diagnostic techniques include chori-onic villi sampling (CVS) and fetoscopy.

On the routinization of ultrasound in obstetrics, see Blume; Kevles; Oakley,

Captured, "History"; Taylor, "Image"; and Wertz and Wertz. On the routinization of amniocentesis, see especially Rapp, "Accounting," "Moral," "Constructing," and *Testing;* on AFP blood screening, see Browner and Press.

2. The research setting was a hospital-based ob/gyn ultrasound clinic in Chicago, where I spent one or two days each week over a period of nearly one year, interviewing women patients and observing medical practice. This sort of clinic is one among a variety of sites at which obstetrical ultrasound may be performed in the United States at this time. Obstetricians or other doctors in private practice may own their equipment, and they either perform scans themselves or hire sonographers to do so; alternatively, obstetrical examinations may be performed by sonographers working in hospital-based radiology clinics, in freestanding medical-imaging clinics (that may also offer CT, MRI, X-ray, and so forth), or in mobile ultrasound services which contract to come regularly and perform ultrasound scans in clinics or private practices. The hospital-based clinic is thus not necessarily typical of the kind of site to which most women would go for ultrasound examinations during pregnancy, and one would naturally expect that medical practice might differ somewhat in various types of settings. There are, however, certain distinctive features of the obstetrical ultrasound examination which I believe remain relatively constant, and it is these which shall concern us here.

As they waited in the clinic waiting room, I explained the nature of my research to women who had come for obstetrical scans, and those who were interested agreed to allow me to accompany them into the examination room during the scan and conduct a semi-structured interview afterward. Participants were thus self-selected, and no attempt was made to limit the study only to women who were pregnant for the first time, or had never had an ultrasound examination before, or were at no known risk for congenital fetal abnormalities, or who had planned their pregnancies jointly with a partner in the context of a stable monogamous relationship. (These and other criteria have, by contrast, been used to select participants in some widely cited studies, suggesting that ultrasound screening, by facilitating maternal "bonding" with the fetus, may encourage women to comply with physicians' directives [to stop smoking, for example] and thereby improve the health of the fetus. I offer a critique of these studies, and of the notion that ultrasound has "psychological benefits," in my "Image").

Over one hundred women eventually took part in my study. This research in the clinic was supplemented by interviews with approximately thirty women residing in the vicinity of the clinic, whom I contacted through personal connections and through the informal circulation of a letter explaining my research. Again, participants were self-selected; some had had obstetrical ultrasound examinations many years ago, some recently, and some on a number of occasions over a period of months or years. I also formally interviewed fifteen sonographers (some practicing clinically, some working in the ultrasound industry, and some working in educational institutions), a few other medical professionals (midwives, radiologists, and obstetricians), and several engineers employed by a company that manufactures ultrasound equipment. This research is discussed at greater length in my dissertation.

3. In her more recent work, Emily Martin has shifted her critical gaze to the topic of immunity, and she has moved beyond the analysis set forth in *The Woman in the Body.* Metaphors drawn from nineteenth-century factory systems of production such as she detected in medical representations of female reproductive processes have, according to Martin, now given way to metaphors of "flexible specialization," reflecting the workings of the "post-Fordist" economy of the United States today (Martin, *Flexible,* "Reproduction"). The feminist argument by analogy

to production which her earlier work helped pioneer has, however, taken on a life of its own and remains influential.

4. Dorothy C. Wertz and Richard W. Wertz make much the same point. "The language of birth is especially telling. Production is still the dominant metaphor. Producing quality children is a job to be contracted out to a team of experts. Doctors speak of the management and progress of labor, of the uterus as an involuntary muscle" (267).

Of course, as Joel D. Howell's (1995) historical study of technology in American hospitals in the early twentieth century shows, medicine modeled itself upon industry in less metaphorical ways as well, and not only in the specialized field of obstetrics and gynecology. One chapter of Howell's fascinating study documents the impact of new technologies and techniques of accounting and record keeping, imported to medicine from the world of business and industry in the years between 1920 and 1925, upon the organization of the hospital in general and in particular upon the practice of surgery. Much remains to be written about the gender dynamics of the conjunction between imaging technologies such as ultrasound and new technologies for the management of information. We might note that most of the sonographers who do the work of producing diagnostic ultrasound imagery are women, and most of the work of scheduling, typing, filing, and otherwise managing the data produced in the course of these examinations is performed by female clerical staff. One woman's labor, it seems, becomes another woman's work.

5. The sonographer takes certain anatomical measurements of the fetus—for example, the diameter of the head (biparietal diameter, or BPD), the length from the top of the head to the rump (crown-rump length, or CRL)—and compares these to standard charts that researchers have developed over the years to obtain an estimate of the "gestational age" of the fetus. These estimates always have a margin of error expressed as a function of time ("plus or minus so-and-so many weeks") which increases as pregnancy proceeds.

6. More data means more work for doctors who interpret the data, but it also entails a great deal of clerical work, most of which is performed by others (see Wertz and Wertz 267).

7. On the experience of abortion following amniocentesis and a "positive" diagnosis of Down's syndrome, see Rapp, "XYLO." On the experience of pregnancy loss in the context of ultrasound diagnosis of fetal death, see Layne, "Fetuses."

8. A growing body of work in science and technology studies documents how technologies and their uses are socially and culturally shaped. For some starting points from which to explore this literature, see, for example, Bijker, Hughes, and Pinch; Bijker and Law; Jasanoff, Markle, Petersen, and Pinch; MacKenzie and Wajcman; and Wajcman.

9. Please see my "Image" for a more detailed discussion of tensions between "medical" and "non-medical" aspects of the "routine" obstetrical ultrasound examination in the United States.

10. One of the features of Storksite, an Internet website geared toward pregnant women and their "pregnant partners," is its collection of "babygrams," a series of one-page bulletins tailored to each week of pregnancy, containing advice and information and including an illustration of "how your baby is growing." The babygram for week 29 of pregnancy (which features not a line drawing, like most other babygrams, but a "29 week sonogram") offers expectant parents a list of items they will "need": crib, crib mattress, bumpers, bassinet, diaper pail, changing space, baby tub, tub seat, bed linens, stroller, umbrella stroller, carriage or pram, infant car seat, child car seat, baby carrier, diaper bag, bottles, disposable bottle liners, nipples, breast pump, bottle brushes, undershirts, drawstring gowns, diaper

wraps and covers, burp pads, booties and socks, pull-on stretch pants, jumpers and one-piece outfits, hooded terry cloth towels, hats, snowsuit/bunting outfit, sweaters, diapers, high chair, playpen, safety gates, drawer and cabinet latches, outlet plug covers, stove knob covers, infant Tylenol, syrup of ipecac, alcohol, cotton balls, sunscreen, calibrated dropper, bulb syringe, nail clippers, thermometer, baby bath soap, diaper rash ointment, petroleum jelly, diaper wipes, baby brush and comb, and Pedialyte.

Storksite also offers "partnergrams" aimed at pregnant partners; the partnergram for month 7 (weeks 28–31) contains another list of recommended items. As the site's authors explain, "For the times you are away, whether at work or travelling, leaving something behind strengthens the Parent and Partner bond." They suggest meals (arrange to have a meal delivered, then "call at mealtime and tell them you wish you were home"), tapes ("breathing tapes" to help the pregnant woman practice her breathing exercises, a tape of the parent singing songs or reading stories), as well as pictures: "Take a clear picture of your face and make a few 8 x 10 prints, perhaps putting them in plastic jackets for wearability. Put one in the crib, and one or two in play areas at eye level. . . . [I]t's a great touch for Parent Partners who have to take business trips or are in the military." It is interesting to note how these material objects are used to create some form of presence for an absent other—tangible and visible evidence of a baby-to-be which is still hidden in the womb, or physical traces of an absent father.

11. Linda Layne has developed this point insightfully, in the context of discussing consumption practices engaged in by women and couples mourning pregnancy loss. She argues that the purchase and giving of consumer goods to, from, or in memory of a lost baby serves to enact a "processual-relational" view of fetal personhood, which differs in important respects from "essentialist" models of personhood also current in American culture ("'I Remember'" 254). Similarly, Margerete Sandelowski discusses the consumption practices of couples awaiting adoption after prior experience of infertility, who struggle between the desire to buy items for the awaited baby and the sense that doing so may intensify the pain of the waiting period (172).

12. Thanks to Paul E. Brodwin for suggesting this final point in his helpful comments on an earlier draft of this essay.

13. See Markens, Browner, and Press for an ethnographically informed discussion of women's dietary practices during pregnancy. The authors examine women's accounts of their dietary practices as a site for the investigation of issues surrounding maternal responsibility and concepts of maternal-fetal conflict. They argue that this "focus on diet is indicative of the extreme individualism in American culture and medicine," and it reflects a growing emphasis on "the exclusive or nearly exclusive role of maternal responsibility for fetal outcome" (369).

14. See Markens, Browner, and Press; and see Daniels (ch. 4) for a discussion of "crack babies" and the criminal prosecution of women for consumption of drugs during pregnancy.

15. Sandelowski notes that ultrasound plays this role for women who have difficulty making the transition from infertility to pregnancy. "One woman recalled having a hard time picturing herself pregnant. Both she and her husband required the first half of pregnancy to 'change gears' from being not-pregnant to pregnant. She had asked an ultrasound technician to convince her she was pregnant: 'they are all saying that I'm pregnant, but I'm not convinced.' By the second trimester, 'the physical manifestations of pregnancy' were there, and everyone 'acted' like a baby was there when she had her amniocentesis, but it was seeing the fetal spine on ultrasound—something 'surefire'—that finally convinced her that she was having a baby" (126).

16. In Fay Weldon's novel *The Cloning of Joanna May,* Alice—one of the sixty-year-old Joanna's four thirty-year-old clones—volunteers to be the one to give birth to a clone of Joanna's dead husband Carl, "on condition she didn't have to rear him." Why would Alice, who has never had any particular fondness for children, be the best candidate among the four for childbearing? Because Alice, as a former fashion model, "didn't smoke, didn't drink, watched her health and her moods—trained as she was in keeping her body well under control" (265).

17. Robbie E. Davis-Floyd makes much the same point (although I believe she could pursue further its implications) in stating that she focused her research "on women who were able to go to private obstetricians and to pay for private (i.e., non-hospital sponsored) childbirth education classes, because of the greater number of options presumably open to such women for exercising individual choice in their childbirth experiences" (4).

18. Much more could be said about "reassurance" and "bonding" (see my "Image"), but we should at least note that the reassurance ultrasound offers is always tentative, since not all problems can be detected, and problems can also arise later in pregnancy.

19. See the work of Rosalind Pollack Petchesky for thoughtful discussions of the fetus as a fetish ("Fetal") and of the liberatory potential of theories of the body as property for feminist movements ("Body").

20. On "reproduction" as a metaphor, and on some other ways in which folk models of procreation remain implicit within anthropological theory, see Delaney.

21. See, for example, Appadurai, *Social, Modernity;* Bourdieu; Douglas and Isherwood; McCracken; Miller, *Acknowledging.* For recent reviews, see Ferguson; and Miller, "Consumption."

22. For more on gender and consumption, see, for example, Spigel and Mann; and de Grazia and Furlough.

23. Analyses of the interplay between consumption and commodification in non-Western contexts appear in, for example, Comaroff; Fisiy and Geschiere; Kopytoff; Scheper-Hughes; Valeri; Weiss, "Northwestern," *Making;* and White.

24. And if, as Arjun Appadurai suggests, consumption is often mistakenly understood to be the "end of the road for goods and services" (*Modernity* 66), this too is perhaps in part due to the subtle workings of the metaphor of eating at the heart of consumption. For after all, if one eats something up, then it is all gone, right?

WORKS CITED

Appadurai, Arjun. *Modernity at Large: Cultural Dimensions of Globalization.* Minneapolis: University of Minnesota Press, 1996.

——, ed. *The Social Life of Things: Commodities in Cultural Perspective.* Cambridge: Cambridge University Press, 1986.

Arditti, Rita, Renate Duelli Klein, and Shelley Minden, eds. *Test-Tube Women: What Future for Motherhood?* London: Pandora, 1984.

Bijker, Wiebe E., Thomas P. Hughes, and Trevor Pinch, eds. *The Social Construction of Technological Systems: New Directions in the Sociology and History of Technology.* Cambridge: MIT Press, 1987.

Bijker, Wiebe E., and John Law, eds. *Shaping Technology/Building Society: Studies in Sociotechnical Change.* Cambridge: MIT Press, 1992.

Blume, Stuart S. *Insight and Industry: On the Dynamics of Technological Change in Medicine*. Cambridge: MIT Press, 1992.

Bourdieu, Pierre. *Distinction: A Social Critique of the Judgment of Taste*. Trans. Richard Nice. Cambridge: Harvard University Press, 1984.

Browner, Carole H., and Nancy Ann Press. "The Normalization of Prenatal Diagnostic Screening." Ginsburg and Rapp 307–22.

Comaroff, Jean. "Consuming Passions: Child Abuse, Fetishism, and 'The New World Order.'" *Culture* 17.1–2 (1995): 7–19.

Comaroff, Jean, and John L. Comaroff. "Goodly Beasts and Beastly Goods: Cattle and Commodities in a South African Context." *American Ethnologist* 17 (1990): 196–216.

Cook, Daniel Thomas. "The Commoditization of Childhood: Personhood, the Children's Wear Industry, and the Moral Dimensions of Consumption, 1917–1967." Diss. University of Chicago, 1998.

———. "The Mother as Consumer: Insights from the Children's Wear Industry, 1917–1929." *The Sociological Quarterly* 36.3 (1995): 505–22.

Corea, Gena, et al. *Man-made Women: How New Reproductive Technologies Affect Women*. Bloomington: Indiana University Press, 1987.

Daniels, Cynthia R. *At Women's Expense: State Power and the Politics of Fetal Rights*. Cambridge: Harvard University Press, 1993.

Davis-Floyd, Robbie E. *Birth as an American Rite of Passage*. Berkeley: University of California Press, 1992.

De Crespigny, Lachlan, and Rhonda Dredge. *Which Tests for My Unborn Baby? A Guide to Prenatal Diagnosis*. Melbourne: Oxford University Press, 1991.

De Grazia, Victoria, and Ellen Furlough. *The Sex of Things: Gender and Consumption in Historical Perspective*. Berkeley: University of California Press, 1996.

Delaney, Carol. *The Seed and the Soil: Gender and Cosmology in Turkish Village Society*. Berkeley: University of California Press, 1991.

Douglas, Mary, and Baron Isherwood. *The World of Goods: Towards an Anthropology of Consumption*. 1979. New York: Routledge, 1996.

Eisenberg, Arlene, Heidi Eisenberg Murkoff, and Sandee Eisenberg Hathaway. *What to Expect When You're Expecting*. New York: Workman, 1986.

Ewigman, Bernard, et al. "Effect of Prenatal Ultrasound Screening on Perinatal Outcome." *New England Journal of Medicine* 329.12 (1993): 821–27.

Ferguson, James. "Cultural Exchange: New Developments in the Anthropology of Commodities." *Cultural Anthropology* 3 (1988): 488–513.

Fisiy, Cyprian F., and Peter Geschiere. "Sorcery, Witchcraft, and Accumulation: Regional Variations in South and West Cameroon." *Critique of Anthropology* 11.3 (1991): 251–78.

Franklin, Sarah, and Helena Ragoné, eds. *Reproducing Reproduction: Kinship, Power, and Technological Innovation*. Philadelphia: University of Pennsylvania Press, 1998.

Gabbe, Steven G. "Routine versus Indicated Scans." *Diagnostic Ultrasound Applied to Obstetrics and Gynecology*. Ed. Rudy E. Sabbagha. 3rd ed. Philadelphia: Lippincott, 1994. 67–76.

Gaskin, Ina May. *Spiritual Midwifery*. 3rd ed. Summertown: Book Publishing, 1993.

Ginsburg, Faye. *Contested Lives: The Abortion Debate in an American Community*. Berkeley: University of California Press, 1989.

Ginsburg, Faye D., and Rayna Rapp, eds. *Conceiving the New World Order: The Global Politics of Reproduction*. Berkeley: University of California Press, 1995.

Howell, Joel D. *Technology in the Hospital: Transforming Patient Care in the Early Twentieth Century*. Baltimore: Johns Hopkins University Press, 1995.

Hubbard, Ruth. *The Politics of Women's Biology.* New Brunswick: Rutgers University Press, 1990.

Jasanoff, Sheila, Gerald E. Markle, James C. Petersen, and Trevor Pinch, eds. *Handbook of Science and Technology Studies.* Thousand Oaks: Sage, 1995.

Kevles, Bettyann Holzmann. *Naked to the Bone: Medical Imaging in the Twentieth Century.* New Brunswick: Rutgers University Press, 1997.

Kopytoff, Igor. "The Cultural Biography of Things: Commoditization as a Process." Appadurai 64–91.

Layne, Linda. "'I Remember the Day I Shopped for Your Layette': Consumer Goods, Fetuses, and Feminism in the Context of Pregnancy Loss." *Fetal Subjects, Feminist Positions.* Ed. Lynn M. Morgan and Meredith W. Michaels. Philadelphia: University of Pennsylvania Press, 1999. 251–78.

———. "Of Fetuses and Angels: Fragmentation and Integration in Narratives of Pregnancy Loss." *Knowledge and Society* 9 (1992): 29–58.

MacKenzie, Donald, and Judy Wajcman, eds. *The Social Shaping of Technology.* Philadelphia: Open University Press, 1985.

Markens, Susan, Carole H. Browner, and Nancy Ann Press. "Feeding the Fetus: On Interrogating the Notion of Maternal-Fetal Conflict." *Feminist Studies* 23.2 (1997): 351–72.

Martin, Emily. "The Egg and the Sperm: How Science Has Constructed a Romance Based on Stereotypical Male-Female Roles." *Signs* 16.3 (1990): 485–501.

———. *Flexible Bodies: Tracking Immunity in American Culture from the Days of Polio to the Age of AIDS.* Boston: Beacon, 1995.

———. "From Reproduction to HIV: Blurring Categories, Shifting Positions." Ginsburg and Rapp 256–69.

———. *The Woman in the Body: A Cultural Analysis of Reproduction.* Boston: Beacon, 1987.

Marx, Karl. *The Marx-Engels Reader.* Ed. Robert C. Tucker. 1867. New York: Norton, 1972.

McCracken, Grant. *Culture and Consumption: New Approaches to the Symbolic Character of Consumer Goods and Activities.* Bloomington: Indiana University Press, 1986.

Merriam-Webster's Collegiate Dictionary. 10th ed. Springfield, MA: Merriam-Webster, 1993.

Miller, Daniel, ed. *Acknowledging Consumption: A Review of New Studies.* London: Routledge, 1995.

———. "Consumption and Commodities." *Annual Review of Anthropology* 24 (1995): 141–61.

———. *A Theory of Shopping.* Ithaca: Cornell University Press, 1998.

Mintz, Sidney. *Sweetness and Power: The Place of Sugar in Modern History.* New York: Penguin, 1985.

Mitchell, Lisa Meryn. "Making Babies: Routine Ultrasound Imaging and the Cultural Construction of the Fetus in Montreal, Canada." Diss. Case Western Reserve University, 1993.

Oakley, Ann. *The Captured Womb: A History of the Medical Care of Pregnant Women.* Oxford: Blackwell, 1984.

———. "A History Lesson: Ultrasound in Obstetrics." *Essays on Women, Medicine and Health.* Edinburgh: Edinburgh University Press, 1993.

Paoletti, Jo B., and Carol L. Kregloh. "The Children's Department." *Men and Women: Dressing the Part.* Ed. Claudia Kidwell and Valerie Steele. Washington, D.C.: Smithsonian Institution, 1989.

Petchesky, Rosalind Pollack. "The Body as Property: A Feminist Re-vision." Ginsburg and Rapp 387–406.

————. "Fetal Images: The Power of Visual Culture in the Politics of Reproduction." *Feminist Studies* 13.2 (1987): 263–92.

Radin, Margaret. *Contested Commodities: Problems with the Trade in Sex, Children, Body Parts, and Other Things.* Cambridge: Harvard University Press, 1996.

Rapp, Rayna. "Accounting for Amniocentesis." *Knowledge, Power, and Practice: The Anthropology of Medicine and Everyday Life.* Ed. Shirley Lindenbaum and Margaret Lock. Berkeley: University of California Press, 1993. 55–76.

————. "Constructing Amniocentesis: Maternal and Medical Discourses." *Uncertain Terms: Negotiating Gender in American Culture.* Ed. Faye Ginsburg and Anna Lowenhaupt Tsing. Boston: Beacon, 1990. 28–42.

————. "Moral Pioneers: Women, Men, and Fetuses on a Frontier of Reproductive Technology." *Gender at the Crossroads of Knowledge: Feminist Anthropology in the Postmodern Era.* Ed. Micaela di Leonardo. Berkeley: University of California Press, 1991. 383–95.

————. "Refusing Prenatal Diagnosis: The Meanings of Bioscience in a Multicultural World." *Science, Technology, and Human Values* 23.1 (1998): 45–70.

————. *Testing Women, Testing the Fetus: The Social Impact of Amniocentesis in America.* New York: Routledge, 1999.

————. "XYLO: A True Story." Arditti, Klein, and Minden 313–28.

Rothman, Barbara Katz. *Recreating Motherhood: Ideology and Technology in a Patriarchal Society.* New York: Norton, 1989.

————. *The Tentative Pregnancy: Prenatal Diagnosis and the Future of Motherhood.* New York: Viking, 1986.

Sahlins, Marshall. "Cosmologies of Capitalism: The Trans-Pacific Sector of 'The World System.'" *Proceedings of the British Academy* 74 (1988): 1–51.

————. *Culture and Practical Reason.* Chicago: University of Chicago Press, 1976.

————. "The Sadness of Sweetness: The Native Anthropology of Western Cosmology." *Current Anthropology* 37.3 (1996): 395–428.

Sandelowski, Margerete. *With Child in Mind: Studies of the Personal Encounter with Infertility.* Philadelphia: University of Pennsylvania Press, 1993.

Scheper-Hughes, Nancy. "Theft of Life: The Globalization of Organ Stealing Rumours." *Anthropology Today* 12.3 (1996): 3–11.

Schmidt, Leigh Eric. *Consumer Rites: The Buying and Selling of American Holidays.* Princeton: Princeton University Press, 1995.

Spigel, Lynn, and Denise Mann. "Women and Consumer Culture: A Selective Bibliography." *Quarterly Review of Film and Video* 11 (1989): 85–105.

Stabile, Carol A. *Feminism and the Technological Fix.* Manchester: Manchester University Press, 1994.

Stanworth, Michelle, ed. *Reproductive Technologies: Gender, Motherhood, and Medicine.* Minneapolis: University of Minnesota Press, 1987.

Taylor, Janelle S. "Image of Contradiction: Obstetrical Ultrasound in American Culture." *Reproducing Reproduction.* Franklin and Ragoné 15–45.

————. "L'Echographie obstétricale aux Etats-Unis: Des Images contradictoires?" *Les Objets de la médicine.* Ed. Madeleine Akrich and Nicholas Dodier. Spec. issue of *Techniques et culture* 25/26 (1995).

Valeri, Valerio. "Buying Women but Not Selling Them: Gift and Commodity Exchange in Huaulu Alliance." *Man* 29 (1994): 1–26.

Wajcman, Judy. *Feminism Confronts Technology.* University Park: Pennsylvania State University Press, 1991.

Weiss, Brad. *The Making and Unmaking of the Haya Lived World.* Durham: Duke University Press, 1996.

————. "Northwestern Tanzania on a Single Shilling: Sociality, Embodiment, Valuation." *Cultural Anthropology* 12.3 (1997): 335–61.

Weldon, Fay. *The Cloning of Joanna May*. New York: Penguin, 1989.

Wertz, Dorothy C., and Richard W. Wertz. *Lying-In: A History of Childbirth in America*. Expanded ed. New Haven: Yale University Press, 1989.

White, Luise. "The Traffic in Heads: Bodies, Borders and the Articulation of Regional Histories." *Journal of Southern African History* 23.2 (1997): 325–38.

Williams, Raymond. *Keywords: A Vocabulary of Culture and Society*. New York: Oxford University Press, 1983.

Part III: Ethics and the Technological Subject

Chapter 7

Computerized Cadavers

Shades of Being and Representation in Virtual Reality

THOMAS J. CSORDAS

One way to address the question of what it means to be human is to begin with the observation that we have a world and inhabit a world. The inquiry unfolds under its own weight from this point, with the next set of questions necessarily having to do with how worlds (for they are always multiple) are constituted, what it means to have them, and how we inhabit them. In contemporary society, biotechnology, one of the central concerns of this volume, is increasingly implicated in transforming the bodily conditions for having and inhabiting any world. This is doubly the case when biotechnology includes sophisticated computer applications, since computers and computer networks are recognized as having enormous transformative potential. Indeed, psychologist Sherry Turkle has suggested important modulations of the self are in the making, and the philosopher Michael Heim has proposed that the computer is leading to a major ontological shift—a modulation in the structure of human reality.

Elaborating the cultural consequences of biotechnology applications of the computer with respect to the having and inhabiting of worlds requires, in my view, what can be called a cultural phenomenology ("Embodiment," *Embodiment, Sacred Self*). For present purposes, the critical feature of such an approach is focus on the interplay between cultural representations and cultural modes of being-in-the-world. Much recent cultural analysis privileges the pole of representation, with culture understood as constituted by symbols, signs, and images. From this standpoint, textuality is the most prominent metaphor guiding the interpretation of culture, and the world is not so much inhabited as represented in a way that can be read. While interpretively powerful, however, the notion of textuality is less apt for specifying cultural modes of being-in-the-world—that is, the kinds of engagement and participation of humans in our worlds—than is the complementary notion of embodiment. This notion places us at once at the most general and limiting condition of our existence. Our bodily existence, or embodiment, is from this viewpoint understood to have a range of potential experiential modalities in relation to features of cultural and historical context.

The interplay between representation and being-in-the-world, and the complementarity between textuality and embodiment, is at issue in biotechnology applications of the computer. First, the human body is the objective target of technology. By being taken up into the technological environment, it is represented and, I would suggest, has its being-in-the-world altered. Second, the computer user is the embodied subjective manipulator of the technology. In this capacity, a person encounters representations of the body and again, I would submit along the lines of Turkle and Heim, has its being-in-the-world altered. The example I offer in this essay involves the use of computer-generated virtual reality to create so-called virtual cadavers that are used for purposes such as the teaching of human anatomy and in computer-assisted surgery. The following discussion takes up these issues as they are being played out in the Visible Human Project, and it concludes with a reflection on their consequences for embodiment with respect to representation and being-in-the-world.

THE CREATION OF COMPUTERIZED CADAVERS

The creation of the Visible Human cadavers was an extraordinary technological feat, achieved with funding from the federal government. National Library of Medicine (NLM) director Donald Lindberg notes that the Visible Human Project originated with his observation in 1987 that the medical school community needed a better way to teach anatomy. In 1991, the NLM awarded a contract for development of the proposed data set to the University of Colorado Center for Human Simulation, with a subcontract for creation of three-dimensional volumetric visualizations of the computerized cadavers to the Visualization Group of the Scientific Computing Division of the National Center for Atmospheric Research in Boulder, Colorado. The first step was to find a suitable cadaver, beginning with a male. In 1993, after two and a half years of searching for a fresh cadaver that was "'normal' and within guidelines of size and age," a qualified thirty-nine-year-old Texas death row inmate named Joseph Paul Jernigan agreed to donate his body to science in exchange for being allowed to die by lethal injection rather than electrocution. This suited the researchers' purposes well, since electrocution modifies the tissues in ways that would defeat the purpose of having as lifelike a body as possible.

The second step was creation of three matching sets of images composed of transverse sections, or slices, of Jernigan's body. The first two sets were obtained by magnetic resonance imaging (MRI) and computed tomography (CT). Next, the body was encased in gelatin, frozen to minus 160 degrees Fahrenheit, and cut into four sections. It was then sectioned into 1,878 transverse slices, each 1 mm thick and corresponding exactly to the MRI and CT images. High-resolution color digital photographs were taken of the block after each slice. By the time a female cadaver—a fifty-nine-year-old woman who died of heart disease and whose family insisted she remain anonymous—was subjected to the same procedure a year later, the researchers decided they could achieve greater detail and higher resolution by making the slices one-third the thickness. Consequently, she was sectioned into 5,189 transverse slices. For both the male and female, the MRI and CT images were created from the whole body prior to freezing. The photographic images were of the face of each section, not as it was planed away by a custom-designed, laser-guided cryogenic macrotome. The physical remains thus became a collection of frozen shavings, which were then cremated, such that their digital remains now have, in some yet-to-be-understood sense that we will explore shortly, a more concrete existence.

Each of the three types of transverse images was digitally captured at 2,048 × 2,048 × 42 bits and aligned with its companion images and images of adjacent slices. The combined data sets are astronomically large: the male takes up 15 gigabytes and the female occupies 39 gigabytes. These data are stored on an FTP site, and with a free license they can be downloaded directly from the Internet.[1] Donald Lindberg states that "With the Visible Human Project, we are returning to the idea that a library holds the knowledge of a profession—not just reprints, journals, and books. The advent of technology gives us the opportunity to store knowledge electronically and distribute it, virtually instantaneously, throughout the world." A NLM project report from 1996 notes that already the Visible Human data "are being applied to a wide range of educational, diagnostic, treatment planning, virtual reality, artistic, mathematical, and industrial uses," and by the end of the year over seven hundred licenses had been issued to users in twenty-seven countries.

Let us explore the capabilities of the two virtual cadavers for producing computer images of the human body. The basic form is the transverse section (each in its own computer file), but because the images are precisely aligned, it is possible as well to produce vertical and horizontal sections through virtually any plane. More sophisticated programs are able to produce three-dimensional representations by stacking slices and isolating sites corresponding to particular internal or external anatomical structures. For example, the accompanying male and female figures (figs. 1 and 2), generated by William Lorenson of the General Electric Corporation, are not photographs but reconstructions of surface features from the slices (215 physical slices for the male, 209 CT slices for the female). Manipulating and combining these images using state-of-the-art visualization programs makes it possible to penetrate the body—giving the sense of walking or flying through (as in walking through walls or using superhero "X-ray vision"). Different levels of depth or systems (skin, muscles, skeleton) can be superimposed to be viewed simultaneously, and discrete anatomical structures can be isolated. Further, these images can be rotated to be viewed from various perspectives, not only in successive still images but in

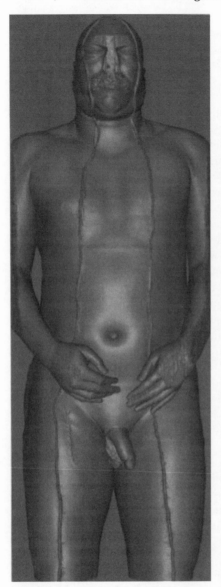

FIGURE 1.
Visible Man, frontal view.
Reproduced by permission of General Electric.

FIGURE 2.
Visible Woman, frontal view.
Reproduced by permission of General Electric.

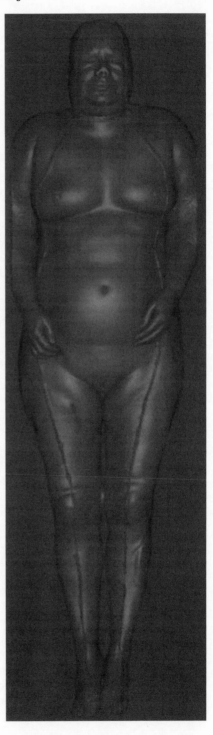

computer animations. Currently, these have advanced to the point of allowing surgical simulations—to be discussed in detail below—similar to flight simulations used in training pilots. In one easily available demonstration from the Center for Human Simulation, one can watch a computerized scalpel make an incision in a thigh sliced off from the body above and below the knee, the incision gradually opening to reveal muscle and fat, and the section then rotating in mid-screen and moving to a close-up to show several views of the incision. Developers of these methods promise that their animations—one is tempted to say reanimations—will eventually include blood coursing through veins and around organs and breath pulsing through lungs. Beyond that is the prospect of transformations that will simulate the processes of aging (as well as, of course, growing young again) and pathology (as well as its reversal). All this occurs in the context of the technical aspiration to render future candidates into ever thinner slices to enter into ever larger databases. One can imagine a race of reanimated virtual cadavers faithful to their human form to the cellular level, infinitely mutable, able to be subject to many simultaneous surgical procedures and healing processes, and capable of reanimation and regeneration in whole or in any part.

SIMULOIDS, AVATARS, AND SHADES

Virtual cadavers are phenomena in and of cyberspace and virtual reality. Let me clarify my understanding of how these terms are related. A person who sends e-mail or peruses sites on the World Wide Web is tapping into cyberspace, but not in a strong sense is this person entering a virtual reality. Here the computer is a communicative tool, a cybernetic enhancement of mail and media. A person outfitted with data glove and data goggles involved in an advanced computerized simulation is entering a virtual reality, but this person is not in cyberspace. Here the computer is an aesthetic tool, a cybernetic enhancement of dramatic and performative techniques by which we create imaginative terrains. To put it more formulaically, cyberspace is an intersubjective medium constituted socially (composed by interaction among participants in the communicative medium), while virtual reality is the subjective sensory presence in that medium (composed by interaction between individual user and technology). The intersection of the two is full sensory and bodily engagement in a virtual reality that is also completely networked and plugged into cyberspace—in other words, the full-blown, Gibsonesque, science fiction version of virtual reality within cyberspace. This intersection is important to identify because, I would argue, it is the cultural locus of the virtual cadavers.

What do we mean by "cultural locus" in this instance? Let us accept the metaphor of cyberspace with sufficient literalness to conceive it not as a cultural domain but as a distinct ethnographic terrain, so that it is legitimate to talk about an ethnography of cyberspace. Indeed, there has recently been a florescence of work along these lines in anthropology, and the number of sessions devoted to related topics at the last several meetings of the American Anthropological Association indicates a groundswell of interest. The most prominent concept in this area is that of virtual communities, or networks of social interaction with fluid boundaries and varying degrees of apparent permanence, composed of actors whose agency is expanded to the point of controlling and manipulating their own identities as pure forms of representation and who interact with ambiguous others whose identities are never certain. While there is thus a clearly articulated concern with the kind of relationships that exist in cyberspace, taking the metaphor of cyberspace literally as an ethnographic terrain allows us to formulate a parallel concern with the kind of beings that inhabit cyberspace. We can then propose a preliminary inventory of such beings, with the caveat that the cyborg is not among them, for a cyborg is by definition a creature of the technological interface. We are cyborgs whenever our bodily capacities are technologically altered or enhanced, including when we have our noses pressed up against the window of cyberspace by virtue of booting up. The goal here is an inventory of beings that exist wholly on the hither side of the interface, in the ethnographic terrain marked by what we just now identified as the intersection of cyberspace and virtual reality.

Accordingly, my preliminary inventory consists of three types of beings: simuloids, avatars, and shades. Simuloids are software-generated entities that have no sentient counterpart in actuality. In the language of the industry, these are referred to variously as humanoid technologies, virtual humans, human-modeling systems, computer-generated humanoids, or autonomous creatures. Simuloids are described as autonomous, not in the sense that they have agency in their own right but in the sense that they are independent of human agency: they are software-controlled rather than human-controlled. They are also autonomous from any necessity to conform to concrete actuality, and thus their features may transcend the human—they can as easily be animal, machine, or monster. However, a great deal of attention is currently being devoted to the development of virtual humans, defined as "computer-generated people that live, work, and play in virtual worlds, standing in for real individuals or carrying out jobs that real people cannot do." The news service story that carries this definition also quotes Sandra Kay Helsel, editor of *VR News*, as pronouncing that "'Virtual Humans will be the growth industry of the 1990s!'" Characters in computer games are simuloids, as are the characters Max Headroom (the hyper-masculine cyborg star of a 1980s television show), the computer HAL from the film *2001, A Space Odyssey*, and the villainous cyberman in the movie *Virtuosity*. The most advanced simuloids include the virtual humans that Nadia Thalmann developed with her Marilyn program, which provide characteristics such as emotional expression, speech, clothes, hair, and the ability to respond to computer users, and which can simulate Marilyn Monroe and Humphrey Bogart as well. Norman Badler's Jack system of human modeling is based on a figure designed to the specifications of the average American male and is capable of articulated motion, including balance modification and collision avoidance, gesture and facial expressions, natural language processing, and transformation in size and color.

The term "avatar" is already in popular usage, and it is sometimes applied to what I have called simuloids, but I want to restrict its meaning to virtual incarnations of human actors on the hither side of the interface that these actors directly control. It is worth playing out some of the cultural connotations of the notion of avatar because of the implications regarding self, agency, and being that we can read into it. The primary meaning of the term, of course, is the incarnation of a Hindu deity in actual human form. By extension, on the one hand, the human computer operator is analogous to a kind of deity that manipulates the computerized avatar in virtual human form. On the other hand, the extension inverts the meaning of the Sanskrit term in a subtle way: whereas the Hindu avatar is an incarnation of a deity into an earthly form, the computer avatar is a virtual apotheosis of an earthly being into an imaginative realm of fantastic powers and shape shifting. A more secularized use of the term, and one that Webster's dictionary offers, suggests that an avatar can be an embodiment, as of a concept or philosophy, usually in human form, a definition which prompts the observation that what the computer avatar embodies is the human form itself.

Another, more thoroughly abstracted definition from Webster's has the avatar as "a variant phase or version of a continuing basic entity." This one prompts the question of whether the avatar is best considered a representation of a person, a cybernetic extension of a person, a projection of a person into cyberspace, or, indeed, a "variant phase or version" of a person; for the avatar is much more than a computerized double or simulation programmed to act like a person. In February 1996, a virtual wedding took place in Los Angeles, California, in which vows were exchanged on-screen via avatars while the human participants remained in separate geographical locations in actuality. What varies from a technical standpoint is the degree of sensory engagement of operator in avatars. In practice, avatars can be imagined forms described to other users by typing in text, as in the interactive computer sex networks where people become animals or creatures endowed with amazing and creative types of sexual organs and multiple genders. Avatars can also be visual "body icons" that can be manipulated and come into the virtual presence of the body icons of other users, but they are now often little more than "grim-looking peg-doll shapes," lacking faces or feet. The most advanced include multisensory feedback in which it is experientially the case that the avatars are not so much representations of the user as they are projections of the user into virtual space. Such projections are customized computer-generated forms, but they may in principle also be computer-animated video images, as is pioneered in the "synthespian" technology that Jeff Kleiser and Diana Waczak developed.

To summarize the distinction between types of beings that I have drawn here, simuloids are computer-generated stand-ins for people, with no connection to any actual person; avatars are projections of living people as digitized persons. To draw on vivid, popular culture examples, the computer-generated villain of *Virtuosity* is a simuloid that crosses the interface from virtuality into actuality and becomes embodied; the character Jobe in the movie *Lawnmower Man* is a human who crosses from actuality into virtuality to literally become his avatar and abandon his body. The purpose of this contrast is to introduce what is a distinct third type of being indigenous to cyberspace. The computerized virtual cadavers the Visible Human Project of the National Library of Medicine produced are what I call "shades," derived from the use of the word to refer to a spirit in the netherworld. There are only two such beings at present, a male and a female, but their existence has profound consequences that we are just now beginning to work out. These shades are "in" cyberspace and virtual reality in a sense distinguishable from either simuloids or avatars. Like a simuloid, the shade can operate as a stand-in for a person, but, unlike a simuloid, it is a distillation of an actual person that can be digitally superimposed on another actual person. Like an avatar, the shade is a projection of a real person into cyberspace, but, unlike an avatar, that person is not only dead but has been dissolved as a physical being. The shade thus exists solely on the hither side of the interface, where it is not an animation but a reanimation—a new kind of being entirely. Let us elaborate this analysis by briefly examining the biotechnological and symbolic structure of shades.

ADAM AND EVE IN THE VIRTUAL WORLD OF THE DEAD

When the first images of the Visible Human male were presented, the regents of the National Library of Medicine broke out in spontaneous applause, prompting a comment by project director Michael Ackerman which was quoted in the *Denver Post:* "'It was sort of like applauding at the end of a movement in a concert. It was inappropriate, and the decorum is not that way. But it was a clear indication of how excited people are'" (Schrader). Given the excitement this biotechnological juggernaut has generated, let us pose the following question: What is the relation between Joseph Jernigan, the housewife from Maryland, and the data sets they have become? In other words, what constitutes their cultural status (being) as shades? For a preliminary answer, I take as data representations of the event in newsletters of the NLM and approximately fifty-five articles from the popular media.

One way of thinking about the change in cultural status is most evident with Jernigan, whose fate was the more public. Here there is a sense of "through the looking glass" with respect to representation of his personhood. The first mention of Jernigan is in retrospect rather eerily incognizant of his subsequent posthumous celebrity. It is a typical article furnished by Reuters in the *New York Times* of 6 August 1993, reporting that he was executed by lethal injection after having admitted to killing a homeowner who surprised him during a burglary. Such deaths by execution are routinely reported in the news media, and the article's significance does not go beyond the observation that because the reinstitutionalization of capital punishment in the United States remains at least a back burner issue of public discussion, executions remain newsworthy. By April 1994, Jernigan's previous existence was a distorted footnote. A bylined article on the forthcoming release of the Visible Human Male data set picked up by several Knight-Ridder newspapers describes the cadaver as a "drug overdose victim" (this remarkably bad article also refers to the film *Fantastic Voyage* as *Incredible Journey*), as do articles in 1994 and 1995 in the *Denver Post*. The *Denver Post* in 1994 mentions the project in an article on "cryonics," the practice of freezing dead diseased individuals in hopes of bringing them back to life when medical technology has advanced sufficiently to cure them, with the concluding comment that "the project doesn't aim to revive the subjects" (Schrader). Nevertheless, one is left with the titillating image of a person, whose personal identity is rather beside the point, who is in some sense capable of being reanimated.

The sense of "through the looking glass" that defines the cultural status of shades is most evident in a series of metaphors invoked to describe them. There are four sets of metaphors: one describing the shades as Adam and Eve, one in images of birth and immortality, another that invokes Leonardo da Vinci, and one as they are a virtual terrain to be mapped. The Adam and Eve metaphor is doubtless the most symbolically charged. This is evident in that it was the original intent of the project directors to use these as the formal titles for their shades instead of the markedly more awkward Visible Human Male and Visible Human Female. The plan had to be abandoned in the face of threatened legal conflict with a company that had already named itself and its interactive anatomy software program "A.D.A.M.," an acronym for Animated Dissection of Anatomy for Medicine. The competition—never mind that the A.D.A.M. company is now licensed to use the Visible Human shades in its product development—indicates that something rather existential might indeed be at stake in the advent of shades.

A sampler of these metaphors will give a flavor of what I mean. The *Philadelphia Inquirer* refers to the shades as a "'perfect couple'—an Adam and Eve for computer immortality" (Kopytoff); the *Denver Post* calls them the "first couple" (Schrader); the *Baltimore Sun* announces that the female shade is a "Partner for 'Visible Man'" (Roylance); and the *New York Times* reports that the NLM was "on the look-out for a female donor to share his life on the wire" ("Computerized"). The *Economist* goes yet further in its reference to "A new family—Adam, Eve, and their embryonic offspring," pairing the Visible Humans with the Visible Embryo Project at the Armed Forces Institute of Pathology ("Adam's" 94). Elsewhere in the piece, they are referred to as "Adam, Eve, and little Cain" (97), and the National Library of Medicine as their "electronic Garden of Eden." Regarding plans to increase the inventory of shades, the *San Diego Union-Tribune* notes that "The future might also bring Visible progeny, Grandpa and Grandma, or a younger pre-menopausal woman" (Fenley).

The Adam and Eve set of metaphors cannot be written off as either tritely cute or opportunistically commercial—it is too good a fit with contemporary gender symbolism (see also Treichler). Although there was no more of a relation between Jernigan and the Maryland housewife than between any two cadavers donated to medical research, there is evident appeal in transforming them into a mythical first couple in a new virtual world, digitally reanimated and capable of necrosexual procreation of a family or species of shades. Conveniently, this time around Adam and Eve are both white. Naturally the male was created first and the female second, albeit this time not from Adam's rib. As is the case generally in contemporary society, the male is guaranteed an identity (although fortuitously because his cause of death was public, court-ordered), while the female remains anonymous (although on the hither side of the looking glass, that anonymity was understood as a right). The image of Adam rendered in 1 mm slices is rougher, while Eve is more refined, or, in a pun spun by one journalist, she "looks sharper" than her male counterpart. Stated otherwise, in the Foucauldian idiom of bodily surveillance, if the male can be scrutinized, the female can be scrutinized more thoroughly. The male was an evil victimizer killed by lethal injection; the woman was an innocent victim who died of a heart (as in "bless her heart") attack. Adam is sufficient in his own right, indeed, a magnificent specimen who pumped iron, while word is that Eve needs to be supplemented by a pre-menopausal counterpart—we can expect polygyny in cyberspace.

The second set of metaphors has to do with birth and immortality. On the one hand, a 1996 World Wide Web self-description of the UCHSC (University of Colorado Health Sciences Center) Center for Human Simulation by its staff observed that their anatomical imaging laboratory has "given birth to the Visible Human—Male and Female," while the *New York Times* reported, "Executed killer reborn as 'visible man' in Internet" ("Computerized"). On the other hand, the *National Library of Medicine News* noted that the visible humans are "immortalized on the Internet"

("Phase II"). The *Independent* announced that "A killer was yesterday let loose on the Internet computer network" ("Executed"); and *Netguide Magazine* declared that "A killer . . . has been immortalized" (Martini). *Life* magazine referred to "electronic afterlife" (Dowling); the *Baltimore Sun* stated that the shade "has won a measure of computerized immortality" (Roylance); and the *Denver Post* published that the project "promises eternal life for the participants" (Schrader). These metaphors are neither idle nor contradictory, but they reflect views from different sides of the looking glass. The humans are in a sense immortal in their new form; their shades are in a sense born again beings of a new space.

Two rather less developed sets of metaphors that yet indicate something of the cultural status of shades can be identified. One is the elevation of the shades in a celebration of the human form that places them alongside the renderings of Leonardo da Vinci. A group at University Hospital in Hamburg, Germany, that has developed an impressive 3-D atlas called VoxelMan draws a direct historical line from Leonardo to the development of X-rays then on to the invention of CT and MRI technology and thence to the Visible Humans. A group of artists in Japan is explicitly juxtaposing Leonardo's representations with representations of the Visible Human shades. Implicit is the ideal of approaching reality via virtuality, a connotation that is also evident in the slightly jarring phrase of the *Baltimore Sun* referring to "real bodies on a computer" (Roylance). The final image, which I found only once in a quote from one of the project coordinators, is telling in its appeal to the inanimate. This was in a reference by project director Ackerman to the need to label each site and segment of the Visible Humans, since " 'Right now, looking at them is like looking at a road map with no street names' " ("Phase II"). Here the shades are understood as a terrain to be traversed—not an unknown virgin territory but a map not yet useful because it is not yet labeled.

Taken together, I would suggest that these sets of metaphors disclose the workings of a deep essentialism that constitutes the cultural status of shades. The Adam and Eve metaphors point to gender essences that the heterosexual reproductive couple defines—the two shades could have been represented as siblings, or, even, given their age difference, as mother and son. The immortality metaphors define a moral essence, whether conceived as an untainted being prior to the Fall or to a redeemed being in the guise of a born-again convict. The Leonardo metaphor outlines an aesthetic essence of the apotheosized ideal man—notably without female counterpart. The map metaphor implies a cosmic essence by assimilating the body to a terrain, but, in particular, to one that still needs to be charted. The positing of essences in these popular metaphors is implicitly a strategy of identity, made all the more compelling by two features. First, positing the essence of the other (the not-me, the shade) is a double-edged act of self-definition either by denial (me as the opposite of not-me) or by desire (the wished-for me). Second, the force of this double-edgedness is enhanced by the paradoxical condition of the shades as virtuality in actuality, their apparent existence as "real bodies on the Internet." But how can all this be so?

THE MEANING OF METAPHOR: VIRTUAL OR ACTUAL?

Two methodological points can help assess the consequences of the foregoing discussion of popular metaphors. First, the discussion assumes that such metaphors offer an interpretive opening to cultural meaning. The analysis can only be valid, for example, if the metaphoric description of the first shades as Adam and Eve is accepted as nontrivial, more than an eye-catching journalistic ploy. It is certainly common enough—only the sober British *Daily Telegraph* consistently reported on the project without recourse to metaphor. Once accepted, this assumption allows the implications of the metaphors to be spun out and regarded as data about the "cultural imaginary," which is further assumed to be as consequential as what we could call the "culturally literal," that is, the language of technology and its application. In this context, the cultural imaginary is the realm of possibility, desire, and fear in which we participate passively insofar as it lurks anxiously beneath conscious awareness (compare Ragland-Sullivan 138–62 on Jacques Lacan's notion of the imaginary) and in which we participate actively by the exercise of imagination, the "capacity to articulate what used to be separate . . . which allows one either to make a new move or change the rules of the game" (Lyotard 52). Looking at a text from the standpoint of the culturally literal, one could argue that a metaphor is only a colorful analogy to help clarify an objective relation or function—the metaphor is discursively subordinate to the function. Regarding metaphor as an opening into the cultural imaginary grants it a far more important role, one in which the cultural imaginary has an equivalent status with—indeed, is in a dialectical relation with—the culturally literal. From the glass-is-half-full perspective, this dialectic is one in which they mutually constitute one another. From the glass-is-half-empty viewpoint, it is one in which they mutually destabilize one another. At the very least, one could say that the cultural imaginary provides the context by means of which the existential implications of the technical innovation can be examined, while the culturally literal leaves discussion at the level of policy implications.

The second point is the significance of identifying the social origins and destinations of the cultural meanings that are spun into the gossamer fabric of a cultural imaginary. In the case of computerized cadavers, the sense-making process that accompanies the technological development originates principally from the Visible Human Project coordinators, from groups developing project applications, and from public media that disseminate information about the project. A fourth source is the metadiscourse of cultural analysts—we must reflexively include, for example, the metaphorical offering of shades to define the computerized cadavers. Members of each group participate both passively and actively in the cultural imaginary, but each has a socially positioned take on the relation between the cultural imaginary and cultural literality. The appeal to the literal through the use of technical and policy language is given greater or lesser weight by project coordinators whose audience is government funding sources, potential database users, and the media; developers of applications whose audience is a potential market for those applications; the media whose audience is the general public; and cultural analysts whose audience is academia. Each is thoroughly ensconced in the dialectic between imaginary and literal, and that dialectic is constituted by the sum of their social positionalities.

This understanding of the relation between the cultural imaginary and the culturally literal remains somewhat strained unless it takes into account what I regard as an orthogonal distinction between representation and being-in-the-world. Cultural analysis will always be subject to suspicions of whether it is dealing with reality if it is cast entirely in terms of representation, or analysis of representation. The popular metaphors surrounding the Visible Human shades, and the images themselves, are necessarily of limited consequence if they are analyzed on the representational level alone. The metaphors are indeed frivolous unless they are interpreted as hinting at or disclosing a subtle shift in our mode of being-in-the-world, and the remarkable images with all their combinatorial possibilities bear no more intimate connection to the original people-cum-cadavers than a photograph torn up and taped together again, unless they do the same. My intent in introducing the notion of a shade is to push us into thinking beyond representation and toward being.

Yet when dealing with such a general notion as being, it is important to think globally in order to avoid essentializing a cultural particularity. Thus, the disposition of the actual bodies might not appear much more radical to a North American than would the disposition of a medical cadaver or a cremated dead person. Consider the integrity of being required in a Buddhist society such as Japan, where one cannot take one's place among the ancestors or expect a higher reincarnation if one goes to the grave lacking a part of one's body. From that standpoint, would one react with complaisance regarding the moral import of creating shades? Final cremation aside, what might be the cosmological status of a human who has been ground into dust? On the other side of the existential looking glass, once transformed into a shade, what might be the status of one who can be divided into chunks repeatedly? On a more mundanely North American ethical level, what about anonymity and privacy? A conventional cadaver used in anatomy training is both anonymous and lacks identity. A shade is not anonymous, for even the woman from Maryland might be recognized by an acquaintance who saw her reconstituted visage. Project director Ackerman has said, "'We're hopeful that if she is recognizable, that people will respect her anonymity. There is nothing we can do'" that would not compromise the data (Roylance). Moreover, a shade retains identity, for even Jernigan at least in some sense stays who he was whether or not he comes to be called Adam instead of Joseph.

All things considered, the argument about the being of shades would remain rather disingenuous if we were not clear that the primary concern was our own being in relation to them, or, in other words, how the technological innovation induces a subtle modulation in our own embodiment and hence in our own culturally situated-being-in-the-world. Subjectivity and intersubjectivity are bodily phenomena, and thus the question becomes the potential transformation of subjectivity on the part of those who use the technology, and especially with regard to physicians, in the intersubjective relation formed with those bodily beings who are their patients. In this light, let us briefly consider the two most immediate sites of impact, namely, anatomy training for medical students, and computer-aided surgery.

"IT EMPOWERS US"

At the fourth annual conference on Medicine and Virtual Reality in January 1996, Michael Ackerman and Victor Spitzer, director of the project at the NLM and director of the contracting group at UCHSC, received the Satava Award, named for Colonel Richard Satava, M.D., a pioneer in virtual reality telesurgery. Helene Hoffman, director of the anatomy curriculum using Visible Human data at the University of California at San Diego Medical School, observed that "This data set has become the new standard for human physiology education. For example, 30 to 40 percent of the papers presented at this year's conference alone relied on this data set." A variety of medical schools are actively developing anatomy curricula based on Visible Human data.[2] The primary debate is over whether these methods are to be used to complement or to replace conventional dissection in anatomy education. Traditionalists are resistant to the idea that medical students would not have the hands-on experience of work with real bodies in what is implicitly sacrosanct as a rite of passage in medical training. Innovators point out that actual cadavers are increasingly short in supply in comparison to the infinitely reusable shades, and that, in any case, most physicians other than surgeons will never have occasion to work on the insides of their patients.

The potential consequences with respect to embodiment for both medical students and their future patients must be understood with respect to the already profound phenomenological transformation wrought in conventional training. In his ethnographic study of medical students, anthropologist Byron J. Good observes that the dimensions of the world of experience built up in their training were "more profoundly different from my everyday world than nearly any of those I have experienced in other field research" (71). Reminiscent of the map without street names metaphor I mentioned above, students learning anatomy are "as geographers moving from gross topography to the detail of microecology" (72). Good repeatedly refers to the intimacy with which medical students come to know the body, and he describes the anatomy laboratory as a kind of ritual space in which the reconstitution of experience takes place. The accustomed body surfaces that define personhood are drawn back, revealing an "interior" that consists not of a person's inner emotional life but a complex three-dimensional space with planes of tissue that are separated to distinguish the boundaries of gross forms and fine structures. As one student said, "'Emotionally a leg has such a different meaning after you get the skin off'" (qtd. in Good 72). The new way of seeing beneath the surface that is central to the medical gaze can usually be turned on and off, but it also bleeds over into the student's everyday perception of other persons, as they are constituted and reconstituted—translated and retranslated—between the perceptual languages of medicine and everyday life. Good observes that this training is profoundly visual, and the profundity of phenomenological transformation can only be enhanced by the new anatomy curricula based on virtual reality. The penetration to increasingly minute levels of biological hierarchy (epidemiological—clinical—histological—cellular—molecular/genetic) will be complemented by a penetration based on transparency, the sense of "X-ray vision."

A preliminary glimpse of the potential change comes from a reflection by a medical student who attended the introduction of the Visible Human Female in 1995. Referring to the Visible Human as a prime example of high performance computing as applied to biomedical science, he writes, "It empowers us. We students know that a world of information is out there at the touch of our fingertips" (Roberts). He was impressed by the fact that brain sections would not fall apart as they sometimes do in dissection of a real cadaver; that one can isolate sections of the body rather than "deal with the whole daunting thing"; that the circulatory system would appear like a real 3-D loop rather than flat as in a textbook; that the database could be reformatted to change body characteristics; that the images could be rotated, dissected, and resected; and that someday he would be able to call up these images in his office to help educate patients about illnesses and procedures.

Several questions arise concerning the ultimate experiential consequences of applying this technology. What will be the effects of isolating body parts for detailed, intensive work? Will it enhance the sense of intimacy that Good notes or will it initiate a more fragmented, objectified sense of the body? What will be the consequences of digital dissection that is both exceedingly neat and comfortably reversible in comparison to actual dissection, in much the way word processing allows easy deletion and substitution in comparison to writing or typing? Will it introduce a sense of arbitrariness of biological process or will it heighten the understanding of meticulous detail? Finally, what will be the consequences of empowerment as it is alluded to by the medical student rapporteur? Will it be the power of humanizing intimacy and compassion or that of apotheosizing omnipotence and objectification of one's fellow beings? Will it refine the sensibilities of physicians as a flower blossom unfolding to reveal its intimate recesses rather than having to be sliced open or peeled apart petal by petal?

"FROM BLOOD AND GUTS TO BITS AND BYTES"

Beyond the training of medical students, shades will increasingly play a role in the development of surgical training and what is called "telepresence surgery." The title of this section is a favorite phrase of Colonel Satava, one of the leading figures in this area, in referring to a major paradigm shift in which the blood and guts of conventional surgery is replaced by the bits and bytes that will facilitate the work of a new generation of "digital physicians" and "Nintendo surgeons." Virtual reality surgical simulations are already available for prostate, eye, leg, and cholycystectomy procedures. Telepresence surgery allows the physician to project himself or herself to a remote location via video and audio monitors, with computer-controlled instruments maneuvered from the remote site with dummy handles able to provide "force feedback" that gives the surgeon—or surgeons collaborating by network from different geographical locations—a sense of tactile immediacy.

Satava distinguishes between artificial and natural virtual reality, the former completely synthetic and imaginary as in the simulation of being inside a molecule, the latter a situation that could physically exist as in surgery on a realistic recreation of a human body ("Robotics" 360–61). Both surgical simulation and telepresence surgery are forms of natural virtual reality, although obviously only the latter is performed on actual patients. Yet Satava says that "the day may come when it would not be possible to determine if an operation were being performed on a real or computer generated patient. . . . [T]he threshold has been crossed; and a new world is forming, half real and half virtual" ("Robotics" 363). He and his colleagues are working on just such a system, in which the operator can fly around the organs and travel through the digestive system, and use of shade data is allowing them to move from a cartoon-like visual display to an increasingly lifelike one. Further, the overlay and enhancement of live CT/MRI data with shade data promises to augment the vividness of telepresence surgery, as the immediacy of the electronic image and remote manipulation come to "dissolve time and space."

The development of shade-enhanced telepresence surgery has consequences for embodiment with respect to the skills it requires of the surgeon—as what Marcel Mauss called a "technique of the body"—and with respect to its applications on the bodies of patients. With respect to the former, the emerging field of Human Interface Technology dictates that a system have sensory intuitiveness—that it "should feel and be used as naturally as possible." As Satava observes, telepresence surgery has the same eye-hand axis as open surgery insofar as the surgeon looks down at a monitor, thus preserving the correspondence of visual with proprioceptive and kinesthetic senses. Contemporary laparoscopy requires visually looking up at a video monitor, while surgical simulation requires wearing a virtual reality helmet such that the surgeon must learn the tool rather than the tool accommodating the surgeon ("Human" 819–20). With respect to patient care, the new technology will allow comparing normal and abnormal organs by substituting images, simulating the biomechanics of muscles and joints to make more effective replacement joints, and demonstrating projected treatment courses for patients. Military applications—one of Colonel Satava's ultimate interests—of shade-enhanced simulation would include plotting the path of a bullet before treating a bullet wound, and applications of telepresence surgery would allow the technology "to metaphorically project a surgeon into every foxhole" ("Modern" 12).

At least two questions are indicated by these developments. The first comes from considering that both surgical simulation and telepresence surgery pose a paradox of simultaneously increased remoteness and enhanced intimacy. Simulation is remote from living persons and telesurgery is geographically remote; both partake of the intimacy afforded by the technologically enhanced medical gaze. What will be the consequences of this paradox; and what will be the limits of access to the inner recesses of biological process? The second inquiry arises in considering Drew Leder's analysis of the typical disappearance of the body from awareness in everyday life as it "not only projects outward in experience but falls back into unexperiencable depths" (53). Leder argues that it is the body's structure that leads to its self-concealment and to a notion of the immateriality of mind and thought that is reified as mind-body dualism. Could it be on the cultural-technological horizon that shade-enhanced virtual reality will make the intimate core of bodily processes accessible in a new way, offering the possibility of transcending this Cartesianism of the natural attitude?

FROZEN REPRESENTATION AND VIRTUAL BEING-IN-THE-WORLD

I want to return to the broader question of the cultural significance of shades not in terms of the relation between the cultural imaginary and cultural practice but in terms of the relation between representation and being-in-the-world. The notion of representation holds a virtual hegemony over contemporary cultural analysis, hand in hand with the associated methodological metaphor of textuality. This extends to cultural analysis of the body, so that scholarly works are filled with phrases such as the body as text, writing on the body, bodies of writing, the inscription of meaning on/in the body, representations of the body, and reading the body. A less prominently articulated tradition understands culture from the standpoint of embodiment as our fundamental and culturally conditional mode of being-in-the-world. As bodily beings we inhabit the world in terms of the space and extension of our bodies; we engage in movement and experience resistance to that movement; we incorporate and explore the world via our senses; and we interact with others or find ourselves in solitude. The modes of representation and being-in-the-world are intimately intertwined in practice, for example, in the way their relation can be superimposed on the relation between subject and object: if the body is conceived as an object, representations of the body are the site of subjectivity; if the body is conceived as subject, representations are objectifications of the body.

I would argue that understanding the interaction between the body as representation and the body as being-in-the-world is critical to cultural analysis in general (see Csordas, *Embodiment*), and, furthermore, that this interaction defines the cultural process that is critically at stake in the existential analysis of the shades that the Visible Human Project created. From the standpoint of Jernigan and the Maryland housewife, are their shades no more than hypertext versions of a photographic representation, no more connected to their particular essences than a snapshot that could be torn to bits then reassembled with tape and glue? Or is there something of the transformation of quantity into quality in the degree of specificity with which their physical beings have been digitalized, some way in which they have gone "through the looking glass"? Indeed, it is possible to indulge a debate about whether even a simple photograph captures something essential about a person (and anthropologists know that in some societies this is thought quite literally to be so) or is better understood as an arbitrary and momentary simulation that can be repeated without limit to the ultimate degradation of meaning, similar to what might happen to the meaning of the word "egg" if it is repeated a hundred times. However, the question of the shades' being-in-the-world is academic insofar as, the tool of science fiction placed aside, there is no question of personal subjectivity for them. What is all the more at issue is the subjectivity of the rest of us—specialist medical students and surgeons to be sure, but also the coming generation. Indeed, UCHSC's Spitzer has said, "I think in the future, kids will grow up with him" (certainly an improvement on Barney). More important, while by the same token there is no question of defining intersubjectivity between shades and users, there is all the more a query of how, given the premise that intersubjectivity is also grounded in our bodily being, it may become transformed, enhanced, or distorted by the existence and application of shades. What will interpersonal relations be like when I can casually visualize your skeleton as we converse, and you can feel your way around inside my brain?

Finally, if the biotechnological innovations in virtual reality of which shades are only one example are indeed pointing toward a modulation of embodiment, it may be so only because of the historical condition in which culture now exists. Daniel J. Boorstin wrote in 1961 that the contemporary world is already one "'where fantasy is more real than reality, where the image has more dignity than its original. We hardly dare face our bewilderment, because the solace of belief in contrived reality is so thoroughly real'" (qtd. in Kearney 252). This is to say that what we are describing is not a technological determinism of embodiment but a highly specific way of incorporating a technological development into the postmodern condition of culture. Understanding this process will require a cultural phenomenology that can capture the essence of the particular in an embodiment constituted in the existential space between virtual and actual, between the cultural imaginary and culturally literal, between remoteness and intimacy, and between representation and being-in-the-world.

NOTES

1. Visible Human images presented here are available on the Internet at the following addresses: General Electric <http://www.crd.ge.com/esl/cgsp/projects>; University of California at San Diego <http://cybermed.ucsd.edu>; and University of Hamburg <http://www.uke.uni-hamburg.de>. Each of these can be accessed via the National Library of Medicine <http://www.nlm.nih.gov>. An impressive array of images based on the Visible Human can be found in Tsiaras.

2. Among medical schools actively developing anatomy curricula are the University of California at San Diego Medical School, Loyola University Stritch School of Medicine, Johns Hopkins in collaboration with the National University of Singapore, SUNY at Stony Brook, University of Pennsylvania Medical Center, Washington University Medical School, the University of Chicago in collaboration with Argonne National Laboratory, and Columbia University Medical School in collaboration with the Stephens Institute of Technology. Outside the United States, projects are underway at the University of Hamburg School of Medicine, the Keio University School of Medicine in Tokyo, Australian National University, and the Queensland University of Technology.

WORKS CITED

"Adam's Family Values." *Economist* 5 Mar. 1994: 94+.
"Computerized Cadaver." *New York Times* 29 Nov. 1994: C8.

Csordas, Thomas J., ed. *Embodiment and Experience: The Existential Ground of Culture and Self.* Cambridge: Cambridge University Press, 1994.

———. "Embodiment as a Paradigm for Anthropology." *Ethos* 18.1 (1990): 5–47.

———. *The Sacred Self: A Cultural Phenomenology of Charismatic Healing.* Berkeley: University of California Press, 1994.

Dowling, Claudia Glenn. "The Visible Man: The Execution and Electric Afterlife of Joseph Paul Jernigan." *Life* Feb. 1997: 40–45.

"Executed Killer Has Piece of Action on Internet." *Independent* 19 Nov. 1994: 3.

Fenley, Leigh. "The Visible Human." *San Diego Union-Tribune* 15 Mar. 1995: E1.

Good, Byron J. *Medicine, Rationality, and Experience: An Anthropological Perspective.* Cambridge: Cambridge University Press, 1994.

Heim, Michael. *The Metaphysics of Virtual Reality.* New York: Oxford University Press, 1993.

Kearney, Richard. *The Wake of Imagination.* Minneapolis: University of Minnesota Press, 1988.

Kopytoff, Verne. "Computer to Profile the 'Perfect Couple': In Death They Will Live On as 3-D Images. A Surgical Simulator for Doctors is the Goal." *Philadelphia Inquirer* 14 Apr. 1994: A2.

Leder, Drew. *The Absent Body.* Chicago: University of Chicago Press, 1990.

Lyotard, Jean-François. *The Postmodern Condition: A Report on Knowledge.* Trans. Geoffrey Bennington and Brian Massumi. Minneapolis: University of Minnesota Press, 1984.

Martini, Adam. "Abracadaver." *Netguide Magazine* 1 Apr. 1995: 29.

Mauss, Marcel. *Sociologie et Anthropologie.* Paris: PUF, 1950.

"Phase II of Visible Human Project Completed: 'Sharper' Visible Woman Joins Male Counterpart on the Internet." *National Library of Medicine News* 50.6 (1995): 1–3.

Ragland-Sullivan, Ellie. *Jacques Lacan and the Philosophy of Psychoanalysis.* Urbana: University of Illinois Press, 1986.

Roberts, Stephen. "From a Different Perspective: A Student's View. Gratefully Yours." *National Library of Medicine Online Newsletter.* (Jan.–Feb. 1996): 3 pag.

Roylance, Frank D. "Md. Gives Partner for 'Visible Man': The National Library of Medicine Has Photographed in Minute Detail the Body of a Maryland Woman and Made the Digitized Data Available to Researchers." *Baltimore Sun* 29 Nov. 1995: 1A.

Satava, Richard. "Human Interface Technology: An Essential Tool for the Modern Surgeon." *Surgical Endoscopy* 8 (1994): 817–20.

———. "The Modern Medical Battlefield: Sequitur on Advanced Medical Technology." Unpublished ms. Advanced Research Projects Agency, Arlington, VA, n.d.

———. "Robotics, Telepresence, and Virtual Reality: A Critical Analysis of the Future of Surgery." *Minimally Invasive Therapy* 1 (1992): 357–63.

Schrader, Ann. "Human Anatomy To Be On-Line Soon: C.U. Work Creating Database for Project." *Denver Post* 6 June 1994: B1.

Treichler, Paula A., ed. *The Visible Woman: Imaging Technologies, Gender, and Science.* New York: New York University Press, 1998.

Tsiaras, Alexander. *Body Voyage: A Three-Dimensional Tour of a Real Human Body.* New York: Warner, 1997.

Turkle, Sherry. *Life on the Screen: Identity in the Age of the Internet.* New York: Simon & Schuster, 1995.

Chapter 8

Chorea/graphing Chorea

The Dancing Body of Huntington's Disease

ALICE RUTH WEXLER

I.

1963: Northwestern Minnesota

"The first time I heard the words Huntington's chorea I was 19 years old and 5 months pregnant with my first child. 'Ma' had the flu very badly. I brought her to the doctor, our elderly family physician. After observing her, he asked me if others in her family had actions [movements] like hers; I said 'no.' He called in a neurologist, who met with us on the hospital steps (my dad, my sisters, my husband and myself—our brother, at 14, was 'too young' to attend). This neurologist told us that our mother had Huntington's chorea, that little was known about it, that there was no cure, and that it had a definite genetic factor. He shook Dad's hand and left us standing on the hospital steps with the elderly doctor, who, after a few minutes, lowered his head and walked to his car. I followed him and said, 'Does genetic mean hereditary?' He told me that after years of depression and illness, my grandfather committed suicide. He then said, 'You don't have to worry, this disease only hits the slow or sickly in the affected family, and it's "dying out" because babies are healthier now.' He told me that my mother was a very sickly child and also that there had been a sister who 'never was quite right' and was committed to an institution as a teen-ager. 'This caused the great depression in their father.' He ended with the statement, 'You're bright and healthy, you and yours will be fine.' I went back to my Dad and told him some of what the doctor had said. My Dad, in his stern German way, said, 'Ya—we better be a lot nicer to your sister so she doesn't end up like ma quite yet.'"

—LaDonna Chouinard, "A Minnesota Story"

Illness, writes the medical philosopher Howard Brody, is a story. And Huntington's disease (HD) is a particularly dramatic story, with its often extreme personality changes, loss of cognitive abilities and emotional control, and, most visibly, the jerky, involuntary, and sometimes dance-like movements which gave the disorder its original nineteenth-century name, Huntington's chorea. Until I was twenty-six, I had never heard of this disease, although it had killed my mother's grandfather and father and all three of her brothers, one by one, and my mother, in her early fifties, was starting to show symptoms. Only when she was diagnosed, in 1968, did I learn of Huntington's in my family's past, and also, possibly, in our future, for my sister and I were at 50 percent risk of inheriting the disease.

I choose to write about Huntington's in the first person in order to acknowledge my position as a person at risk vis-à-vis questions the new technologies of predictive testing pose. My arguments come partly from personal experience, my particular version of the at-risk narrative. They also come from my professional identity as a historian, a writer, and a trustee of the Hereditary Disease Foundation, a private, nonprofit organization dedicated to stimulating and funding basic research on Huntington's and related genetic and neurological disorders. Like my sister, who is president of this foundation, I have a strong stake in the outcome of debates about the meanings of the knowledge and power the new predictive tests produce. In this essay, I situate my arguments—for these unfold in two different voices—alongside the narratives of several other people at risk for Huntington's, in order to echo within the text the dialogue that is unfolding in the Huntington's community.

The revelation of Huntington's disease in the family had for me a numbing effect. I did not want to be drawn back into the family dramas I had been struggling to escape. This is what Huntington's meant to me: a biological metaphor for the identification with my mother that I had fought all my life. I did not want to think of the lingering death that awaited her, a decline typically lasting ten to twenty years, as Huntington's first slowly robbed her of her identity and then took her life.

In the summer of 1983, researchers identified a genetic marker for Huntington's, that is, a small piece of genetic material that, in families affected by the illness, traveled together with the HD gene as a sort of constant companion. Knowing the whereabouts of the companion greatly aided researchers in narrowing their search for the Huntington's gene to a specific locale of the humane genome, in this case the small arm of chromosome four. This advance came in part through the efforts of my father, Milton Wexler, who had started the Hereditary Disease Foundation, and through the work of my sister, Nancy Wexler. She was the critical link between the molecular biologists in the laboratory and the communities of affected families, organizing and leading annual expeditions to the shores of Lake Maracaibo in Venezuela to work with the families who provided the crucial pedigrees and DNA samples that made the marker discovery possible.

It was during the euphoric summer of 1983, when we could fantasize that a cure lay around the corner, that I began to conceptualize my project of exploring the impact of the illness in my family and of documenting the ongoing research on Huntington's disease. I decided to write a book, envisioning the narrative as a sort of double helix, one strand being the science story, the other strand a personal one. I called the book *Mapping Fate* because it was, in part, a story about gene mapping. But I also envisioned the book as a kind of map, and my project as one of mapping my fate as well as the course of HD research, although recently I have begun to think in terms of the metaphor of chorea/graphy, in homage to the earlier name of this disorder, Huntington's chorea, but also to suggest a more active orchestration of the changing meanings of this disease, the possibility of a new kind of dance.

The most dramatic new dance so far has been the predictive gene test, which, while it clearly has many Foucauldian potentialities for marginalization and exclusion of those found to be gene-positive, has made possible new kinds of knowledge (Foucault 184; see also Nelkin and Tancredi; Quaid 5–6). To grasp the implications of the predictive test it is essential to understand the Mendelian pattern of inheritance in Huntington's disease. Huntington's is an autosomal dominant disorder, which means each child of a parent with Huntington's faces a fifty-fifty chance of inheriting the disease; men and women are afflicted equally. Unlike the pattern with recessive disorders such as sickle cell anemia or cystic fibrosis, only one parent need have the disease to pass it on to a child. And until recently, the Huntington's disease gene was thought to be 100 percent penetrant, meaning that if one had the abnormal gene, one would definitely develop symptoms if one lived a normal life span, considered to be seventy-five for men, eighty-one for women (Brinkman et al. 1208). However, the identification of the gene is 1993 led to the realization that this situation did not always apply. The abnormal HD gene is characterized by an excessive number of nucleotide triplets, in this case CAG, or Cytosine-Adenine-Guanine, a triplet which codes for the amino acid glutamine. Everyone has a string of these CAG triplets at the Huntington's disease gene locus, or address, on chromosome four. Individuals possessing twenty-eight or fewer of these CAG repeats will not, within a normal life span, develop Huntington's disease. Nor are their children at risk for the illness. Within the normal life span, someone with forty or more CAG repeats will almost certainly develop the symptoms. However, a gene with between thirty-five and thirty-nine CAG repeats is considered to lie in the zone of reduced penetrance. The individual with this number of CAG repeats may or may not develop the disease, depending upon other factors which are not yet understood. But even if this person remains healthy, his or her children run a risk of developing the disease, since the number of CAG repeats sometimes expands in the parental germ-line—egg or sperm forming the basis of the next generation. (For reasons that are unclear, such expansion seems to occur more frequently in sperm than in eggs.) Similarly, individuals with between twenty-nine and thirty-four CAG repeats will not develop Huntington's, but their children are at increased risk of inheriting an expanded number of CAG repeats, thereby putting them into the range of the disease (National 13).

Because of the severity of this disease and its inexorable pattern of transmission, geneticists have long argued for the value of a predictive test to identify in advance of symptoms who will develop the disease. For example, in 1966, Ntinos Myrianthopoulos, a geneticist at the National Institute of Health whose early interest would help spark research on the disease in the 1970s, wrote that the absence of a predictive test was "downright tragic . . . as anyone who has followed the course of the disease in an individual or family will readily admit, especially since onset is usually after reproduction has been completed" (312). But if the geneticists' interest in developing a presymptomatic test had a definite eugenic dimension, people at risk and their relatives also shared a desire for some mode of prediction, even if there was no treatment and nothing medically one could do with that information. The opportunity to escape the awful uncertainty, especially the chance of learning one did not carry the disease gene and would not pass it on, or had not already passed it on to one's children, made the gamble seem worthwhile to many. About three-fourths of those surveyed in the 1970s said they would take such a test if it became available. Indeed, when a test did become available in 1986, as part of a clinical research protocol at a few academic testing centers, there was considerable anger by many people at risk at the difficulty of getting the test. There was also a lot of anger at the protocol required for those going through testing, even though this protocol, which included extensive counseling, had been set up with participation from lay people in the voluntary organization, the Huntington's Disease Society of America, or HDSA ("Guidelines"). The testing process was time consuming and stressful, in part because linkage testing, the kind of test that was offered initially, required blood samples from several family members, including at least one affected person. From 1986 to 1993, the test identified not the Huntington's gene, which was still unknown, but the next-door-neighbor genetic marker, which almost always traveled together with that disease gene in stricken families but was not itself a cause of disease. The marker was a visible sign of the presence of the as-yet-invisible gene: an indication, as some commentators have quipped, of nearby "gene trouble."[1]

Of course Huntington's is in certain ways an extreme, worst case scenario that should not be uncritically projected onto other disorders. With many of the more common disorders such as Alzheimer's disease or breast cancer, for example, a gene mutation may be associated with *greater susceptibility* to a disease but not the kind of straightforward either/or situation which is generally the case with Huntington's and other autosomal dominant diseases. While the media pose the sensational question, "Do you really want to know how you will die?" the more vexing and complex questions involve the interpretation of gene mutations whose medical meanings are far from clear. The human genome is full of variation which has nothing to do with pathogenesis.

II.

October 1989: University of Minnesota Hospital

"Carl and I were seating ourselves in a small examining room in the neurology department. A psychiatrist, a neurologist-geneticist, and a social worker entered the room and jockeyed around for chairs with us. Before handshakes and greetings had been completed, before all persons present had settled comfortably in their seats, the neurologist-geneticist—a young woman with long, straight, blonde hair—focused a serious gaze on my face and spoke.

'I'm afraid we have bad news,' she said. 'You have the gene for Huntington's. That answer is 92% accurate.'

I began to sob. Someone passed me a box of tissue. Carl put his arm around my shoulders, and that made me cry more. Three pairs of professional eyes watched my every action intently.

When I could catch a breath, I asked questions about the testing procedure, and they gave me patient, detailed answers. None of it really mattered. None of it changed the fact that I had inherited the gene for Huntington's disease from my father, that one day I would begin that long, slow, process of mental and physical deterioration. I began to regain my composure, and then I lost it again with just a few words.

'My girls. What about—?' Sobs choked my throat, and I couldn't finish.

I don't know why I asked the question when I already knew the answer, which was that my two daughters were now at 50% risk, though they couldn't be tested until they became 18 and legal adults to choose for themselves. I knew that, but I asked anyway, hoping for a different response, hoping somehow they were safe.

At the end of the hour, the three professionals gave me an appointment to come back in three weeks. They wanted to be sure telling me had been a positive thing. They wanted to be sure this whole testing business was a good idea when there wasn't a damn thing medical science could do for a person in my shoes. No medication, no treatment, nothing could stop the onset of the disease.

I was their sixth patient to be given test results at the University of Minnesota, the fourth to get bad news. The truth was this was all virgin territory at the testing centers across the country. These three people didn't know for certain what I would do with the knowledge they had given me. They didn't have a stack of medical journals or a pile of statistics to tell them what to expect, and it worried them.

I saw fear in their eyes that day as I left them. I also saw frustration. They couldn't even tell me when the symptoms would first appear. . . . They wanted to do something medically for me and couldn't. They wanted to follow me home and watch how I incorporated this sad news into my life, and couldn't do that either.

> They wanted to be absolutely certain I would show up for that next appointment, but they knew there were no guarantees.
> So I left to go home that day, still firmly convinced that knowledge was a good thing, but wishing for the briefest of moments that I could give back just this one little piece of personal information."
>
> —Sally Spaulding, "Autumn Patience"

At chapter meetings as well as at the national convention of HDSA, the number of people who have taken the HD gene test has increased over the last few years. There is now a significant group of asymptomatic people who have "tested positive." (As of the middle of 1996, some 5,800 people had taken the predictive test worldwide, 40 percent of them male and 60 percent female. About 2,320 people, some 40 percent of all those tested, now knew they were at increased risk.) At HD meetings, even around many people familiar with Huntington's, the people who are gene-positive—at least those who have been public about their status—possess a kind of aura, a transparency, as if it were possible to see inside their bodies to the gene itself. They are visible in a way that the rest of us are not. In spite of ourselves, we look at them more intensely, scrutinizing them for any telltale choreic movements.

Keep in mind the critical distinction between a *diagnostic test*—that is, a gene test taken by someone who has begun to show symptoms—and a *predictive or presymptomatic test* taken by an individual who shows no symptoms clinically. In the latter case a gene test can indicate only whether the individual will become ill *in the future;* it is *not* a diagnosis in the present. In discussions of gene testing, the two are sometimes conflated; they are both gene tests, but the contexts for testing—in the presence of symptoms or without symptoms—makes the experience very distinct. The people I am talking about here are healthy people with no symptoms who have learned they will develop the illness sometime in the future, nobody knows when.

> "There were about 25 of us. . . . We had the chairs in a circle and kept having to make the circle bigger and bigger as more people kept coming in. First off there was a huge sense of relief at being in a room with so many others who were in the same tragic, weird shoes as each one of us. . . . The biggest concern expressed by members of the group was the uncertainty regarding age of onset and what the first symptoms are going to look like. There was lots of variation in what the first symptoms had looked like in parents and family members. One woman's mother didn't have any chorea until almost the end of her life. What is it going to be like to lose short-term memory first? How can you tell if it's happening to you?
>
> One theme running through everything was family members' reaction to the knowledge about testing positive for the gene. Some families still had not even gotten to the point of accepting that HD was present in their families. Other families were very supportive and open; some had known for generations that the risk of HD was there.

This group looked pretty strong emotionally and psychologically. There were tears and lots of sad stories, but these people seemed insightful and knowledgeable about HD. I would say that being at the convention is a sign of someone who is dealing with things in a positive way. What got repeated again and again was that people were focusing on the present and on enjoying what they had right now." (Spaulding, Letter)

III.

The largest center of predictive HD testing, that is, the center which has tested the most people, is at the University of British Columbia in Vancouver. This center has also been the most enthusiastic in affirming the psychological benefits of genetic testing. According to recent studies emanating from Vancouver, predictive genetic testing for Huntington's has long-term "significant psychological benefits" for people who have a *decreased or an increased risk*. In the Vancouver stories, people who go through genetic testing for Huntington's are less depressed, less anxious, and more hopeful than those who do not, whether they get an increased or a decreased risk of Huntington's in the future (Wiggins et al.).

Since the Vancouver claims are being cited in the development of genetic testing for a variety of other conditions, for example breast cancer, it is worth noting how such claims were made (Letter). First, the evaluations of psychological responses to testing were carried out as part of the testing protocol itself. There was no independently conducted evaluation of people's responses to the testing process. Moreover, the evaluations took place within a research protocol where, along with the genetic information, people received extensive counseling—a service which is becoming increasingly difficult to sustain in the current context of the commercialization of testing. Indeed, all the published Vancouver studies acknowledge the possibility that favorable responses may be due as much to counseling as to the actual information. In addition, the studies relied significantly, though not exclusively, on paper and pencil, self-reporting surveys, with no information cited from spouses or other family members or friends for comparison or confirmation. And except for acknowledging that about two-thirds of those seeking predictive testing for Huntington's are women, these studies did not evaluate gender as a possible variable in shaping responses. It is significant, too, that the Vancouver studies used as a control group two distinct sets of people, some of whom received inconclusive gene results, and others who did not choose to learn their genetic status but were willing to complete the psychological tests: a control group which may not be appropriate for comparison.

Most important, a significant number of the initial research population did not complete the follow-up studies. At the American Society for Human Genetics meeting in November of 1996, the Vancouver researchers published an Abstract of their most current results from some 200 individuals whom they claimed gained "significant psychological benefit" from testing, whether they had an increased or a decreased risk. When researchers presented their results orally at one of the sessions, however, they acknowledged, as the Abstract did not, that they had begun the study with about 300 individuals, but that nearly one-third of their research population, or close to 100 people, had failed to complete the five year followup, possibly the very individuals who did not experience "significant psychological benefit."[2]

The question I want to raise here, however, is not so much methodological as epistemological, that is, whose voices count in these testing stories? Who gets to represent the "experience" of testing? Who will define the meanings of predictive testing for Huntington's? How is "benefit" defined and for whom? What counts as a "scientific" study of testing? How do those of us who believe in "situated knowledges," who practice narrative, ethnographic, or clinical approaches, confront the powerful claims of quantification and "objectivity"?

The new claims for the "significant psychological benefits" of predictive gene testing for Huntington's have important implications for other genetic tests. These assertions are not only statements about past responses to testing, they are recommendations for future practices. Moreover, they are being used to promote genetic testing for other late-onset diseases for which the genetics is much less straightforward than in Huntington's. A 1996 study, published in the *Journal of the American Medical Association* (*JAMA*), of testing for the BRACA1 gene associated with certain forms of breast cancer cited the Vancouver HD studies as evidence that predictive testing yielded long-term psychological benefits, but it did so without including any of the qualifications and reservations even the Vancouver researchers acknowledged (Lerman et al.).

Today within the Huntington's community, we are seeing growing pressures on people at risk to get tested, even though there is still no medical benefit from such a test. Increasingly, pressures to test are coming from doctors, including neurologists, who are accustomed in their clinical practice to ordering "tests," and who may suggest adding an HD gene test to a battery of otherwise routine tests. Studies have shown that a sizable minority of medical geneticists would give people information about their genetic risk even when they did not request it and did not want it (see Codori, Hanson, and Brandt 172; see also Nelkin and Tancredi). Pressures also come from insurers, adoption agencies, the courts, the police, and from other family members, including parents who want their minor children tested. Often, subtle psychological pressures come from genetic counselors and psychologists, whose language implicitly conveys the idea that going through testing is "courageous" and truth seeking, while deciding against testing is denial, a wish to "avoid" the truth, or even a "preference for ignorance." Some counselors have referred to the "want-to-knowers" and the "avoiders." The informational literature that genetic counselors and testing centers dispense oftentimes communicates the strong desirability of testing and the undesirability of not getting tested. As one center's brochure put it, "Without predictive testing, people at risk are consigned to a waiting list, left to anticipate or ignore the arrival of symptoms that may never come" (Siebert 52; "Predictive").

But even with so devastating a disorder as Huntington's, there are questions about the "need" for certainty and the "need" to escape from ambiguity that are worth pondering. In this regard, I think it is useful to consider the analysis of Abby Lippman, a Canadian feminist epidemiologist and biostatistician, whose arguments about prenatal testing have relevance for other kinds of genetic tests. According to Lippman, the language of reassurance offered to women for prenatal testing often evades questions of why reassurance is sought in the first place and how the "need" for reassurance is constructed, since it is an acquired rather than an inherent characteristic of testing. As she points out, "We must first identify the concept of need as itself a problem and acknowledge that needs do not have intrinsic reality. Rather, needs are socially constructed and culture-bound, grounded in current history, dependent on context and, therefore, not universal" (149). In Lippman's view, the anxiety and fear women experience and seek to escape through prenatal testing are in part created by cultural values and biomedical institutions more interested in a quick fix and individualized remedies than in social and environmental change that might accommodate a wider range of disabilities. Lippman asks, "What does it mean . . . to 'need' prenatal diagnosis, to 'choose' to be tested?" (149).

Certainly there are important differences between predictive testing for a biomedical catastrophe such as Huntington's disease and prenatal testing for disabilities which are not life-threatening or progressively physically and mentally disabling. Yet substituting predictive for prenatal can offer useful perspectives for comparison. What does it mean to "need" predictive testing and to "choose" to be tested, for disorders that have no available treatment? (Once a treatment becomes available, the entire landscape changes.) HDSA, speaking for the affected families, has taken the position that predictive testing must be an entirely voluntary decision. However, what does voluntary mean in this regard? Is the anxiety surrounding at-risk status for Huntington's, and the need for testing that many people at risk experience, intensified by larger cultural anxieties about all intermediate and ambiguous categories and the general emphasis in Western culture on clear-cut, binary, either/or classifications, whether they are categories of race, sexuality, gender, or, in this case, genetics? Many in the Huntington's community use the language of individual rights to argue for easier access to testing, figuring the body as a kind of private property over which the individual exercises sole rights of ownership. But others emphasize the family context in which the individual necessarily makes his or her choices, and the implications for one's children, siblings, spouses or partners, and parents.

Genetic testing is currently represented as a means of gaining control over one's life and over oneself, acquiring an ability to plan for the future. Control and planning for the future are powerful cultural ideals. As one psychologist wrote in an ostensibly "informational" article published in the newsletter of HDSA: "After years of uncertainty, at-risk individuals may make plans for the future. . . . An increased sense of control will now be possible both for gene-carriers and those who find that they are no longer at risk" (Jones 6–7). However, the speaker here, rather than merely responding to the need for increased control she perceives in her anxious at-risk clients, may be helping to create the "need" for an increased sense of control. She conveys the idea that the uncertainty of being at risk is unbearable. This discourse, common in the literature on testing, suggests that people's "need" to escape from uncertainty and ambiguity is not only a response to the terrifying possibility of disease but may also be produced through interactions with health professionals who express anxiety about the ambiguities of being at risk. Moreover, the discourse on testing often implies that the only way to assuage the anxiety of being at risk for Huntington's is to learn one's genetic status. Yet testing may short-circuit other possible strategies for reducing anxieties, such as counseling without the genetic test. Indeed, the availability of a predictive test may make genetic uncertainty even more intolerable.

But the "certainty" offered by genetic testing is ambiguous. If one is lucky and gets a decreased risk, that is, a negative result, then there is obviously one kind of certainty. But an increased risk, a positive result, may exchange one form of uncertainty for another, since the age of onset is unpredictable at this point. Still, those who have tested positive emphasize various aspects of this outcome. Prior to testing, "'Tina'" believed she lived "'an airy-fairy life,'" unable to make commitments in love or work. "'Since I got the test results and did the [motivational] course, I've made all these decisions,'" she says. "'I've bought a house, I've got married, I'm having a child and I've written a book. It's so exciting. I'd say that Huntington's is one of the motivators'" (Gray 161). Sally Spaulding, a law school graduate who was already married with two young children, had a different perspective. For her, "the worst part about testing positive is the uncertainty: when will the symptoms start and what will they look like?" ("Autumn").

IV.

Partly out of my dissatisfaction with the testing stories coming out in the genetics and medical literature, and partly because of my experience writing an HD narrative, I have been collecting, and in some cases eliciting, the stories of people at risk or affected or care giving in the Huntington's community. Because testing is a finite event, one with a beginning, middle, and end, it lends itself to narrativization, particularly by those who get good news at the end of the long ordeal. Eliciting the stories of those who test positive has been more difficult, even for the genetic counselors and geneticists involved in clinical research on responses to testing. The fact of a bad outcome—that is, testing positive—thwarts the effort at telling the story. People do not want to revisit the scene of the crime, even in memory. Often people find it painful to come back to the office where they received such devastating information, and this has been one of the big problems in evaluating reactions to testing, because evidence suggests that those having the greatest difficulty coping with the results are also more likely to drop out of studies or fail to return for future visits.[3]

I hold no brief for raw experience. Nor am I searching for "authentic" voices to tell the stories of predictive testing. The testing stories and risk narratives we have been eliciting within the Huntington's community are highly mediated constructions of experience which people craft from a variety of sources, including medical discourse, popular culture, and religious discourse. People at risk for a disease such as Huntington's belong to ethnic, racial, class, religious, and geographical communities which shape their identities and discourse as well. Some stories may even be mediated by the brain destruction that the gene portends. Indeed, Emily Martin has recently spoken about the dilemma of establishing one's authority in speaking of a psychological disability or illness in others when one has oneself been diagnosed with or is at risk for that condition. Yet these autobiographical accounts offer a valuable kind of witnessing, not least because they are constructed outside of the therapeutic-medical sites where such stories are usually elicited, that is, the neurological exam, the counseling session, the psychiatric interview, or the psychological survey. At minimum, such personal narratives convey the cultural meanings circulating around a disease such as Huntington's, inside as well as outside the medical milieu. LaDonna Chouinard's eloquent account of the elderly family physician standing on the hospital steps, who "lowered his head and walked to his car," conveys powerfully the shame and stigma which even doctors associated with Huntington's, as well as the culturally inscribed ways in which medicine becomes a vehicle for morality. Sally Spaulding's brave account of a positive gene test result suggests the Foucauldian sense of heightened visibility which she experienced and her awareness of herself as an object of the professionals' intensified medical gaze. Her description of a support group meeting of people who had tested positive also acknowledges the resistance strategies, including the sharing of stories, that people develop to deal with devastating information. Spaulding's emphasis on the uncertainty which follows a positive gene test subverts the popular notion, conveyed also by many testers, that testing can offer a reassuring certainty even if one does test positive. And finally, there is "Tina's" very different account of how testing positive (and taking a motivational course!) changed her from someone living an indecisive, "airy-fairy" life to a woman who made decisions and accepted commitment, including getting married and having a baby. Her narrative suggests how the ambiguity of being at 50 percent risk can haunt a person's entire life; it also speaks to the range of choices a positive gene test can motivate.

I often ask myself whether I cannot really hear the voices of those who say that they are glad they got the information, because of my own stake in not getting tested. I am aware that my preference not to get tested undoubtedly shapes my interpretation of these stories and colors my view of which voices command the greatest authority. If my sister and I have difficulty believing the reports that people gain "significant psychological benefits" from testing even if they test positive, is this because of our critical evaluation of the data, our observations of people living with this knowledge, or our emotional stake in rejecting that claim? There is no neutral ground in this landscape, and I want to be clear about where I stand: although I choose not to take the test myself, I have come to appreciate the compelling reasons why people may decide to learn their genetic status, including older people who wish to inform their adult children whether they are at 50 percent risk and young people considering whether to start a family. I have talked with many people who have been through testing, including many who have gotten a positive result, and who are now living, and, I believe, living well with this "toxic knowledge." These are people with amazing courage and strength. Yet my conviction is that, as one young woman who is gene-positive put it after an overly optimistic support group discussion, "this [gene-positive] situation really sucks!"

V.

In a recent essay, "Artificiality and Enlightenment: From Sociobiology to Biosociality," the anthropologist Paul Rabinow evokes a scenario in which identity terms will circulate increasingly around the distribution of particular polymorphisms at specific gene loci, a scenario which is already a reality for many in families affected by Huntington's disease. Certainly predictive testing has heightened intensities and given new precision to the concept of risk. Now we are even seeing people in HD chat rooms on the Internet comparing the number of their CAG repeats, as if this number were somehow definitive of identity, a symbol of the time allotted their lives, although scientists insist that such numbers are not definitive of age of onset except at the highest extremes. Yet being at 50 percent risk for Huntington's has always been a powerful vector of identity for those of us in affected families, particularly for those growing up with an ill parent. Moreover, discussions within the HD community, as well as published studies, suggest the tenacity of pre-test identities, especially for those who receive a negative result. Testing negative for the HD gene does not necessarily free people from their worry and anxiety, presumably because of other family members who test positive for the gene or are currently suffering from the disease. Indeed, family identifications may remain relatively unchanged by a negative test: this is so much the case that a former president of the HDSA, who went through testing herself and got a negative result, declared in an influential article in the *New England Journal of Medicine* that "genetic testing must be viewed as a family issue, not an individual one" (Hayes 1450). This emphasis, now widely accepted in the Huntington's community, directly challenges the individualistic ethos of American medical practice, while also possibly reinforcing the traditional family identifications of women, who form the majority of the tested population so far.

For those who receive a gene-positive result, the implications are more powerful and potentially devastating, but, even here, prior expectations sometimes persist. Some people translated their risk back to pre-test levels, that is 50 percent, even though they had been told their risk was now close to 100 percent (Codori and Brandt 177). And at least one person reported that getting a positive test result had less emotional impact than learning for the first time that she was at 50 percent risk. Others, however, affirmed that a positive result was harder to cope with than they had anticipated and that they worried more about Huntington's after getting this information, even though the test resolved their uncertainty about whether or not they carried the gene.[4]

In the context of growing numbers of people at risk for HD who are choosing to learn their genetic status, support groups have proliferated, addressing the concerns both of those considering testing and those who have already been through the process. Yet, it is worth recalling that these support groups, lobbying groups, and other networks concerned with Huntington's disease preceded the development of the predictive tests, growing, rather, out of the health activism of the 1970s and 1980s. The predecessor to HDSA, the Committee to Combat Huntington's Disease, was formed in 1968. Indeed, the technologies of predictive HD testing, while part of the general advance in molecular genetics of the 1970s and 1980s, were also partly a response to the demands of the new HD activists, not so much for a predictive test but for research focused on the gene. Thus Rabinow's allusion to the "new group and individual identities and practices," in my view, describes a situation that preceded rather than followed the rapid advance of the new genetic technologies.

Whether predictive testing for Huntington's will in the long run genuinely serve the interests of people at risk or the interests of the insurers, health providers, employers, and biotechnology companies who stand to profit from the proliferation of tests remains to be seen. Now that a test is available to prospective parents, how will people regard the birth of babies who are at risk or gene-positive? Will those who develop symptoms in the context of a prior positive gene test experience them differently from those who have never been tested? Although predictive testing for the Huntington's gene has been offered since 1986, many questions about the impact and meanings of this information remain unanswered; and they are possibly unanswerable, given the powerful political and economic agendas that swirl around the practice of testing and the difficulties of independent evaluation.

Certainly with the expansion of genetic testing for other more common disorders for which the genetics is much less straightforward, such as Alzheimer's and various forms of cancer, many more people will find themselves labeled "at increased risk" or "gene-positive" for something. One's genetic "profile" will form an increasingly salient dimension of personal identity, as twenty-first-century medicine comes to rely more and more on genetics, although genetic diagnosis currently remains far in advance of gene therapies. In the absence of treatments, this proliferation of testing, as many critics have pointed out, has the potential to reinforce old hierarchies and inequalities based on race, class, gender, and sexual orientation, opening what Troy Duster calls "the backdoor to eugenics."

Yet the Huntington's story suggests more hopeful potentials as well. The number of those who are choosing to get tested remains low, around 10 to 15 percent of all those at risk as of late 1998. Clearly, people carefully consider their choices and resist pressures to take the test in the absence of any medical benefit. Many of those who do get the test remain, or even become, activists in the HD community, whether they receive a positive or negative result. And the Huntington's story, like that of AIDS, also prompts the ability of those at risk or otherwise directly affected by an illness to help shape research and testing agendas. Researchers studying other diseases, health professionals, and individuals and families afflicted by other disorders have looked to the HD networks as a model of collaborative science, effective advocacy, and the creation of an educated community. Our experience also subverts claims for a simplistic genetic determinism, showing that even a straightforward Mendelian disease such as Huntington's is patterned by more than just the gene.

Hopefully, the new genetic technologies will soon lead to effective treatments for some of the most devastating diseases, such as Huntington's and Alzheimer's. But with or without such "magic bullets," one challenge certainly remains: how to turn the growing activism of patients and people at risk into an effective political movement, not only for greater citizen involvement in establishing the priorities and practices of biomedical science but also for universal health insurance so that all of us may reap the benefits.

NOTES

1. For an account of the identification and use of the marker, see Wexler.
2. See Quaid and Wesson 50; Taylor and Myers 368.
3. At a meeting of testing centers in November 1996 at the American Society for Human Genetics convention, genetic counselors talked at length of the difficulties of follow-up and of clients whom they could not contact soon after giving them their positive test results.
4. The HDSA distributed a survey of attitudes toward predictive testing in the fall of 1996, and results were tabulated in the spring of 1997 at the University of Pittsburgh's Graduate School of Public Health (see Kallenborn). Two hundred women, sixty-six men, and four others (who did not indicate gender) completed the survey. Of these 274 individuals, sixty-seven had received an increased risk or positive result, and sixty-two had received a negative result. Four were indeterminate. The rest had chosen not to be tested.

WORKS CITED

Brinkman, R. R., et al. "The Likelihood of Being Affected with Huntington's Disease by a Particular Age, for a Specific CAG Size." *American Journal of Human Genetics* 60 (1997): 1202–8.
Chouinard, LaDonna. "A Minnesota Story." Unpublished ms., 1982, 87–95.
Codori, Ann-Marie, and Jason Brandt. "Psychological Costs and Benefits of Predictive Testing for Huntington's Disease." *American Journal of Medical Genetics* 54 (1994): 174–84.
Codori, Ann-Marie, Rebecca Hanson, and Jason Brandt. "Self-Selection in Predictive Testing for Huntington's Disease." *American Journal of Medical Genetics* 54 (1994): 167–73.
Duster, Troy. *Backdoor to Eugenics*. New York: Routledge, 1990.
Foucault, Michel. *Discipline and Punish: The Birth of the Prison*. Trans. Alan Sheridan. New York: Vintage, 1979.
Gray, Alison. *Genes and Generations*. Wellington, New Zealand: Wellington Huntington Disease Association, 1995.
"Guidelines for Genetic Testing for Huntington's Disease." Rev. ed. New York: Huntington's Disease Society of America, 1994.
Hayes, Catherine. "Genetic Testing for Huntington's Disease—A Family Issue." *New England Journal of Medicine* 327 (1992): 1449–51.
Jones, Randi. "Genetic Testing for Huntington's Disease—What's New?" *The Marker* (Huntington's Disease Society of America) (spring 1994): 6–7.
Kallenborn, Melissa Meg. "A Retrospective Survey of Attitudes toward Presymptomatic Testing for Huntington Disease." Master's thesis. University of Pittsburgh, 1997.
Lerman, C., et al. "BRCA1 Testing in Families with Hereditary Breast-Ovarian Cancer." *Journal of the American Medical Association* 275 (1996): 1885–92.
Letter. *Journal of the American Medical Association* 276 (1996): 1139–40.
Lippman, Abby. "Prenatal Genetic Testing and Screening." *Genetic Counseling: Practice and Principles*. Ed. Angus Clarke. London: Routledge, 1994.
Martin, Emily. "Canons of Rationality in Categories of the Mental." American Ethnographical Society. San Francisco. Mar. 1996.

Myrianthopoulos, Ntinos. "Huntington's Chorea." *Journal of Medical Genetics* 3 (1966): 298–314.

National Institute of Neurological Disorders and Stroke. *Huntington's Disease.* Bethesda: NIH Publication No. 98–49, Mar. 1998.

Nelkin, Dorothy, and Laurence Tancredi. *Dangerous Diagnostics: The Social Power of Biological Information.* New York: Basic, 1989.

"Predictive Testing for Huntington's Disease: If You Want It, an Answer is Probably Waiting for You." Wichita: Donald J. Allen Memorial Huntington's Disease Clinic, 1988.

Quaid, Kimberly. "A Few Words from a 'Wise' Woman." *Genes and Human Self-Knowledge.* Ed. Robert F. Weir, Susan C. Lawrence, and Evan Fales. Iowa City: University of Iowa Press, 1994. 3–17.

Quaid, Kimberly A., and Melissa K. Wesson. "Exploration of the Effects of Predictive Testing for Huntington Disease on Intimate Relationships." *American Journal of Medical Genetics* 57 (1995): 46–51.

Rabinow, Paul. "Artificiality and Enlightenment: From Sociobiology to Biosociality." *Essays on the Anthropology of Reason.* Princeton: Princeton University Press, 1996. 91–111.

Siebert, Charles. "Living with Toxic Knowledge." *New York Times Magazine* 17 Sept. 1995: 50+.

Spaulding, Sally. "Autumn Patience." Unpublished essay, 1992.

———. Letter to the author. 23 July 1996.

Taylor, Catherine A., and Richard H. Myers. "Long-Term Impact of Huntington's Disease Linkage Testing." *American Journal of Medical Genetics* 70 (1997): 365–70.

Wexler, Alice Ruth. *Mapping Fate: A Memoir of Family, Risk, and Genetic Research.* 1995. Berkeley: University of California Press, 1996.

Wiggins, S., et al. "A Long Term (= 5 years) Prospective Assessment of Psychological Consequences of Predictive Testing for Huntington's Disease (HD)." Abstract #28. American Society for Human Genetics. Nov. 1996.

Chapter 9

The Ventilator/Baby as Cyborg

A Case Study in Technology and Medical Ethics

ROBERT M. NELSON

INTRODUCING MICHAEL

I first met Michael when he was admitted to the pediatric intensive care unit as a transition to our home ventilation program. The diagnosis was nemaline rod myopathy—a rare and slowly progressive neuromuscular disease that renders a person immobile (Iannacone and Guilfoile 30–32; Connolly, Roland, and Hill 285–88). Unable to move and unable to breathe, Michael had undergone a tracheostomy procedure a month earlier so that his airway was secure and he could be ventilated more easily. A tube had also been placed into his stomach through his abdominal wall so that he could be fed without placing a temporary feeding tube through his nose. Michael was then five and a half months old. His mother appeared tense and protective, yet no more so than other parents in similar circumstances. Michael stayed in the intensive care unit for ten days. After being switched to our "low-tech" home ventilator, he was transferred out of the intensive care unit to the ward of the hospital where all our chronic home ventilator patients are admitted. All that remained was to train the family and discharge Michael home.

Three and a half months later, I received a telephone call from the physician responsible for taking care of Michael. At the time, I was on call for ethics consultations and he was seeking our help in trying to resolve an apparent impasse in discharging Michael. I knew there were problems, since part of my clinical responsibilities also involved covering this particular ward off and on for a week at a time. There were two broad areas of potential conflict. The first issue was trying to decide where Michael should stay after discharge. The options included his mother's home, the home of a relative, or the home of a foster parent skilled in taking care of so-called technology-dependent children.[1] At his mother's request, plans had been developed for temporary foster placement. The second issue was his mother's continued insistence that she had the right to take Michael off the ventilator and let him die. It was the fear that this right might be threatened that caused her to scuttle plans for foster placement and precipitate the ethics consultation. The concrete result of the consultation was to reaffirm the plan to discharge Michael home—an event that did not take place until after Michael's first birthday (three months later), when the hospital administration warned that they would seek involuntary foster placement.

The most compelling aspect of this case is how Michael's mother progressively lost control over the one decision she cared about—the determination of whether Michael would or would not be subject to the technology of home ventilation. To begin to understand this apparent shift in decision-making power, we also need to pay attention to the organizational and cultural context within which these decisions were made. In the course of Michael's treatment, the ventilator appeared to take on a life of its own and began to define and redefine the boundary of Michael's body. We need to explore this apparent agency of our technology in the process of social production, that is, in how the ventilator produced the subject of Michael as both body and machine. However, we should not reify the agency of technology in a way that obscures and thus privileges the agency of the designer—as in the *Wizard of Oz* when we are told, "don't pay attention to that man behind the curtain." This essay is less an examination of bioethics than an exercise in cyborg anthropology, for Michael truly is a machine-human/human-machine whose body and machine components are mutually interdependent (Downey, Dumit, and Williams 264–69).[2]

Let me first situate myself with respect to the story of Michael and his mother. I am a pediatrician trained in neonatal and pediatric intensive care. I have a degree in religious studies, and I teach bioethics in a medical school setting. I participated in Michael's medical care and bear some responsibility for the events as they unfolded. The physicians involved I count among my close colleagues. As I believe that the opportunity for understanding arises from an initial awareness of difference, I am acutely aware that my participation in and thus acceptance of many of the presuppositions of the medical and bioethical environment may constrain my imagination.[3]

THE MEANING OF TEARS

During the ethics consultation, Michael's mother mentioned her belief that his constant tearing indicated emotional and physical distress. In her eyes, Michael was fragile and unable to tolerate any activities such as being propped up in a wheelchair or taken out of the hospital on field trips to the local park. Rather than a reminder of his suffering, the staff interpreted Michael's tears as a natural result of the inability to close his eyelids. The nurse told of his taking pleasure in simple things and of his ability to sit contentedly in a chair for hours. The staff believed that Michael would cry when his mother arrived and "suffocated" him during her brief visits. The image of an overprotective mother who fails to appreciate the strength and ability of her child is brought to mind; however, the use of the metaphor of suffocation takes on a more literal and provocative meaning given the nurse's knowledge of the mother's stated desire to stop the ventilator at some point in the future. Fundamentally, Michael's mother continued to see the ventilator as something other than Michael: as a threat, as an invasion of his body, as something foreign. In the eyes of the staff, the threat to Michael was not the ventilator, but his mother. In reflecting on the ethics consultation, I began to wonder whether the staff saw Michael not as a body on a machine but as both body and machine. If so, to ask the staff to participate in turning off the ventilator was to ask them to amputate part of Michael's body.

Was there a "fact of the matter" which could settle this dispute? Was there a third point of view, acceptable to both parties, from which the relationship between Michael's tears and his suffering could be "objectively" determined? I certainly did not believe that the ethics consultant sat in that other point of view. In spite of my sympathy for the mother's dilemma, I found myself secretly agreeing with the staff's interpretation of Michael's tears. Glossing the difference between pain and suffering, the absence of other signs of pain such as sweating and a rapid heart rate supported the staff's position. This was an argument that Michael's mother could only lose; the best she could hope for was for everyone to agree to disagree.

There was no apparent disagreement over the moral principle that Michael's treatment could be stopped if the burden of such treatment outweighed the potential benefit (Nelson and Nelson 427–33). What was in dispute was the description of Michael as either suffering or as simply a child who could not close his eyelids. The staff was able to substitute a physiologic argument about the presence of pain for the existential question of whether Michael was suffering—in effect, shifting a moral argument about the worth of living life on a ventilator to a technical dispute about the interpretation of a physical sign. In addition, the staff had the political power to threaten Michael's mother with a charge of failing to provide necessary medical treatment, that is, "medical neglect," thus throwing the matter into court. The moral and political discourse about what to do hereby assumed the form of a rational technical discourse that in the minds of the staff was uncontested. As such, the dependence of knowledge on subject position was implicitly denied; the power to determine objectivity was invisibly exercised. Rather than "moral and political discourse [serving as] the paradigm of rational discourse," the technical discourse set limits around what was morally and politically possible (Haraway 193–94).

KNOWLEDGE AND COMMUNICATION

An appeal to the shared "sense experience" of Michael's tears failed to attain the staff's agreement to the mother's factual claim that Michael was suffering. In addition, the staff failed to convince his mother that Michael's tears simply meant that he could not close his eyelids. The staff believed that his mother's claim to know that Michael was suffering simply reflected her subjective experience. The staff's insistence that Michael was not suffering reinforced their professional interest in continuing treatment. In answer to the question, "What protects knowledge from being the arbitrary expression of subjective desires or the tool of social and personal interests?" Helen E. Longino offers an approach she refers to as "contextual empiricism." Longino, as does Charles Peirce and Karl-Otto Apel, grounds "objectivity" or the truth of a statement concerning a sense experience in an "inter-subjective" or "social" process which should assure "the inclusion of all socially relevant perspectives in the community engaged in the critical construction of knowledge" (200, 202–3). A necessary part of this communicative process is a critical examination of the implicit assumptions which establish the relevance and interpretation of observational data. However, although the natural world cannot impose one interpretation, power differentials in this social process may constrain the potential plurality of interpretations to the one interpretation that is consistent with the dominant discourse.

Such power may be manifest either as the imposition of one interpretation through constraining the freedom of expression and diversity of legitimate knowledge or through structuring or restricting the community of discourse in such a way as to predetermine the outcome. For example, other than the final meeting at which she had legal representation, Michael's mother was effectively alone and isolated in all of the conversations in which I was involved. In my estimation, we thus failed to establish a meaningful community of inquiry concerning the question of Michael's suffering. Langdon Winner, in discussing contemporary policy debates about technology, observes that the lack of a coherent community of discourse "contributes to two distinctive features . . . : (1) futile rituals of expert advice and (2) interminable disagreements about which choices are morally justified" (75–77). The moral uncertainty involved in the application of ventilator technology to the indefinite support of patients such as Michael cannot be resolved by an appeal to the technical advice and expertise of the physician. Moreover, the lack of an appropriate community of discourse resulted in the privileging of the physician's technical expertise, and thus interpretation of Michael's experience, and in a conflict of interpretations which in the end served to reinforce the power and interests of the physician.

KNOWLEDGE AND POWER

How do we empower a parent to make decisions concerning his or her child's medical care? The notion of personal autonomy or self-rule has resulted in a significant shift of power from the physician to the patient over the past two decades. Once we abandoned the concept of the child as property, the notion of parental autonomy as a justification for the right of a parent to direct a child's medical care became problematic. Each one of us may have an absolute right to determine our own medical care. A parent has, at most, a prima facie right which is limited by the child's right to life and freedom from serious bodily injury or disability (Committee 279–81). Within this constraint, we expect that a parent will make decisions that benefit the child or, in other words, are in the child's "best interest." Thus, the parent's vision of the good is imposed on or becomes the child's vision—an imposition we accept given the diversity and, at times, incompatibility of competing visions of the good within our society. We nevertheless expect a parent to express a decision concerning his or her child not as "what is good for the parent" but rather as "what is good for the child." However, to the extent that our concept of childhood is simply the absence of adulthood, we give "voice to the voiceless" by articulating adult values (Archard 23). "If I were in this condition, I would want . . . ." If we seek to avoid this imposition of adult values by supporting the child until he or she is capable of self-expression, we reinforce the physician's tendency for the relentless application of life-sustaining technology. Consequently, the concept of a child's best interest appears to be the arena for an unavoidable expression of adult power. The stakes are high, for if the parent understands the child's best interest in such a way as to refuse what the physician otherwise believes to be necessary medical care, the parent may find himself or herself in court facing a charge of medical neglect.

The past two decades have seen a lively debate in the bioethics literature and the courts concerning the withholding and withdrawal of life-sustaining technology (for example, see President's Commission). Some have argued that removing a person from a ventilator is to choose death based on the judgment that the anticipated quality of life is not worth living. Others, concerned about the potential abuse of quality-of-life judgments, have posited that such decisions are better understood as the choice of how to live while dying (Dyck 529–35) or as simply the decision to remove technology that is no longer medically indicated (Ramsey 145–88; see also President's Commission 60–90). The first argument, that of how to live while dying, makes the decision to remove a ventilator dependent upon a prior determination that the patient is dying—a determination that the technology itself makes more difficult. A child with nemaline rod myopathy who is on a ventilator may not die for years in the absence of an intervening complication. Thus, once one has put the child on the ventilator, one cannot remove it unless the child is dying, and he or she is not dying unless one removes the ventilator. The second argument, that technology can be removed when it is no longer medically indicated, either restricts the removal of technology to those situations where more narrow technical goals cannot be achieved or obscures the physician's determination of an acceptable quality of life behind the veil of professional technical competence. A ventilator is medically indicated when a patient has respiratory failure; it is not indicated when either the patient recovers or the ventilator fails to correct the respiratory failure. Thus, Michael's ventilator is medically indicated. If a physician argues that the ventilator is not medically indicated since correcting the patient's respiratory failure does not contribute to the overall good of the patient, we necessarily must engage the question of what is or is not in the patient's best interest—a discussion that cannot avoid assessments of the patient's quality of life. The problem, then, of trying to avoid an explicit debate of a child's anticipated quality of life is that the physician's power and authority are inadvertently reinforced.

The physicians in Michael's case appeared to impose their power by establishing what counts as legitimate and credible knowledge, rather than by forcing a choice for one of either two plausible options. Michael's mother and the health care team disagreed over the description of Michael's life, not over the moral evaluation of an agreed-upon description (Hauerwas and Burrell). It was simply not credible to the medical staff that Michael was suffering. In discussing the problem of technology as ideology, Robert B. Pippin asks whether we have "been so influenced by technical instruments . . . that our basic sense of the natural world has changed . . . so fundamentally that . . . possibilities for social existence are seen only . . . in terms of such technical imperatives" (46). The physician's reliance on technology "reaches a point where what ought to be understood as contingent, an option among others, open to political discussion, is instead falsely understood as necessary; what serves particular interests is seen, without reflection, as of universal interest; what is a contingent, historical experience is regarded as natural" (46). Physicians have lost any sense of the natural or the contingent as a moral category. Rather, the natural serves to mark the domain that resists the physician's intervention, as in "let nature take its course." The natural becomes that which cannot be technically overcome, rather than that which should not be overcome. The natural is subservient to the technical, which in turn resists the explicit introduction of moral and political questions.

THE DIFFERENCE BETWEEN STARTING AND STOPPING

The belief that technology is a neutral means to whatever ends are selected on either moral, political, or more narrow physiologic grounds is a fundamental conviction and ideology of medical practice. The meeting at which it was decided, with the agreement of his mother, to perform a tracheostomy on Michael was held without the presence of the physician or nurse who work with children who are on home ventilators and before the diagnosis of nemaline rod myopathy was confirmed by a muscle biopsy. Any decision to limit ventilatory support was deferred for another six to eight weeks in hopes of potential improvement given the uncertainty of Michael's diagnosis and prognosis at the time of the meeting. In retrospect, the decision to perform a tracheostomy was not intended as a decision for home ventilation in spite of the fact that Michael was transferred to the home ventilation program prior to the end of this eight-week probationary period. Michael's mother insisted that she had been told that a decision to perform a tracheostomy did not preclude a determination at some point in the future to remove Michael from the ventilator—"what was done could be undone." Although she could not remember who told her this, I assumed her report was trustworthy given the widely endorsed bioethical teaching that there is no significant moral or legal difference between withholding and withdrawing treatment (President's Commission 77).

There are a number of important assumptions behind the use of this bioethical maxim. First, it assumes a symmetry in the application and removal of medical technology consistent with the prejudice that technological means are value-neutral. It also supposes a symmetry between an endotracheal tube and a tracheostomy by reducing each to its essential function of establishing an airway for the purpose of mechanical ventilation. However, as an endotracheal tube is inserted either through the mouth or nose, the tape required to hold it in place covers a major portion of an infant's face. A tracheostomy surgically inserted through the front of an infant's neck results in the entire face being visible and thus capable of expression. Second, the maxim appears to ignore any relevant differences that may occur between the moments of application and removal of the technology apart from any changes in the medical indications. Third, and related to this historically naive stance, is the view that the organizational context in which these decisions are being made is apparently unimportant. An appreciation of the value-laden nature of a tracheostomy, along with the significance of time and context, counsel against a premature surgical procedure and then transfer to a home ventilation program. One wonders whether the use of this bioethical maxim was based on a reasoned ethical stance or if it was employed as a rhetorical device to finesse the discussion of more difficult ethical issues to a later date. The interpretation that this maxim served merely as a rhetorical device unfortunately is supported by the fact that the physician who was in conflict with Michael's mother was unable to find any other physician willing to assume Michael's ongoing medical care, including any of the physicians who had cared for him prior to the tracheostomy.

IS TECHNOLOGY VALUE-NEUTRAL?

The bias that our medical technology is simply a collection of devices emphasizes the functional aspects of technology and obscures an understanding of its social context. As a result of this dichotomy between function and context, our technology appears value-neutral while only the application of this technology becomes morally problematic. Andrew Feenberg, reflecting on the relationship between technology and power, points out that this "dichotomy of goal [function] and meaning [context] is a [contingent] product of functionalist professional culture" rather than a necessary component of technology (9). Echoing Herbert Marcuse's critique of Max Weber, Feenberg asserts that this ideology of technology as value-neutral reinforces the dominant forms of power which compose the cultural horizon and social meaning of technology. Feenberg refers to this as the "bias of technology" by which "apparently neutral, functional rationality is enlisted in support of a hegemony." As a result of this bias, to obscure the fact that our application of technology carries with it the values and presuppositions that shape this technology serves as an effective technique by which the professional control of medical technology is perpetuated (12).

The professional control of medical technology is also reinforced by the perception that it should be used when it can be used, that is, by the so-called technological imperative. Barbara A. Koenig suggests that this moral imperative for the use of medical technology develops through a social process by which the technology becomes habitual or routine (466, 469–70). Her field research involved the use of therapeutic plasma exchange, a procedure which entails the removal and then replacement of blood plasma. The first of three features that Koenig identified in the process by which plasma exchange became a routine therapy for certain conditions was a transformation in roles and responsibilities. The physician-nurse relationship shifted from egalitarian to hierarchical while, at the same time, the physicians moved from being closely involved to delegating many of the routine tasks to nurses (476–79). Similarly, the physicians participating in the home ventilation program maintain close control while shifting many of the routine tasks from in-hospital nurses and respiratory therapists to parents and visiting home nurses. The second feature that Koenig ascertained was the use of treatment rituals which appeared to reduce uncertainty, anxiety, and disorder and thus established the meaning of the technology as standard therapy for both patients and staff (479–80, 482–83). When a patient is placed on a home ventilator after a tracheostomy, there is an orderly and nearly invariable process of parental training, arranging for nursing services, equipment purchase, and so forth that must take place prior to discharge from the hospital. Any deviation from this process generally results in hesitancy, inefficiency, omissions, and the like. The third feature that Koenig identified was the generation of research data. Noting the enthusiasm with which the physicians engaged in plasma exchange collected clinical data as part of their ongoing research, she speculated that the machine's capability of producing research data supported the physician's tendency to use the technology (483–85). Although this may be true with therapeutic plasma exchange, it does not appear that the development of home ventilation programs was driven by a research imperative. The physician's use of home ventilator technology is more likely driven by factors such as the need to find alternative placements for children who otherwise would survive intensive care but remain dependent on ventilator technology. Also, the immediate efficacy of the ventilator when compared to plasma exchange is obvious, for otherwise the child would die. In spite of these differences, I believe that Koenig's conclusion remains essentially correct. The "technological imperative" is transformed into a "moral imperative" through the development of a "sense of social certainty experienced by health professionals" (485–86). The technology simply begins to feel routine.

The decision to transfer Michael to a unit where the use of chronic home ventilation is considered routine thus placed him and his mother under a different set of moral imperatives. The unit is organized so that home ventilator technology is accepted as standard therapy. Within this social context, it becomes difficult if not impossible to question whether this technical standard of care ought to be used for any particular child. The moral question of what is in a particular child's best interest is then understood more generally as "usually in these circumstances we provide the standard technology." The moral meaning of our medical technology is thus created and sustained by the professional culture of the hospital. Since the statutory definition of medical neglect in Wisconsin is simply failure to provide necessary medical care, the technological and moral imperatives that medical and nursing professionals experience have "the potential to wrest control of decisions about the use of technology" from parents and patients (Koenig 486, 489–90).

Feenberg proposes that one of the assumptions behind our modern image of technology is that social institutions must adapt to the technological imperative. Noting that "the economic significance of technical change often pales besides its wider human implications in framing a way of life," Feenberg encourages us to study the "social role of the technical object and the lifestyles it makes possible" through defining "major portions of the social environment, such as . . . medical activities and expectations" (6, 9, 15–16). This assumption that we must adapt to technology is readily apparent over the past two decades with the development of home care programs for so-called technology-dependent children. The family is explicitly expected to change in response to the demands of caring for a child who is to be discharged from the hospital on a home ventilator. The only other available option is foster care, as was the case with Michael and his mother. The foster care option is problematic for two reasons. While the child is in foster care, a parent may lose control over any decisions to either withhold additional medical treatment or withdraw existing medical treatment. It was the fear of loss of control that led Michael's mother to cancel plans for foster care placement in spite of the obvious strain on her personal and family resources and the presence of a capable foster parent. The foster care option also is problematic given the often unspoken assumption that to choose this option reflects poorly on the ability of a parent to provide for his or her child. Although many parents choose to take their ventilator-dependent child home out of a sincere concern for their continued life and well-being, the normative pressures against choosing otherwise are enormous once the child is within the context of the home ventilation program. This assumption, then, that social institutions such as the family must adapt to the technological imperative is another manifestation of the extension of professional power implicit in the ideology of value-free technology.

The apparent inevitability of the technological imperative is rejected by both Koenig and Feenberg. Consistent with Koenig's thesis, Feenberg asserts that "technology is just another dependent social variable" and the "scene of social struggle" (8). Contrary to the claim that technology itself requires professional control, Feenberg argues that technology has been used to block the extension of public or democratic control to "technically mediated domains of social life" (20). Thus, the professional medical culture seeks to reinforce the image of technology as both value-neutral and complex in order to maintain control in spite of the routinization process of placing this same technology in the home.

CONTESTING POWER OVER TECHNOLOGY

What power do we as a community have over medical technology? Our discussion of technology as socially constructed may give the impression that we have complete power over technology and that power differentials are based solely on social and political considerations independent of the constraints of technology. The opposite extreme is that we have no power over technology and that our moral and cultural response is determined by the concrete and independent reality of technology.[4] My intent is not to argue for one or the other extreme but rather to show how the assumption of the neutrality of technology serves to reinforce the professional dominance of physicians. By failing to recognize the extent to which technical knowledge is constructed by and for the interests of a particular community, we are likely to respond to technology in ways that strengthen the power of that particular community (Vogel 28–29).[5] The technological determinism which constrains our social and moral choice in the application of this technology risks "making humans subservient to things" (Vogel 37).[6] Who gives life to whom? The ventilator to the child or the child to the ventilator?

Controlling the technical mediation of social activities such as medical care is a major source of public power within our society. The ability to manage or expand this technical mediation results in the concentration of power in an elite group of experts, the narrowing of acceptable options for public discussion, and an increase in the extent of administrative control over aspects of daily life (Pippin 43–44). Changes in the way medical technology is delivered or applied to a specific problem will require a shift in this expert control of technology. As Feenberg writes: "If authoritarian social hierarchy is truly a contingent dimension of technical progress . . . and not a technical necessity, then there must be an alternative way of rationalizing society that democratizes rather than centralizes control" (5). Is the apparent link between the physician's social role and the control of medical technology necessary or contingent? For example, one approach to the issue of physician-assisted suicide is to allow for assisted suicide while preserving the traditional social role of the physician by making available to the general public the technical tools that to date remain under the physician's prescriptive authority.

If we move the control of medical technology into the public domain, we will need to create an appropriate community of discourse concerning the development and application of this technology. Such a task may be difficult given the diversity of our current communities. Although the reform of technology may appear to be a better option than simply resistance, it is not clear that those of us who have been socialized in the modern medical ethos possess the resources to sustain such a reform while avoiding the danger of new forms of professional control (Pippin 51). The creation of a community for the reform of medical technology should include those who anticipate needing or who may resist medical technology and thus will require abandoning the notion of professional expertise. In addition, such a community of discourse would need to begin by questioning the assumption that technology is a value-free instrument—a supposition that serves to reinforce professional control and hinder rational debate. Or should we simply recognize the legitimate existence of disparate communities and thus reframe the question of the appropriate application of medical technology as a choice of which community to belong to?

THE BOUNDARY OF BODY

One of the original questions that stimulated my interest in Michael and his mother was the perception that the medical and nursing staff saw Michael differently than his mother saw him. As the disease progresses, a child with nemaline rod myopathy cannot move, is unable to breathe, cannot express emotion, and, indeed, is not able to make any facial expressions; communication at best can occur through the movement of an eye in response to a question. Consequently, it would be difficult if not impossible to get any indication of what tearing meant to Michael. As the passive object of our application of ventilator technology, Michael was reduced to

either a resource for our instrumentalist projects or a mask for our dominant interest in maintaining control (Haraway 197).[7] Modern medical
technology, as we have seen, clearly includes the feature of the technical
control of some human beings by others. Donna J. Haraway attributes this
modern tendency toward technical domination to the dualism between
objective nature and subjective culture so that the projects or interests that
shape our determination of natural objects are hidden. As an alternative,
she offers us a view of "objectivity as positioned rationality" (196). To capture a notion of the object as active and not passive, Haraway asserts that
"bodies as objects of knowledge . . . materialize in social interaction.
Boundaries are drawn by mapping practices; 'objects' do not pre-exist as
such" (200–201). The issue then is the various positions from which each
one of us, including his mother, viewed Michael—a question that necessarily draws us back into an explicit discussion of power in determining the
boundaries of Michael as the object of our attention.

How are we to understand who Michael is, this body attached to a ventilator? Through an exploration of the "semiotic use of the body" among
the Kayapo of the Brazilian Amazon, Terence Turner illustrates how "the
body is at once a material object and a living and acting organism possessing rudimentary forms of subjectivity that becomes, through a process of
social appropriation, both a social identity and a cultural subject" (145).
For example, the Kayapo use various modifications of their body surface to
define and redefine their social identity, as in the use of ear piercing to
create and indicate the capacity for social communication and participation. The individual Kayapo, as both a social body and an embodied subject, assumes the dual role as product and producer (167). In our case, the
body called Michael as a material object of our technical interventions took
on the social identity of a patient in the home ventilation program. Although his mother later tried to resist this medical appropriation of Michael's body, the tracheostomy and attached ventilator tubing were key
modifications of his body which produced his social identity as a patient in
the home ventilation program (see Turner 146). The ventilator-infant as
"embodied subject" appears to be the "socially patterned" product of our
technical activity, rather than the "producer" of its own activity. Similar to
the ideological consequences of the view of technology as value-neutral,
the misrepresentation of the "cultural subject" of the ventilator-infant as
an "objective (natural) feature existing independently" of our social production further reinforces the dominant power of the physician (167).[8] In
infancy, it is unclear to me that there is any content to the notion of the
subject existing prior to and independently of the social production of the
embodied subject by others. In other words, what meaning can we give to
the notion of the best interest of the child apart from the specific interests
of a particular embodied social subject? Following Turner, the "specific
representations of bodiliness and . . . embodied subjectivity are produced

as integral components of specific social organizations of productive relations" (169). Once Michael underwent a tracheostomy and was placed on a home ventilator, he was and would remain a patient in the home ventilation program. Thus, we come full circle to the notion of the ventilator-infant as cyborg, the machine-human as embodied subjectivity rather than the machine as external to the body. The social identity of the patient named Michael is a product of being a machine-human hybrid, that is, the ventilator gives life to the body and the body gives life to the ventilator. To contemplate taking the patient named Michael off of the ventilator would be to contemplate amputation—a request that the medical and nursing staff could not and would not honor.

CONCLUDING REMARKS

What have I learned from the story of Michael and his mother? I have come to doubt the universality of the classic teaching of the symmetry between withholding and withdrawing technology. I have a renewed understanding of the insight that our medical technology is not value-neutral, and it often serves to reinforce the professional dominance of physicians. Although I was aware of different units having disparate cultures, the impact of the organizational context on the ability of patients and parents to control the application of medical technology is greater, on reflection, than I had previously appreciated. This effect is not simply through the imposition of a distinct set of moral values, but it is also through fundamental shifts in how we see and thus know our patients. Finally, I am disheartened how we as physicians constrain what is admitted as knowledge within the medical setting.

Six months after he was discharged from the hospital, Michael died at home. I was told that two weeks before his death, Michael's mother had asked that all of the monitoring equipment be removed from her home. There had been little contact between Michael's mother and the physicians and nurses of the home ventilation program, with arrangements having been made for Michael to be cared for by a local physician. I know little about the circumstances surrounding Michael's death. It seems that in our technological enthusiasm we lost touch with both Michael and his mother.

NOTES

1. The range of conditions to which the term "technology-dependent children" may refer includes children with a tracheostomy either with or without a ventilator, children who require home oxygen, and children who require parental hyperalimentation for nutritional support.

2. My purpose is not to discuss, at least directly, the ethics of the various decisions that were made during the course of Michael's hospitalization.

3. "Cyborg anthropology is a dangerous activity because it accepts the positions it theorizes for itself as a participant in the constructed realms of science and technology.... One danger of participation in institutionalized science and technology, even if retheorized, is co-optation. That is, accepting participation can shade into the acceptance of presuppositions that constrain the imagination of alternate worlds and undermine the critical edge of ethnographic investigations" (Downey, Dumit, and Williams 268–69).

4. Technology as applied science appears to bridge the natural and the social as it seeks to apply our knowledge of the natural world to the resolution of human problems. However, both interpretations of the metaphor of application are untenable given the shared assumption that one can clearly distinguish the natural and the social. See Vogel 23–24.

5. Vogel relies on Jürgen Habermas's critique of Herbert Marcuse to argue that "it is the conflation of the technical interest (means-ends) with a communicative interest (mutual understanding) that reinforces domination" (29).

6. "The mistake is to try to ground social theory in nature, which is to say in the extra-social—a move that always ends up making humans subservient to things, and that [Karl] Marx had criticized as fetishism and [Georg] Lukács as reification" (37).

7. Michael's mother did not have a privileged access to Michael as subject, and thus Michael was positioned as object to her as well; however, I am less comfortable speculating about the nature of her instrumental projects or interests.

8. Turner refers to this as a "fetishistic inversion."

WORKS CITED

Archard, David. *Children: Rights and Childhood.* New York: Routledge, 1993.
Committee on Bioethics, American Academy of Pediatrics. "Religious Objections to Medical Care." *Pediatrics* 99.2 (1997): 279–81.
Connolly, Mary B., Elke H. Roland, and Alan Hill. "Clinical Features for Prediction of Survival in Neonatal Muscle Disease." *Pediatric Neurology* 8.4 (1992): 285–88.
Downey, Gary Lee, Joseph Dumit, and Sarah Williams. "Cyborg Anthropology." *Cultural Anthropology* 10.2 (1995): 264–69.
Dyck, Arthur J. "An Alternative to the Ethics of Euthanasia." *Ethics in Medicine: Historical Perspectives and Contemporary Concerns.* Ed. Stanley Joel Reiser, Arthur J. Dyck, and William J. Curran. Cambridge: MIT Press, 1977. 529–35.
Feenberg, Andrew. "Subversive Rationalization: Technology, Power, and Democracy." Feenberg and Hannay 3–22.
Feenberg, Andrew, and Alastair Hannay, eds. *Technology and the Politics of Knowledge.* Bloomington: Indiana University Press, 1995.
Haraway, Donna J. *Simians, Cyborgs and Women: The Reinvention of Nature.* New York: Routledge, 1991.
Hauerwas, Stanley, and David B. Burrell. "From System to Story: An Alternative Pattern for Rationality in Ethics." *Truthfulness and Tragedy: Further Investigations in Christian Ethics.* Ed. Stanley Hauerwas. South Bend: University of Notre Dame Press, 1977. 15–39.
Iannaccone, Susan T., and Timothy Guilfoile. "Long-term Mechanical Ventilation

in Infants with Neuromuscular Disease." *Journal of Child Neurology* 3 (1988): 30–32.

Koenig, Barbara A. "The Technological Imperative in Medical Practice: The Social Creation of a 'Routine' Treatment." *Biomedicine Examined*. Ed. Margaret Lock and Deborah Gordon. Dordrecht: Kluwer, 1988. 465–96.

Longino, Helen E. "Knowledge, Bodies, and Values: Reproductive Technologies and Their Scientific Context." Feenberg and Hannay 195–210.

Nelson, Lawrence J., and Robert M. Nelson. "Ethics and the Provision of Burdensome, Harmful, or Futile Therapy to Children." *Critical Care Medicine* 20.3 (1992): 427–33.

Pippin, Robert B. "On the Notion of Technology as Ideology." Feenberg and Hannay 43–61.

President's Commission for the Study of Ethical Problems in Medicine and Biomedical and Behavioral Research. *Deciding to Forego Life-Sustaining Treatment: A Report on the Ethical, Medical, and Legal Issues in Treatment Decisions*. Washington: U.S. Government Printing Office, 1983.

Ramsey, Paul. *Ethics at the Edges of Life: Medical and Legal Intersections*. New Haven: Yale University Press, 1978.

Turner, Terence. "Social Body and Embodied Subject: Bodiliness, Subjectivity, and Sociality among the Kayapo." *Cultural Anthropology* 10.2 (1995): 143–70.

Vogel, Steven. "New Science, New Nature: The Habermas-Marcuse Debate Revisited." Feenberg and Hannay 23–42.

Winner, Langdon. "Citizen Virtues in a Technological Order." Feenberg and Hannay 65–84.

Chapter 10

The Ethics of the Organ Market

Lloyd R. Cohen and the Free Marketeers

DONALD JORALEMON

The U.S. Department of Health and Human Services' Division of Transplantation (DOT) organized its 1996 annual meeting around the theme Toward the Year 2000: Concepts and Considerations in the Consent Process (Washington, D.C., 21–23 March). The goal of the conference was an exploration of the psychosocial and socioeconomic factors influencing whether people agree to donate their own or loved ones' organs. The meeting was motivated by the frustration of the transplantation profession over a growing gap between organ supply and demand.[1]

Citing the "tragedy of lives lost due to organ scarcity," conference speakers declared that it was time to "push the limits" of current procurement practices for "the good of the cause." Physician Michael Rohr, director of transplantation services at Bowman Gray School of Medicine at Wake Forest University, related the challenge faced by the profession of transplantation medicine to the space program and urged a new "science of donation" aimed at a better understanding of the process of donation requests. Rohr also advised his audience to reconsider the ethical objections and "legal impediments" that have stood as obstacles to two alternative approaches to organ procurement: taking organs unless the deceased had registered an objection (that is, presumed consent) and offering financial incentives for permission to take organs.

Rohr's plenary address set the stage for a smaller afternoon session on the topic of alternatives to procurement. Among the speakers was Lloyd R. Cohen, a jurist and professor of economics at George Mason University. Cohen has long been an outspoken advocate of market forces as the solution to the present organ shortage. He summarized the argument of his recent book, *Increasing the Supply of Transplant Organs: The Virtues of an Options Market,* which favors a "futures market" in which a per-organ fee of $5,000 would be paid to the estate of those who agree to sign contracts for the removal of their organs, should they die in a fashion that permits transplantation.[2]

After defending his proposal and urging that it be given a "fair test,"[3] Cohen responded to questions from the audience. An elderly, white-haired woman explained that not long ago she and her husband had donated their daughter's organs after a fatal traffic accident. She asked Cohen why he insisted on using terms such as "salvage" and "harvest" instead of "donation" and "gift" when referring to the process by which organs are made available for transplantation. Cohen responded bluntly: he wanted to be precise and to avoid language that obscures the fact that everyone but the "supplier" is profiting from the transaction. The woman, stunned by Cohen's dismissive tone, promptly left the room.

This exchange led me to consider several currents in American thinking about death and treatment of the dead. In the case at hand, the donor's mother clearly held the notion that the physical and spiritual aspects of a person are not quickly separated by a declaration of death. Deciding to agree to the removal of her daughter's organs was, in her mind, a symbolic act celebrating the generosity of the deceased. Like so many who make the same decision, she talked about how the sudden death was made more bearable by the knowledge that some parts of her daughter were still alive in others. Flesh and self remain connected postmortem.

By contrast, Cohen's argument is predicated on a libertarian understanding of personal autonomy and property rights and on a purely material view of the cadaver—flesh is stripped of self by death. One's body is comparable to the family china; the disposition of both after death should be the prerogative of the previous owner. Cohen asks, If we do not own our bodies, who does?

It would be wrong to imagine that these two views of the meanings of death and of the dead represent mutually exclusive alternatives. We need only look at the history of English common law regarding disposal of the dead to see that materialist and non-materialist priorities have long commingled (Richardson; Scott). However, it is useful to separate them and to ask if there has been a shift in the balance between them as a result of medical interventions such as transplantation. Are we prepared to think about the cadaver in more material terms because of its increasing medical utility for others? What significance does this have for our understanding of death?

In this essay, I use Cohen's market argument and moral objections to it as a way to explore these questions. My reflections are part of an emerging anthropology of organ transplantation in which the cultural and socioeconomic ramifications of this high-tech medical specialty are scrutinized (for example, Joralemon; Sharp; Lock and Honde; Marshall, Thomasma and Daar; Hogle). My discussion also shows how anthropology can contribute to bioethical debates "by calling attention to the cultural underpinnings that sustain and reinforce ethical constructs in matters of health and illness" (Marshall, "Anthropology" 62).

Let me review in detail what Cohen proposes, and then I will survey published criticisms of the general idea of financial incentives for organ retrieval. I am especially interested in the ethical arguments both sides make. Do they reflect or portend important changes in the view of the public at large? What does this debate indicate about our ideas of when death occurs, about what happens at death to the "self" or "person" that was once associated with the physical entity, and about the obligation(s) we have for the survival of strangers?

COHEN'S OPTIONS MARKET

The full explication and defense of Cohen's proposal is found in his 1995 book, *Increasing the Supply of Transplant Organs: The Virtues of an Options Market*. He argues that organs for transplant are in short supply not because they are "scarce" but because present law and practice provide insufficient incentives for people to agree to give up their own or loved ones' organs. He insists that resistance to donation is not based on any "deeply felt antipathy" (49), but that even superficial objections are not overcome by the current altruism-based system. Cohen considers financial incentives to be the best way to stimulate organ supply because he believes that "the most direct, efficient, and least costly way to induce [someone to do something] is to pay them" (51).

In Lloyd Cohen's world, people would be able to enter into contractual obligations for the postmortem sale of their vital organs and tissues. The proceeds would go to their estate or to named beneficiaries. Cohen argues that since the "seller" would receive no direct payment, the market would not coerce poor people into selling their organs ("the poor man's sale").[4] Cohen also insists that the proposal need not be vulnerable to the objection that only the wealthy would be able to buy organs ("the rich man's purchase"). It could be paired with legislated allocation schemes that would assure equal access to the organs for rich and poor alike, perhaps using a combination of fixed prices and subsidies.

Cohen does not defend his proposal solely on his conviction that it will increase the number of organs made available for transplantation. Instead, he claims the moral high ground by associating his market with the priorities of individual autonomy and the saving of lives. He invokes the "moral canon that a person's body is in the first instance his private property, and that, like other property, he has the right to provide for its disposition after his death" (83). Cohen declares that those who reject his plan on the grounds of "pseudo-intellectual moral posturings" (76) are insensitive to the loss of lives that results from an insistence on uncompensated donation. His opponents offer only "the empty moral pieties of armchair philosophers incapable of a reasonable balancing of human needs" (90).

After baiting his critics for their moral shortcomings, Cohen offers two suggestions that he knows will be viewed as reprehensible: encouraging people who are considering suicide to take their lives in a fashion that will facilitate organ retrieval, and taking organs from executed prisoners (108, 112). Cohen admits that the first idea may sound "heartless and ghoulish" (108). Making the case for the second, he urges his reader not to be "overly prissy about where we get the organs" (112). Like others in the politically conservative tradition of the study of law and economics (for example, Richard Posner), Cohen seems to enjoy staking out positions that offend public standards of decency, even as he claims clear vision and moral superiority.

Apart from his debating style, what is especially striking about Cohen's proposal is the willingness of ethicists and specialists in transplantation to treat it as an idea worthy of discussion. When financial incentives were proposed earlier in the history of transplantation medicine, they were met with moral indignation and legal prohibitions (for example, the National Organ Transplantation Act of 1984, Public Law 98–507). Less than a decade later, versions of Cohen's proposal are presented and discussed at conferences (for example, the Fourteenth International Congress of the Transplantation Society, Paris, August 1992) and in academic and professional journals: to name a few, *Transplantation Proceedings* (Cohen, "Futures"); *Journal of the American Medical Association* (Evans); *Issues in Science and Technology* (Fentiman); *Technology Review* (Reed); *Journal of Medical Ethics* (Harvey); *The Public Interest* (Kass); the *Review of Industrial Organization* (Barnett, Beard, and Kaserman); and the *Hastings Center Report* (C. Campbell).

This shift in the perceived respectability of market options for organ procurement stems from the desperation of those who most directly confront the loss of lives on the transplant waiting lists: fatally ill patients and their families, transplant surgeons, organ procurement specialists, et cetera. The same desperation has resulted in other proposals, some of which have been tested, that would have been unthinkable a few short years ago. These include: changing the definition of "brain death" to encompass those with functioning brain stems, taking organs from anencephalic infants, transplanting across species, permitting medical interventions that preserve organs for transplant from "non-heart beating cadaver donors" even before receiving consent from the next of kin,[5] passing "mandated choice" legislation which would require all adults to state their preference regarding donation, and allowing "routine salvaging" of organs in the absence of a recorded objection (that is, a "presumed consent" system).[6]

In considering Cohen's proposal, then, we have to keep in mind that it is part of a larger collection of schemes which transplantation medicine has floated for its own benefit, but which have profound implications for our understanding of death and the dead. These suggestions are all premised on the assumptions that transplantation is an unqualified good, and that those on transplant waiting lists have a moral claim on the rest of society to support their survival.[7] These proposals have a common focus on the "supply" side of the organ shortage; with rare exceptions, there is no mention of lowering the "demand" for organs by reducing the waiting list and/or managing better the existing supply.[8]

Sociologist Renée C. Fox quotes the theologian Paul Ramsey in arguing that these proposals move us closer to "'the reduction of persons to an ensemble of . . . interchangeable . . . spare parts in which everyone [becomes] a useful pre-cadaver'" ("Afterthoughts" 265). Critics such as Fox urge against starting down this slippery slope.

MORAL GROUNDS FOR OPPOSING ORGAN SALES

Critics of financial incentive plans such as Cohen's (for example, Fox, "An Ignoble Form"; Fox and Swazey; Kass; Anderson; C. Campbell; Murray, "Organ Vendors") have raised both moral and practical objections. Certainly, a sound moral argument against organ markets would make consideration of the feasibility of market solutions unnecessary.[9] There are two common moral arguments opponents offer. The first objects to organ sales on the grounds that it would threaten the social cohesion and sense of community which the present altruism-based system promotes. The second focuses on the moral consequences of commodifying the body.

Legal scholar Mark F. Anderson provides a good example of the argument based on altruism. In his recent article, "The Future of Organ Transplantation: From Where Will New Donors Come, To Whom Will Their Organs Go?" he concludes that compensation for organs would destroy an altruistic institution which plays "an important role in the integration of a society increasingly estranged from itself" (299).

> Organ donations are an important part of the voluntary contributions we make as citizens to help improve our community. If money enters the picture, the element of community is lost. We are no longer giving of ourselves to help those in need of the most basic of human resources. (299)

Cohen answers that, because requests for organs are relatively rare occasions compared to other opportunities for charitable acts, the impact of introducing financial incentives could not have much of an effect on the role of altruism in public life. He also points out that "permitting sale does not mandate it" (*Increasing* 74); indeed, if compensation for organs were available, the decision to donate without payment would be an even greater act of altruism. Besides, Cohen insists, his proposal calls for an act of altruism, one which is directed toward the decedent's heirs.

The objection to organ sales based on the moral implications of commodifying the body—treating the body as marketable property—has been made by ethicists (for example, Murray, "Human Body"), legal scholars (for example, Radin), philosophers (for example, Chadwick), and social scientists (for example, Fox and Swazey). The commodification argument is of particular interest to the current discussion because it confronts directly the question of how we conceive of the human body, dead or alive.

The most persuasive and carefully argued version I have found of the commodification objection comes from University of California at Los Angeles legal scholar and philosopher Stephen R. Munzer, who turns to Immanuel Kant's notion of human dignity to "make an uneasy case against property rights in body parts":

> The argument from dignity runs as follows. Human beings have dignity (*Wurde*). Dignity is an unconditioned and incomparable worth. Entities with dignity differ sharply from entities that have a price on the market. If human beings had property rights in body parts and exercised those rights, they would treat parts of their bodies in ways that conflict with their dignity. They would move from the level of entities with dignity to the level of things with a price. (266)

Munzer summarizes the link between this conception of human dignity and Kant's effort to provide a foundation for universally applied equal rights:

> Dignity is an attribute not mainly of isolated individuals but of persons as "ends in themselves" who are members of a "kingdom of ends." Persons so understood belong to a moral community in which dignity and autonomy must be ascribed to every rational person with a will. Accordingly, the construction of a legal system for these persons must observe the dignity of each individual. (267)

Munzer's case against commodifying the body is "uneasy" in part because he sees different degrees of offense to human dignity in various kinds of sales of body parts. He asks us to consider the motivation of the seller relative to the centrality of the particular body part to be sold to the "normal biological functioning of the person" (269).[10] A body part that is integral to biological functioning can only be removed at some risk to the person. The seller, to accept the risk, must be motivated primarily by profit. This reason is incommensurate with the significance of the part to the human dignity of the person. Thus, the sale demeans and degrades the moral status of a human being.

By this logic, the sale of replenishable or naturally shed parts that can be removed at little or no risk to the person (for example, hair, blood, and sperm) pose the least threat to the person's dignity. Other body parts, those which can be removed with minimal risk (for example, ova and bone marrow) occasion more concern, because a monetary value on such parts might pressure people into selling and thereby diminish their dignity (that is, transform their body into a means rather than an end).

Munzer sees the most serious potential offense against human dignity in the sale by a living person of a nonreplenishable solid organ (for example, kidney or cornea). Given the risks, only the severest financial exigency would motivate someone to engage in such a sale. To the degree that the seller is thereby demeaning his/her moral worth for reasons that are not commensurate with the centrality of the part to the whole person, the sale demeans human dignity.

Interestingly, Munzer places the sorts of sales Cohen recommends in an intermediate position on a scale of potential offense against human dignity. Munzer seems to accept the idea that the cadaver has no moral significance relative to human worth (278). The only threat he sees in a futures market is the possibility that a well-publicized secondary market, one which buys and sells organ contracts, might have the effect of encouraging people to be viewed "as repositories of body parts with a market worth rather than as entities with a Kantian dignity" (286).[11]

It is the invocation of ideas such as "human dignity" (or "human flourishing"—Radin) that so infuriates Lloyd Cohen. Characterizing arguments such as Munzer's as perverse, Cohen asks, "By what leap of moral logic could it be that the very preciousness and sacredness of human life is affirmed by condemning innocents to death, and that saving human life through market transactions degrades life?" (*Increasing* 75).

LIVES AND MORALS

Cohen rejects these moral arguments against his proposal by individualizing the issue. He casts himself as the savior for identifiable people—those on waiting lists—while his opponents have chosen to make their case at a far more abstract level (for example, harm to American society at large or to humanity in general).[12] Cohen advocates on behalf of "innocent lives"; his opponents can only speak to "offended moral principles!"

For the moment, let me take sides by suggesting that Cohen's critics need not yield the rhetorical advantage of speaking on behalf of people who would be harmed by the failure to institute a futures market. It is also possible to follow out the potential consequences of violating moral principles in a way that makes clear the disservice that could occur to real people if a futures market in organs were established.[13] Imagine the following scenarios in a Lloyd Cohen world:

Scenario 1: The elderly woman mentioned at the outset of this essay is approached by a doctor in the emergency room. She is informed that her daughter is dead and that the body has been taken to the operating room where the organs will be harvested for transplantation. "There is a silver lining here," says the doctor, "You'll get $5,000 for each organ we can use."

Scenario 2: An African-American family arrives at the emergency room because the eldest son has been critically wounded in a drive-by shooting. A white doctor enters and informs the family that the injury was fatal; the young man has been declared brain dead. The doctor also tells them that there is a contract for the sale of the young man's organs. The family registers strong opposition to the idea of having the organs taken; they do not trust the doctor and ask if he let their son die so that he could get the organs. The doctor explains that he has no choice but to proceed with the removal of the organs because he is legally liable for their prompt delivery (see Cohen, *Increasing* 82). A heated argument ensues and the police have to be called to remove the distraught family.

Scenario 3: A forty-year-old woman, having explored a New Age religion, becomes convinced that body and soul remain connected for some time after the cessation of physical life. Unfortunately, at age twenty she had signed a contract for the removal of her organs upon her death. She now finds that she has to compensate an investor who purchased that contract on a secondary market (see Cohen, *Increasing* 114). In effect, she has to buy back her own organs.

Scenario 4: An eighteen-year-old male buys a motorcycle. The family immediately begins to receive postal and telephone solicitations from companies eager to have this likely candidate for brain death sign a contract for his organs (see Cohen, *Increasing* 115). One caller asks the mother, "Wouldn't you rather get something back if he's sent through a windshield?"

Each of these scenes is consistent with the policy Cohen promotes and each would represent real harm to real people. Is the disservice justified by the "good" of saving more lives from the transplant waiting list? It is only if one accepts Cohen's implicit argument that people who are suffering from organ failure have a strong moral claim on strangers, and on society as a whole, to provide them with the means to survive. But then, were there agreement about the moral responsibility owed to strangers, would financial incentives be necessary in the first place?[14]

ANTHROPOLOGICAL REFLECTIONS

I noted above the apparent agreement between Stephen Munzer and Lloyd Cohen on the question of the moral worth of the cadaver. Cohen is especially fond of the image of bodies being "fed to worms" (*Increasing* 16, 138) as a way of conveying the materiality of the corpse. Munzer sees less risk of an offense to human dignity in a futures market because it is not a living person whose body is ultimately commodified.

Bioethicist Thomas H. Murray, among the most eloquent opponents of financial incentives for organ procurement, is less willing to view the cadaver as nothing more than rotting flesh. He points to the "great moral concern and social meaning" with which the dead human body is invested by all cultures (Murray, "Organ Vendors" 116). Murray writes:

> In this light, financial incentives for organ recovery can be seen as an attempt to transform the relationship that a family has to the remains of its newly deceased relative. The relationship is transmuted from one of intimate social and moral connectedness to something more like the relationship the owners of old cars have with their vehicles, which are now being stripped of still-useful parts in a salvage yard: impersonal, with the hope that a little utility and a little money can be extracted from the now lifeless hulk. (117)

Murray is on solid anthropological footing here. The simple social fact about death, so well documented in my discipline (Huntingdon and Metcalf), is that it represents a transition that disrupts the social order and requires ritual markings which give primary attention to the significance of the deceased to the survivors. In evolutionary perspective, death rituals are among the most important early signs of the human capacity for culture. They evidence a characteristic human concern about the consequence of continuity in human relationships, even in the face of death.

Whatever reductionistic science might have to say about the physiological locus of consciousness and selfhood, it is the social memory of the person that remains attached to its embodied form for a prescribed time after death. In some cases (for example, Egyptian pharaohs, Inca rulers), the linkage between identity and physical form is considered eternal; flesh and self are never separated. In other cases, powerful symbolic means (for example, cremation, endocannibalism) are required to disaggregate identity from its physical referent. But in all instances, the dissociation of body and self after death proceeds according to symbolically laden scripts because of the difficulty humans experience in making a memory of one who was recently alive.

Cohen and his fellow free marketeers have trouble seeing any serious value in the symbolic significance of the dead body to the living, especially in comparison to its utility as a collection of recyclable parts. However, it would be hard for these advocates to deny that death rituals built on the continuing association of the person and the physical form could be as universal as they are without some important social purpose behind them. That purpose is to minimize the impact of an individual's death on the community of which he/she was a part by structuring a process of disassociation wherein the person becomes memory. This is the foundation of mourning rituals of all varieties, of the idea that combatants have a right to retrieve their dead from the battlefield, and of the widespread notion that one must show proper respect for the cadaver.

CONCLUSION

In a review of the relationship between anthropology and bioethics, Patricia Marshall writes, "What has distinguished anthropological from philosophical examinations of prescriptive norms and values is a decided absence of moral judgment. The anthropological perspective has not measured ethical problems against a definitive standard of moral rectitude but instead has viewed them as culturally constituted and continually evolving" (Marshall, "Anthropology" 54).

I am not sure that the anthropological presumption of moral neutrality on the grounds of cultural relativism has ever been more than a convenient smoke screen for researchers' firmly held views of right and wrong. In any case, my position regarding financial incentives as a means by which to increase the supply of organs for transplantation should be clear. I oppose it, and in this essay I have sought to use anthropological evidence to defend my contention.

In this concluding section, I want to ask if my strong reaction to proposals such as Lloyd R. Cohen's represents an anthropological and/or personal blind spot. By focusing on past cultural precedents regarding the treatment of the dead, am I missing evidence of a transformation of attitudes in American society?

There is reason to believe, at the very least, that Americans are headed in a more materialist direction as pertains to the treatment of the dead. Historian Ruth Richardson points to the increasing acceptance of cremation and the reports of out-of-body experiences as indicators that cultural views of body and self are not as closely linked as they once were (80). The growing separation of the elderly and terminally ill from the living community (for example, warehousing the old in nursing homes and medicalizing the last hours of life in hospitals) may be additional signs that contemporary society wishes to disassociate itself from the soon-to-be deceased even before a declaration of death.

We might also note the variety of ways that the human body, as well as the "American Way of Death" (Mitford), has been commercialized. Just think of the capital costs of funerals, the marketing of cell lines, the sale of sperm and ova, the purchase of bodily alterations (for example, plastic surgery); do these not indicate that American culture has embraced a monetized body and a retail approach to death? Could a cadaver with a price tag be far away?

At the same time, there are signposts pointing to a general anxiety about the treatment of the body as a reservoir of spare parts.[15] I was recently asked to comment for the *Philadelphia Inquirer* (D. Campbell) on the proliferation of stories about business travelers who find that a casual drink with a stranger can lead to drugging and removal of a kidney. Similar plots have appeared in movies (for example, *The Harvest*) and on television programs (for example, *ER*). Sensational headlines surface, condemning the kidney trade in southern India (Moore and Anderson) and the extraction of organs from executed prisoners in China (Field). When celebrities seem to get organs immediately, the media asks whether status buys privilege on the transplant waiting list (Belkin).

Whether media accounts and movie plot lines reflect and/or shape public attitudes and concerns, it is clear that transplantation professionals are quick to assume they might. When ethical issues are raised, organizations such as the National Kidney Foundation and the United Network for Organ Sharing (UNOS), as well as hospital public relations specialists, have at the ready complete press packets defending their procedures. Similarly, when changes are envisioned in the policies governing procurement practices or allocation rules, there are efforts made to gauge public response via surveys, focus groups, members forums, debates in hospital ethics committees, and professional conferences (for example, DeVita and Snyder).

All of this invested energy on the part of transplantation professionals suggests an uneasiness about how prepared Americans are to revise their understanding of death and to reconsider the meaning of cadavers. These specialists worry that a misstep could turn public sentiment against transplantation and hinder its continued expansion. At the same time, they are impatient with what they consider cultural barriers to medical progress. "Pushing the envelope" with proposals such as Lloyd Cohen's is the order of the day.

I will not venture to predict the outcome of this effort. I do hope that there is a bedrock of opposition in the United States to the idea of paying for body parts. I also hope that this essay contributes to buttressing that opposition.

NOTES

I am indebted to Phillip Cox for philosophical guidance and to Kim Fujinaga for expert research assistance. Gail Joralemon and Brian Hansen deserve special thanks for their combined editorial work.

1. Cadaveric donation rates in the United States have remained between 4,500 and 5,000 per year while the number of patients on national transplantation waiting lists continues to increase dramatically (over 44,000 at the time of the meeting, UNOS). This situation has developed despite new federal and state legislation (for example, "required request" rules) and public information campaigns designed to increase the number of organs available for transplantation.

2. It is worth noting that on the day of the conference, *USA Today* asked readers whether Cohen's proposal was acceptable; three of four respondents quoted in the newspaper reacted negatively on moral grounds.

3. Proponents of financial incentives frequently ask to have their proposals tested in limited trials, assuming that ethical objections will fade away as large num-

bers of organs are made available. This suggestion appeals to two American cultural priorities: the notion that new ideas deserve a chance and the faith in progress by experimentation.

4. Cohen does not object to direct payments in advance of donation, but he acknowledges that such plans raise concerns that make them a hard sell. He has set out to propose financial incentives that avoid as many objections as possible.

5. This extraordinary proposal, now instituted to some degree at transplant centers around the country, is the subject of a collection of essays originally published in the *Kennedy Institute of Ethics Journal* (3.2 [1993]) and then in a book (Arnold et al.). It is worthy of more attention than I can give it in this essay.

6. Renée C. Fox, the social scientist with the longest history of research on transplantation, has noted these developments. She finds them so ethically troubling that she has decided she can no longer be involved in research on the topic.

7. Both assumptions may be challenged. For instance, the quality of life for organ recipients has been greatly exaggerated (Joralemon and Fujinaga); the financial costs of transplantation and the allocation of organs have been questioned on equity grounds (for example, Caplan, "Obtaining"; Evans; Koch); and the illegal sale of kidneys by living "donors" in poor countries has been the subject of heated debate (for example, Moore and Anderson). As regards the "innocence" of lives saved by transplantation, public media have asked whether recipients such as Mickey Mantle, whose alcoholism caused liver failure, are deserving of transplantation (for example, Kolata).

8. The recent suggestion that priority on the waiting list be given to those most likely to survive rather than to those in the most critical condition is an exception. Other "supply side" solutions include: eliminating re-transplants (that is, each person gets one chance); favoring younger candidates; and ending multiple organ transplants for a single patient (see Anderson).

9. I do, however, find some feasibility arguments persuasive. For instance, the historical case Ruth Richardson made, comparing current financial proposals to those offered 150 years ago to increase the supply of cadavers for dissection in anatomy schools, is convincing.

10. Munzer calls this line of argument the "integration strategy." The idea is that the more biologically irreplaceable a body part, the more important it is to the whole person. He also offers a "derived-status strategy" to assess the degree to which sales of particular body parts risk offending human dignity (275). Both are designed to avoid the "fallacy of division," by which one mistakenly assumes that what is true of the whole must also be true of all its parts (275).

11. Leon R. Kass expressed this risk in a memorable way: "If we come to think about ourselves as pork bellies, pork bellies we will become" (83).

12. Ethicist Arthur L. Caplan has made this point as well ("Telltale" 210).

13. Note that I am only addressing the implications of Cohen's proposal; other plans for paid donations have their own set of potential consequences.

14. A good argument can be made that society is already doing more to promote the survival of this small segment of the population than it does for others whose lives are also endangered. Consider the financial cost of each transplant (for example, hundreds of thousands of dollars for livers); in most cases the cost is borne by taxpayers and/or participants in health insurance plans. Consider the thousands of families that presently, without compensation, agree to give the organs of loved ones. If we add to these considerations the fact that transplantation does not save lives so much as prolong them, then I think even a self-proclaimed utilitarian such as Cohen has to take a second look at where the moral balance rests.

15. In an earlier essay, I used an analogy to the biological processes of rejection to talk about "cultural rejection" of materialist views of the body (Joralemon).

WORKS CITED

Anderson, Mark F. "The Future of Organ Transplantation: From Where Will New Donors Come, To Whom Will Their Organs Go?" *Health Matrix: Journal of Law-Medicine* 5 (1995): 249–310.

Arnold, Robert M., et al., eds. *Procuring Organs for Transplant: The Debate over Non-Heart-Beating Cadaver Protocols.* Baltimore: Johns Hopkins University Press, 1995.

Barnett, Andy H., R. T. Beard, and D. L. Kaserman. "The Medical Community's Opposition to Organ Markets: Ethics or Economics?" *Review of Industrial Organization* 8.6 (1993): 669–78.

Belkin, Lisa. "Fairness Debated in Quick Transplant." *New York Times* 16 June 1993: A16.

Campbell, Courtney S. "Body, Self, and the Property Paradigm." *Hastings Center Report* Sept.-Oct. 1992: 34–42.

Campbell, Douglas A. "Kidney Snatchers? Tale on Internet Dismissed as Hoax." *Philadelphia Inquirer* 27 Feb. 1997: A3.

Caplan, Arthur L. "Obtaining and Allocating Organs for Transplantation." *Human Organ Transplantation.* Ed. D. H. Cowan et al. Ann Arbor: Health Administration Press, 1987. 5–17.

———. "The Telltale Heart: Public Policy and the Utilization of Non-Heart-Beating Donors." Arnold et al. 207–18.

Chadwick, Ruth F. "The Market for Bodily Parts: Kant and Duties to Oneself." *Journal of Applied Philosophy* 6.2 (1989): 129–39.

Cohen, Lloyd R. "A Futures Market in Cadaveric Organs: Would it Work?" *Transplantation Proceedings* 25.1 (1993): 60–61.

———. *Increasing the Supply of Transplant Organs: The Virtues of an Options Market.* New York: Springer, 1995.

DeVita, Michael A., and James V. Snyder. "Development of the University of Pittsburgh Medical Center Policy for the Care of Terminally Ill Patients Who May Become Organ Donors after Death following the Removal of Life Support." Arnold et al. 55–67.

Evans, Roger W. "Organ Procurement Expenditures and the Role of Financial Incentives." *Journal of the American Medical Association* 269 (1993): 3113–3318.

Fentiman, Linda C. "Organ Donations: The Failure of Altruism." *Issues in Science and Technology* 11 (1994): 43–48.

Field, Catherine. "China's Vampire Gangs." *Marie Claire* (Dec. 1996): 62–63.

Fox, Renée C. "Afterthoughts: Continuing Reflections on Organ Transplantation." Youngner, Fox, and O'Connell 252–72.

———. " 'An Ignoble Form of Cannibalism': Reflections on the Pittsburgh Protocol for Procuring Organs from Non-Heart-Beating Cadavers." Arnold et al. 155–63.

Fox, Renée C., and Judith P. Swazey. *Spare Parts: Organ Replacement in American Society.* New York: Oxford University Press, 1992.

Harvey, J. "Paying Organ Donors." *Journal of Medical Ethics* 16.3 (1990): 117–19.

Hogle, Linda F. "Transforming 'Body Parts' into Therapeutic Tools: A Report from Germany." *Medical Anthropology Quarterly* 10.4 (1996): 675–82.

Huntingdon, R., and P. Metcalf. *Celebrations of Death: The Anthropology of Mortuary Ritual.* New York: Cambridge University Press, 1979.

Joralemon, Donald. "Organ Wars: The Battle for Body Parts." *Medical Anthropology Quarterly* 9.3 (1995): 335–56.
Joralemon, Donald, and Kim Fujinaga. "Studying the Quality of Life after Organ Transplantation: Problems and Solutions." *Social Science and Medicine* 44.9 (1997): 1259–69.
Kass, Leon R. "Organs for Sale? Propriety, Property, and the Price of Progress." *The Public Interest* 107 (1991): 65–86.
Koch, Tom. "Normative and Prescriptive Criteria: The Efficacy of Organ Transplantation Protocols." *Theoretical Medicine* 17.1 (1996): 75–93.
Kolata, Gina. "Transplants, Morality, and Mickey." *New York Times* 11 June 1995, sec. 4: 5.
Lock, Margaret, and C. Honde. "Reaching Consensus about Death: Heart Transplants and Cultural Identity in Japan." *Social Science Perspectives on Medical Ethics.* Ed. G. Weisz. Dordrecht: Kluwer, 1990. 99–120.
Marshall, Patricia. "Anthropology and Bioethics." *Medical Anthropology Quarterly* 6 (1992): 49–73.
Marshall, Patricia, D. Thomasma, and A. Daar. "Marketing Human Organs: The Autonomy Paradox." *Theoretical Medicine* 17.1 (1996): 1–18.
Mitford, Jessica. *The American Way of Death.* New York: Simon, 1963.
Moore, Molly, and John W. Anderson. "Kidney Racket Riles Indians." *Washington Post* 30 Apr. 1995: A25.
Munzer, Stephen R. "An Uneasy Case against Property Rights in Body Parts." *Social Philosophy and Policy* 11.2 (1994): 259–86.
Murray, Thomas H. "On the Human Body as Property: The Meanings of Embodiment, Markets, and the Meaning of Strangers." *Journal of Law Reform* 20 (1987): 1055–88.
———. "Organ Vendors, Families, and the Gift of Life." Youngner, Fox, and O'Connell 101–25.
Radin, Margaret J. "Market-Inalienability." *Harvard Law Review* 100 (1987): 1849–1937.
Reed, Susan. "Toward Remedying the Organ Shortage." *Technology Review* 97 (1994): 38–45.
Richardson, Ruth. *Death, Dissection, and the Destitute.* New York: Viking Penguin, 1988.
Scott, Russell. *The Body as Property.* New York: Viking, 1981.
Sharp, Leslie A. "Organ Transplantation as Transformative Experience: Anthropological Insights into the Restructuring of the Self." *Medical Anthropology Quarterly* 9.3 (1995): 357–89.
Youngner, S. J., R. C. Fox, and L. J. O'Connell, eds. *Organ Transplantation: Meanings and Realities.* Madison: University of Wisconsin Press, 1996.

Part IV: Biotechnology and Globalization

Chapter 11

Reach Out and Heal Someone

*Rural Telemedicine and the Globalization
of U.S. Health Care*

LISA CARTWRIGHT

Here is a "vision of the future" offered by the National Information Infrastructure (NII) Task Force Committee in 1994:

> In a rural area, a child awakens with severe coughing, fever, and a rash on her chest. Her mother dials the interactive telecommunication connection to access medical care support and describes her child. The nurse at the other end asks for the mother to connect special probes that monitor the child's temperature, blood, pulse. She then listens through an electronic stethoscope to the child's breathing. She examines the rash through the high resolution telecommunications viewer. After consulting information through the NII about recent health events reported in the community, such as the incidence of measles, bacterial and viral infections, she recommends action to the mother.[1] ("Health Care")

Osteopaths Christopher Simpson and Martha A. Simpson note that rural residents tend to have a higher unemployment rate, are more likely to be poor, are apt to have higher rates of chronic disease, and are less likely to have health insurance when compared to their urban counterparts.[2] These facts, combined with knowledge that many rural areas in the United States do not have strong enough economies to sustain hospitals and medical practices, much less well-equipped ones, make it difficult to conjure up an image of a rural home equipped with an electronic stethoscope and high-resolution telecommunications viewer (Simpson and Simpson 502–8; McCarthy 111–30). Unsurprisingly, a recent U.S. Department of Commerce survey of telecommunications technology access in American households reports that rural homes are least likely to have computers. Within this group, poor households rank lowest in terms of computer penetration (4.5 percent) and—among those with computers—modem penetration (23.6 percent), even as compared to poor households in central cities (7.6 percent and 43.9 percent). Moreover, if the rural family in question happens to be Native American or black, chances of their having a computer and modem are further diminished (U.S. Department of Commerce). These findings suggest that the NII vision of telecommunications as an answer to the problem of providing health care for rural populations deserves some serious discussion.[3]

In his summary of the First National Conference on Consumer Health Informatics in 1993, journalist Steven Carrell was as optimistic as the NII Task Force about the future of telecommunications in health care: he estimated that in the United States roughly 95 percent of health care problems were resolved by lay management without professional involvement. "Consumers conduct online searches, join online medical self-help groups, make decisions about their own care, and read and write in their own medical records"—a situation that, for Carrell, suggests a more democratic model of health care in the making. Indeed, he predicted that by the year 2000, "half of the U.S. population will provide a substantial portion of their own health care, working as partners with their health professionals" through the combined use of phones, faxes, interactive multimedia, and personal computers (Carrell).

These visions are supported by the recent emergence of telemedicine, or medical telecommunications, as a lucrative specialty in health care delivery. Telemedicine involves the use of imaging and telecommunications modalities, including the Internet, satellites, phone lines, and video and digital imaging systems, to exchange medical information and images and to conduct video teleconferencing. A concept that first appeared in the medical literature in 1959, during some of the earliest federally and privately funded experiments using television technology in health and education, telemedicine was introduced to make it possible for doctors to consult, diagnose, and even treat patients at a distance. The commercial success of this field can be gauged by the fact that, by January 1995, telemedicine had been the subject of well over one hundred articles on the computerized medical literature database, Medline. *Telemedicine,* a monthly journal, was launched; and university departments devoted to its practice now exist in medical institutions throughout Europe and the United States. Indeed, Queens University of Belfast has instituted a chair in the specialty ("Telemedicine: Fad").[4] Nationally, government support for this field is surprisingly strong in a period of cost-cutting in health care: according to the *Journal of the American Medical Association,* federal and state allocations for telemedicine programs were likely to exceed $100 million in fiscal 1994–1995. At least thirteen federal agencies, including the Department of Commerce, the Health Care Finance Administration, the Office of Rural Health Policy, and the Department of Defense, have begun telemedicine research and demonstration programs (Perednia and Allen).

The vision of virtual health care illustrated above begs the questions of who will benefit from medical treatment at a distance and which half of the U.S. population (if we accept Carrell's estimates) will acquire knowledge and agency in medical treatment and care via new communications systems. Numerous advocates of the field extol telemedicine's benefits, especially for underserved rural and remote populations in the United States and abroad. Programs linking such communities to distant medical research centers have proliferated in recent years, suggesting that rural poor communities may be the field's primary beneficiaries. Gerhard Brauer, a Canadian health information specialist, argues that the recent emergence of telemedicine as a burgeoning specialty is a long overdue outcome of work since the 1970s by avant-garde physicians and health educators bent on delivering more effective health services to rural and underserved populations globally (in developing countries) as well as domestically (151–63).[5] The early projects Brauer describes demonstrate telemedicine's basis in the liberal ideals of a more democratic flow of medical information, treatment, and care at the domestic level and a benevolent transfer of medical knowledge and technology globally.

Brauer's optimism about telemedicine's progressive roots notwithstanding, it is important to ask whether it is likely that a more democratic model of health care via new technologies can materialize in the U.S. health care climate of the mid to late 1990s. Telemedicine has emerged as a medical specialty during a period marked by GOP-led assaults on federal funding of health care programs for the poor, public hospital staff cuts and closings, and the development of private health maintenance organizations (HMOs) as the specious new leaders in national health care. As David Keepnews of the American Nurses Association writes, "the end of the national health-care reform debate . . . represented a defeat for efforts to provide a regulatory framework for the transformation of the U.S. health care system. As a result, 'market-based' changes in the health care system have accelerated" (280–81). Such a climate is unlikely to support the growth of domestic health care programs geared toward improving medical services for the populations whose hospitals and clinics are facing cutbacks and elimination, and which historically have been underserved.[6]

Telemedicine, I argue, is thriving in the domestic health care climate partly because it enables a kind of cost-effective population and resource management appropriate to the changing health care economy. Diagnosis and even treatment at a distance can eliminate physician and patient travel time and costs, as well as the expenses that would be incurred by duplicating health facilities, technology, and staff throughout a region. Second, as public health facilities are scaled back or eliminated, the market for HMOs has grown. As competition for clients intensifies, these corporations are likely to find telemedicine a useful means of expanding their catchment area, or regional client base, beyond a single facility's geographic scope.[7] Finally, those specialists seeking markets beyond the United States may find telemedicine a viable means of managing a catchment area no less expansive than the world. The promise of lucrative overseas markets for medical services without the investment of time and money that physical travel requires and without the need for creating fully outfitted facilities overseas may be appealing to some medical corporations, especially at a moment when the demand for specialists "at home" is predicted to decline.

Certain questions and problems arise with these changes in health care delivery. Paradoxically, the argument that state-based public telemedicine programs may facilitate a reduction in domestic health care costs by reducing staff, facilities, and technology is frequently made on the basis of projects that were established to extend more and better medical services to those underserved populations whose services are now targeted for reduction. Programs such as the Medical College of Georgia's rural telemedicine project and the Center for Telemedicine at the University of Oklahoma Health Sciences Center were launched in order to bring advanced medicine to poor rural communities. Publicly supported programs such as these in states where medical care has not been readily available to the rural population were started through public funds designated for improving rural telecommunications and medical services. This objective echoes the mission of private sources that have also been tapped to support research in telemedicine, such as the Robert Wood Johnson Foundation's program Reach Out: Physician's Initiative to Expand Care to Underserved Americans, 1995.[8] As telemedicine shifts from public and private charity grant-based pilot projects to fully functional networks operating in the mainstream medical market, the field's managers are turning their attention from the mission of fostering more and better care for underserved populations to market and logistical questions. While it would be unrealistic to expect any field to survive without addressing market concerns, it is nonetheless important to consider the consequences of this shift for the communities these projects initially targeted. This consideration is especially crucial in light of the privatization of both the telecommunications and health care industries in the 1990s. As private managed-care systems in partnerships with private telecommunications companies take over many of the health programs previously under state and federal operation, we must ask whether their technologically sophisticated services will improve quality of care and whether telemedicine will grant the target communities in particular the degree of patient agency and autonomy predicted by Carrell, the advocate of consumer health informatics quoted at the outset of this essay.

The relationship between these goals and profit motives vis-à-vis telemedicine deserves further elaboration. John Patterson, past vice president and chief information officer of New England Medical Center in Boston, offers some insight into telemedicine's relation to the growth of managed care. He describes a hypothetical scene wherein "a pregnant patient presents herself to a primary-care physician's office, and an ultrasound scan shows there is a problem." Rather than referring the woman to a specialist, Patterson suggests, the physician can use a telemedicine network to get a consultation on the spot from a specialist located at a tertiary center that is part of the same delivery system. This statement implies that telemedicine is grounded in a mission to bring the benefits of specialized diagnosis to a broader population. However, Patterson emphasizes instead the potential economic benefits this hypothetical solution might offer a health care delivery system laboring under the pressures of managed-care structures: "The cost implications [of this solution] are considerable. Ultimately, those efficiencies and cost savings will be what drives telemedicine to the heart of health care systems" ("Medical" 1, 7).

In order to understand the cost implications to which Patterson refers, and how telemedicine might emerge at "the heart" of newly configured health care systems, it is necessary to summarize some issues regarding managed-care systems. Within a managed-care capitation system, physicians are paid to provide service for a fixed number of plan members rather than receiving a flat fee for services rendered. This system has been widely criticized: physicians Steffie Woolhandler and David Himmelstein, for example, observe that by linking salaries to numbers of patients treated, HMOs pressure doctors to put a smaller percentage of their patients in the hospital and give them less outpatient treatment. Under this type of managed-care system, they argue, some doctors will "boost their incomes by suppressing the use of services" (1706–8).[9] Given this widely shared perspective, it is likely that Patterson sees telemedicine at the heart of a system designed to ratchet down services in order to accommodate an expanding client base. Viewed in this context, telemedicine is unlikely to facilitate the establishment of a more democratic and decentralized network of medical knowledge and service.

EXPANDING THE CLIENT BASE BY CREATING MEDICAL "COMMUNITIES" AT A DISTANCE

Bruno Latour describes historical cycles of accumulation of knowledge and artifact that allow a particular geopolitical point (from a laboratory to "a puny little company in a garage") to become a dominating center of knowledge and power by acting on other points at a distance:

> How to act at a distance on unfamiliar events, places, and people? Answer: by somehow bringing home these events, places, and people. . . . By inventing means that render them (a) mobile so that they can be brought back; (b) keep them stable so that they can be brought back and forth without corruption or decay, and (c) are combinable so that whatever stuff they are made of, they can be cumulated, aggregated, or shuffled like a pack of cards. (223; ch. 6)

One of the goals of telemedicine is to expand a given medical facility's reach, to "bring home" to the research center distant medical cases, clinics, and patients. Telemedicine renders patients mobile in the sense that they can be electronically transported to a central medical facility via tele-conferencing and other communications modalities. At the same time, tele-medicine keeps patients stable, safe from the threat of "decay" or physical stress that actual travel between the remote community and the medical center poses. Most important, telemedicine keeps distant patients readily combinable, making possible an aggregation of geographically dispersed patients within a single medical network. This technique of assembling a dispersed pool of clients is one of the chief characteristics of rural telemed-icine. The rural communities brought together under the shelter of the medical center via telemedicine constitute a kind of virtual community whose status as such is imposed top-down by the aggregating force of the telemedicine process.

State-based telemedicine facilities consolidate a client base from among low-income and poor populations in designated rural and remote areas, people whose lifestyles and labor are determined in part by their existence in rural and remote environs. This dispersed client base is not a community in the conventional sense—that is, its members do not necessarily share economic status, geographical space, political agendas, or a sense of cul-tural identity. The term "rural community" conjures up a national patch-work of small family farms; yet less than 10 percent of the U.S. rural population lives on a farm (many communities are dependent on mining and manufacturing, for instance). Although dramatic economic diversity exists among rural areas, many factors are common to most rural popula-tions. These include less ethnically diverse populations, a higher percent-age of elderly residents, higher rates of chronic disease, less education, a higher unemployment rate, and a greater likelihood of being poor. More-over, employed rural people are less likely to have employer health insur-ance benefits, and families living below the poverty line are more likely to be two-parent families, which do not qualify for Aid to Families with Depen-dent Children (and hence Medicaid) (Simpson and Simpson 502–8).

The term "community" is also evoked to generate a sense that the dis-persed individuals serviced by telemedicine share in common a bond with the distant medical center that serves them, much like the geographically local community served on site by the medical center. In other words, the community a telemedicine facility serves is a virtual community linked by factors other than geographical proximity. Obviously, the nature of treat-ment and care the hospital's dispersed virtual community offers is different from that provided to the geographically local community. The former constituency's members may never see the inside of the medical center, or the face of the treating physician, except on a video monitor. Moreover, the distant medical patient may not interact with a community of patients experiencing similar conditions and treatments, as he or she might during a clinic visit or a hospital stay. The practice of telemedicine thus potentially creates a class of patients who share a lack of experience with the routine and embedded components of social interaction that are typical of more conventional medical care. However, the experiences of the constituencies serviced by distance medicine can also be described in positive terms. The experiences of telemedicine's culture that members of this community or class share include having as their immediate caregivers a team of intermediaries (computer technicians in addition to nurse practitioners and physician assistants) rather than a single physician, an arrangement that may decenter the authority of the physician in the more conventional one-on-one patient-doctor relation, as well as having one's body diagnosed and perhaps even treated by a distant physician with whom one never comes into physical contact. How will consultation, diagnosis, and treatment at a distance shape the self-understanding of the physical body among recipients of this sort of health care? How will it change the value and meaning of exchanges in real life—physical contact, presence—as a privileged mode of patient-doctor communication? How will the rural poor populations targeted by telemedicine programs respond to this top-down administration of classification, and general medical authority, from a distance? Although these questions are difficult to answer this early in telemedicine's short history, they are crucial to our understanding of how these virtual communities may function as something other than simply market groups or artificial aggregations falsely promoted through the cozy term "community." This new way of defining community is not to be dismissed: the fact that virtual communities are forged top-down does not preclude their members' subsequent functioning as active agents in the shaping of telemedicine's future and the future of their own health decisions and health care cultures.

GLOBAL TELEMEDICINE

Rural telemedicine also has a global component, suggesting that the communities telemedicine facilities forge are sometimes transnational in scope. Since the 1970s, a number of Western medical research centers have conducted pilot telemedicine programs designed to bring fresh technologies and specialized services to developing regions, such as Nairobi in Kenya and Kampala in Uganda (see House et al. 399). Many of these programs operate on the principle of benevolent technology transfer. Much of the current literature on telemedicine, however, suggests that there is a thin line between this liberal politics of technology transfer and the generation of new markets. A relationship that is becoming more typical of telemedicine in the late 1990s is that between the Mayo TeleHealth Center and hospitals in the United Arab Emirates, where the UAE government has paid for a state-of-the-art telemedicine system linking their hospitals to specialists in cardiology at Mayo.[10] Mayo does not do telemedicine for profit.

However, as HMOs scramble for enrollment among potential customers, telemedicine can extend an organization's geographical reach, allowing it to dip into more distant markets while concentrating labor, costs, and revenue in localized centers of control. In cities such as Boston, where an unusually high concentration of health care providers vie for the business of a relatively limited catchment area, telemedicine has proven an especially helpful resource. According to Patterson, "In Boston there are several tertiary hospitals and we all compete like junkyard dogs for patients. The only way we can make a long-term impact is to change our service area" ("Medical" 1).

Apparently the "junkyard" is the world—or, more precisely, the developing world. In the late 1990s, New England Medical Center, which deploys telemedicine in applications ranging from distance fetal monitoring to telepsychiatry in nursing homes, extended its services to parts of Latin America, providing, for example, second opinion consultation services to a pool of four hundred thousand paying patients in Argentina ("Medical" 1). Similarly, Massachusetts General Hospital formed a for-profit subsidiary that reads X-rays transmitted over phone lines from hospitals and clinics in the Middle East.[11] Perhaps the most stunning example of the potential for telemedicine to construct a global catchment area is the failed Glasgow-based multinational, Health Care International Hospital, a system that was designed to service clients around the globe. This forging of a global "Glasgow" medical community raises major problems around issues such as differences of politics, national law, cultural identity, and medical training and practice across this virtual space of medical treatment—the very questions that have yet to be resolved domestically, within federal or state jurisdictions. It is inevitable that this transnational telemedicine marketplace will be a source of intense legal, political, and ethical debates as distinct national legal policies and laws come into conflict. As the lines between consultation—the transferal of medical information—and actual treatment become blurred, further ethical and political dilemmas will emerge. For along with advice transmits the ideology of superior Western knowledge and the greater legitimacy of scientific medicine. Who can we say is most responsible for actual treatment outcomes when American medical personnel are behind crucial treatment decisions in Saudi Arabia and Italy? And how will those developing countries who do not have the oil-based economy of, say, Saudi Arabia figure into this newly market-focused economy of global telemedicine? These questions remain open as the field continues to take shape.

Global ventures such as these, successful or not, create a transnational and virtual Third World community of clients linked by a common experience with distance medicine in the absence of adequate local services, a community more far-flung than the rural communities aggregated within domestic medical systems by state-based telemedicine programs. In the case of transnational telemedicine, however, what is at stake is not more cost-effective local management of basic health care but increased dependency on distant centers through global expansion of the market for specialized services, primarily in Third World markets able to afford such services (oil-producing and newly industrializing countries, for example). Whereas U.S. medical industries have until recently been able to generate markets for goods (equipment) in these "remote" settings, now medical practitioners will be able to market diagnosis and treatment services as well, without leaving their home institutions, with one outcome that technological development takes the form of dependency on advanced communications systems controlled from a distance. These systems do not simply threaten to take the place of "imported" medical experts; they foreclose on the potential for training and jobs for health care professionals from within the local community.

How are we critically to address the formation of these national and transnational aggregations of health care consumers bound by shared ties to health communications services? Earlier I stated that telemedicine is thriving in part because it enables a kind of cost-effective population and resource management. As I explain below, telemedicine is not a new idea in community-based health care; it has been a central component of rural health monitoring schemes since the 1970s. Its future must be considered from the perspective of earlier uses of the technology—uses that suggest a telemedical future in which the definition of "local" health care and "community" health change in dramatic ways.

BEYOND GLOBAL TELEMEDICINE: NASA'S SPACE MEDICINE PROGRAM

In 1972, in conjunction with the space program and during the early years of the Internet's formation as a military communications system, the Life Sciences Directorate of NASA's Johnson Space Center announced its plans to use space communications technologies to provide health services to rural communities. In what surely must have been a public relations ploy along the lines of the Teacher in Space Program, NASA posted advertisements soliciting rural communities interested in becoming a demonstration site for the project. From among the respondents they chose the Papago, a Native American community situated less than one hundred miles from Tucson, Arizona, and they titled the program Space Technology Applied to Rural Papago Advanced Health Care (STARPAHC).[12] The Papago were not an arbitrary choice. Their health problems, which included diabetes, hypertension, otitis media, dysentery, cirrhosis of the liver, and a high infant mortality rate, were regarded as representative of conditions on most Native American reservations. The alleged representativeness, however, was not the only rationale for NASA's selection of this community. As an officer of the Indian Health Service (IHS, an entity brought in to manage this program) explains, NASA selected the Papago because much public health information had already been collected about them since 1967, when the federal Office of Research and Development (ORD) happened to be housed, with tribal council approval, in an abandoned tuberculosis hospital on the reservation. The ORD had used its proximity to the Papago to experiment with health monitoring and management, setting up an early, computerized health care information system. ORD personnel designed a computer simulation model for reordering and predicting the pattern of patient flow through a given health facility. This system allowed the ORD to accumulate extensive epidemiological data to "bring home" to federal health research centers, while also facilitating the management of the community's health matters from medical centers outside reservation boundaries. This precedent suggests that the Papago were already seasoned federal health research test subjects when they were selected for the STARPAHC program.

Another factor influenced NASA in their selection of the Papago. The main (although unadvertised) purpose of STARPAHC was to provide a test site for new data systems and equipment for the remote diagnosis and treatment of astronauts living in space. For NASA, the culturally "remote" Papago were apt surrogates for Americans traveling in the geographically remote reaches of outer space. The Papago, living as they did dispersed across large tracts of undeveloped reservation land—"no man's land" in the sense that it did not belong to the U.S. government—without benefit of advanced communications and transportation systems, must have seemed to NASA the perfect test subjects for their planned simulation of medical communications systems designed to serve hypothetical astronauts drifting around that more distant uncharted no man's land, outer space.

On the most obvious level, STARPAHC was a multilocational, networked health services system whose decentered structure reflected the geographically dispersed organization of the reservation community. Health officials, rather than requiring people to travel distances to get treatment, dispensed their services via a mobile clinic—a huge trailer with a satellite dish mounted to its roof. Those wishing to be treated could enter the trailer and be seen by a traveling nurse or consult with a distant medical professional via satellite link. Homes, schools, and other community buildings served as treatment sites along with the existing clinics and hospital staffed by public health nurses, disease control workers, and health personnel under tribal administration. The hospital housed an interactive television system linked to a medical research center in Phoenix. This blanketing of the reservation with vans, personnel, and makeshift facilities was an imposition upon the culture of the reservation, bringing federal health policy and practice into every corner of the region. What was telemedical about this program, ultimately, was not so much the electronic delivery of services to the Papago but the transmission of data about the Papago outward, to the Phoenix medical center, the regional "centre of calculation."

Writings on STARPAHC are more than clear about the project's innovative method of information-gathering, but they are less than forthcoming about changes in the health of the Papago resulting from this new system. The little documentation of Papago responses to STARPAHC that health workers provided suggests that the community was less than compliant with the program. This noncompliance may be read as a form of community mobilization against the threat STARPAHC posed to indigenous medical practice. Indeed, evidence adduces that noncompliance was a point of great concern for Indian Health Service workers, causing them to question the success of the technology as a whole. As two IHS workers explain, the transfer of sophisticated technology to meet the needs of minority groups does not need to result in the disruption of traditions or lifestyles. However, implementation methods and operational policies may well have a potential for disruption, resulting in an eventual discrediting of the technology (Justice and Decker 170). This enigmatic statement suggests a nascent awareness on behalf of STARPAHC's advocates that communities have the capacity to resist technological interventions when these interventions threaten to displace traditional modes of health care. Interestingly, noncompliance is regarded as grounds for discrediting a technology's overall viability. The health workers' statement indicates that what was ultimately at stake in the STARPAHC project was not the health of the Papago but the future of the technology—specifically, its promise as a distance medicine technology for the space program.

There is irony to the STARPAHC story. How often do we see capital-intensive, elite space technology used to support chronically underfunded areas such as Native American health and social services? It is more often the case that sophisticated medical technologies with ties to military research initiatives (ultrasound or nuclear medicine, for example) are developed for specialized diagnosis and treatment, not for the implementation of programs designed to advance general health among an underserved population. The idea of NASA selecting a Native American community living on the low end of the economic and technological spectrum as a model for a heavily funded, high-tech community of astronauts is a striking paradox. But perhaps the most significant aspect of this narrative is that, during the 1970s, mortality rates among the Navajo were highest in areas where dependence on welfare and involvement in the wage economy tended to be highest. Unfortunately, there are no statistics telling us STARPAHC's impact on the population. STARPAHC's documentation does indicate, though, that the issue of community agency in the determination of medical service systems—particularly in the case of those systems designed and managed from a distance—cannot be underestimated in the consideration of telemedicine's potential for community medicine. As the IHS workers' statement suggests, the technology ultimately stands or falls on the basis of its rejection or acceptance within a community.

This brief recounting of NASA's foray into—and fantastic redefinition of—community in medicine demonstrates how a liberal position advocating federal support for health education, technology transfer, and community health service programs in underserved areas can be a front for far less humanitarian goals, specifically, the development of a space medicine program. The Papago's conscription into service as test subjects potentially benefitted a much more privileged community—in this case, the imaginary astronauts living in space, a group that existed only in NASA's futuristic fantasy. Significantly, though, in the STARPAHC narrative, the Papago and the hypothetical astronauts living in space are rendered members of the same virtual community, insofar as both groups exist, within the terms of the program, as remote participants in the same class of health management and care delivery. For the treating physicians in the urban medical center, the locations of the Arizona desert and outer space are equally remote inasmuch as both appear on the same screen, as both bring home to the center patients who otherwise would remain peripheral to such "local" care. Thus telemedicine confounds not only geographical distance but also cultural difference for the purpose of creating a "community" out of those disparate remote individuals brought within reach of the medical center.

STARPAHC is not representative of all telemedicine pilot projects purporting to bring medicine to underserved communities. Brauer describes the use of the jointly operated U.S.-Canadian Hermes satellite in 1976 to deliver diagnosis and treatment at a distance to northern Canada's medically underserved and poor rural communities.[13] A more recent development along these lines is the Canadian Northern Hepatitis Liver Project, a University of Alberta telemedicine system that links northern Inuit and Dene hepatitis patients with southern consultants. Typically, the communities the Project serves are composed of fifty or sixty people whose care providers are usually nurse practitioners and in some cases a doctor who flies in periodically. It is estimated that up to 30 percent of the Inuit and Dene people will contract hepatitis B in their lifetimes. The Project, funded by the Canadian Liver Foundation, supports a telemedical consultation system that allows the person with hepatitis to get specialist consultation and a recommendation for drug therapy to be administered at home. According to the doctor who instituted this project, "everybody benefits."[14] The person with hepatitis is treated—and the southern consultant gets pertinent information to pass along to the university for epidemiological study. The purpose this program serves for public health management at a distance is clear. Yet, who would object to such epidemiological management techniques when the benefits to the communities treated are so evident? The Canadian project benefits everybody, I would suggest, for a number of reasons. First, its foundation funding and situation within a more socialized medical system (compared to that of the United States) allow the program to exist in a state of relative autonomy from larger market concerns. Problems arise in cases where remote communities like the Inuit and Dene are targeted as viable models or as future markets for a more profitable use of telemedicine health delivery systems, or where such a system is brought in to curtail existing local services (to replace extant facilities and practitioners, for instance). Second, the project may be a less invasive alternative to the possibility of bringing a southern specialist physically into the community. By limiting outside specialist presence to the consultation process, and by encouraging the administration of treatment by the patient or by local health workers, the project fosters a degree of patient autonomy and community agency lacking in the STARPAHC model, where outside health workers and new facilities invaded the physical space of the reservation.

REACH OUT AND HEAL SOMEONE:
"CARE" AT A DISTANCE

The problems of community coherence and integrity raised in the example of the hepatitis project among the Inuit and Dene are far from resolved. They are at the center of most rationales for the continuation of this practice. In a 1995 presentation to the U.S. House of Representatives, physician Helen L. Smits, then deputy administrator of the Health Care Finance Administration (HCFA), emphasized the pertinent role telemedicine is certain to play in advancing the sense of community and connectedness among rural health providers. She noted that while the technology might siphon patients away from rural doctors, more likely it would allow these doctors to stay in touch with their distant colleagues, remain current on research, and recruit and retain coworkers while also improving the health of the community by importing more up-to-date information, consultation, and treatment. In other words, far from undercutting the agency and autonomy of the local practitioners within his or her community, distance medicine would, paradoxically, enhance his or her sense of belonging within the local community while also giving him or her a place within the imaginary global community of advanced medical research. A similar local-global paradox inheres in writings about projects such as the Georgia telemedicine program, where care is telemedically administered to patients in rural areas of the state.[15] But for the telemedicine program, it is asserted, patients might otherwise have been dislocated from their home towns, transferred out of their small local hospitals for specialized care in distant, alienating high-tech hospitals—or worse, might have been denied access to advanced medicine altogether. Like the telemedically supported rural doctor, the telemedically served patient can maintain his or her place within the safe and familiar world of the local while also being transported into the realm of world-class medical treatment, without ever leaving home.

It is important to consider the specific geographical and institutional coordinates of this multilocational local-global schema. Smits stressed the technology's potential within the framework of public health.[16] Currently, the state is the larger unit within which the idea of a public, a locality, or a community is conceived, in the case of both the medical community and the patient community. The state (and not the nation, the university, or the "universe" of medical knowledge) is the primary unit for the administration of domestic telemedicine programs, partly due to the inconveniences of existing regulatory structures. Prevailing U.S. state licensing laws prevent a doctor practicing in one state from diagnosing or treating residents of another state (although use of the ambiguous category of "consultation" allows some finessing here). Policies on accreditation and liability are equally limiting. (An exception is the establishment of universal cross-state licensure within the Veterans Administration and the Indian Health Service.) Inconsistencies among different states' laws also create obstacles to market expansion. In 1997, legislation was passed requiring the Health Care Financing Administration to reimburse rural providers for telemedicine consultations. This bill is predicted to impact the 745 rural counties designated as Health Professional Shortage Areas (HPSAs) and to combine with legislation passed or pending in several states regarding Medicaid coverage of telemedicine for recipients, including the estimated three million beneficiaries in designated HPSAs ("Telemedicine Gets").

Equally important to the establishment of a multilocational, state-centered model is the existence of an adequate communications system throughout rural regions to effect such programs in HPSAs. The Georgia system provides a useful example of the relationship between telecommunications systems development and the expansion of this model. In 1993, the state of Georgia passed the Distance Learning and Telemedicine Act to enable a $75 million project that allotted one third of this sum to the task of upgrading the smaller telephone companies throughout the state. The project was designed to effect the establishment of telemedicine systems in those rural, low-income communities where telecommunications and other aspects of development lagged behind the rest of the state. The technologically advanced care offered to clients at research hospitals such as the Medical College of Georgia is now delivered to low-income communities without the costs of travel or new facilities, technology, and staff (Perednia and Allen 486). As I mentioned above, Republican-backed changes in federal policy have facilitated market conditions favorable to the expansion of telecommunications services to previously underserved communities, and it is tempting to read this extension of high-tech communications and medical technology to underserved regions of the United States to indicate a recent acknowledgment by medical professionals about the class politics of technology and treatment access. But, as I have argued already, such a reading is contradicted by the fact that telemedicine has emerged along with a broader managed-care mentality that entails private corporate expansion of markets along with cost cutting through techniques of population management and economizing on personnel, time, and facilities. Telemedicine, along with burgeoning new industries such as home health care, is a practice that aids the trend toward decentralizing care at the federal level while centralizing oversight of local (home or community) health management at the state and regional levels, whether through multilocational "local" managed-care facilities or through multilocational state-centered public health programs. For health managers in either sector, treatment from a distant high-tech facility through a telemedicine system may emerge as an economical solution to the problem of managing more, and more dispersed, people with fewer resources. Whether through state-operated or HMO-centered health management strategies, the creation and maintenance of models of local health care within a global framework will remain central to cost-effective market expansion, and telemedicine will stay "at the heart" of this ambiguously local-global formulation of community.

TELEMEDICINE'S DISPLAY OF THE BODY

Confidentiality and privacy are the concerns most widely discussed among those who are less than sanguine about the degree to which telemedicine's mode of reaching out constitutes adequate or appropriate care. As one assessment cautions, "the nature of the doctor-patient relationship changes dramatically when the open airwaves carry the personal histories, images, and concerns of a patient" (Norton, Lindborg, and Delaplain 340; Hopper, TenHave, and Hartzel 493–99). Online charts and medical records also are subject to questions about confidentiality when those being treated are already denied many of the privileges of privacy that come with middle-class citizenry through experiences with myriad institutions, including loan institutions, food banks, and welfare agencies. While some practitioners using videoteleconsulting over the airwaves or by direct satellite link make a point of scrambling their broadcasts to ensure confidentiality, issues of privacy during consultation are still contentious. During a conventional office visit, clients usually may see who is present in the room; there is little reason to suspect that an unseen observer might be lurking in the shadows. Teleconsulting often requires technical staff at both ends of the system, introducing off-screen observers, including technicians, into the doctor-patient relationship. At Kwajalein Hospital on the Marshall Islands, a remote site linked by satellite for teleconsultation to Tripler Army Medical Center in Honolulu in the early 1990s, patients were asked to read and sign a "'statement of understanding'" that includes the line: "'I understand that participation in the medical VTC [videoteleconsultation] constitutes a waiver of the usual rights to doctor-patient privacy'" (Norton, Lindborg, and Delaplain 340). The extent of privacy lost is indicated by the design of the Tripler facility adapted for the program: a Defense Commercial Telecommunications Network studio was equipped with three 17-inch television monitors and seats ten to twenty people. Privacy was also an issue in the Marshall Islands' examination facility—a slightly larger but similarly equipped studio staffed with a physician and various technical staff members (Delaplain et al.). In small communities such as this one, the patient sometimes knows the technical staff socially, further eroding his or her sense of privacy. The consultation consists in a session in which the patient sits before a monitor on which he or she views the consulting physician, first providing a narrative account of his or her complaint and then displaying areas of the body at the request of the physician. The physician at the medical center is framed in medium close-up, sitting at a wooden desk affixed with a plate that displays his or her name, degree, and specialty. According to the authors of this study, this framing prevents the patient from being aware of the disturbing presence of technical personnel, avoiding a "fish bowl atmosphere," while the mise-en-scène reassures the patient that this situation is similar to a conventional consultation (Norton, Lindborg, and Delaplain 340). Technical staff at the remote site record footage of the patient's body, ranging from macroscopic close-ups (for example, of skin lesions) to full-body images, shot as per the physician's directives.

This scenario suggests a kind of public performance in which the patient must actively speak about and display to local and distant medical-technical personnel his or her symptoms, rather than allowing an individual doctor to discern signs of illness by means in addition to sight, such as palpation and auscultation. (Interestingly, a 1994 survey of patients' responses to the telemedicine process indicated that women tended to note feeling self-conscious during the interview, and they also identified touch as an important but missing part of the medical encounter [Allen and Hayes 1693].) Along with photographic images, radiographs, lab sheets, electrocardiograms (EKGs), and other physical data are transmitted via a graphics stand situated within the studio. By signing the consent form, the patient agrees to the possibility that this videotaped consultation, which includes varied footage of bodily regions in addition to the kind of talking-head shots more typically evoked by the term "consultation," may be displayed later by "various personnel" (a group that might include technical as well as medical personnel) for training and administrative purposes.

This scopic relation has important implications for knowledge and power in telemedicine's dispersed network. Whereas the Marshall Islands' patient cannot see all of the Hawaiian personnel who document and later observe a record of his or her body and personal narrative, the Hawaiian personnel have unrestricted access to this spectrum of images and informa-tion—and rights to their display and reproduction in unforeseen contexts. This situation raises the issue of patients' rights in relation to visual docu-mentation, an area of medical regulation fraught with ambiguity. While the issue of imaging privacy has been posed historically on the model of the doctor-patient relationship, the introduction of multiple personnel (technical and medical) in the Marshall Islands' telemedicine scenario poses new problems vis-à-vis patient privacy. The privacy issue pertains not only to the body but to the body image, as the media artifacts of care at a distance are quickly transformed into the raw material for medical training and simulation.

This situation begs the question of how the telemedicine doctor's novel position as distant observer and manager of health will impact his or her regard for the patient. Another "vision of the future" from the National Information Infrastructure helps to illustrate my point:

> In the hospital of a major medical university in the state, Dr. Jones visits a virtual reality learning center to review procedures for surgical removal of the prostate (prostatectomy). As she sits in the virtual reality clinical educa-tion room, she takes the electronic scalpel and feels the sensation of cutting into the patient, the texture of the skin, the hardness of the prostate as she is guided in making the proper decisions. ("Health Care")

Touch, the sense missing from the distance medicine experience and one that women patients surveyed noted as an unfortunate absence, is ironi-cally highlighted in the simulated experience of "healing" described in this scenario. As patient records in the form of electronic images become raw material for such programs, will physicians exhibit the same degree of "care" for those patients whose distance treatment may be identical to the treatment performed on such medical models? And how will the absent sensation of touch, accented in the above account to emphasize its central-ity to the realism and humanity of the surgical experience, impact the pa-tient's experience of "care"? This situation suggests that telemedicine may function as a kind of sensory anesthetic to the body (and hence the "human" feelings) of the physician—which is not to say that other sensory experiences may not take the place of those sensory pathways the technol-ogy blocks.

CONCLUSION: THE MEDICAL CONSUMER AS MEDIA CONSUMER

Clearly, telemedicine is integral to the management of newly configured local public health systems and the expansion of local, private health care markets to include a virtual global market. Both aspects—local medical management and global market expansion—depend in no small part on developments in the politics and economy of the telecommunications industry, with current trends within U.S. government and industry toward deregulation and privatization of telecommunications systems at the crux of these developments. The most significant product telemedicine research centers provide to remote sites is not better treatment facilities and equipment but a better communications system—technology that can receive and send high-resolution graphics, video, and other kinds of data. The remote site health care provider becomes on some level a communications facilitator, someone who mediates between patients and a disembodied team of medical experts. The extent of disembodiment, and the subsequent need for a critical focus on these absent players, must be underscored: one account of the Massachusetts General Hospital project closes by noting that "the system demonstrated in the Middle East emphasizes the need to focus on the local patient and physician"—a conclusion that is perplexing given the complete absence of mention of these people in the text that precedes this statement (Goldberg et al. 1500). Other absent players in this media context include the computer specialist. At telemedicine hubs, doctors rely on the skill and knowledge of information specialists, the computer personnel who are often without formal medical training but who encode and transmit medical images and texts. The dynamics of this relationship suggest that authority at the telemedicine hub is not wholly embodied in faceless corporations and their specialists, but it is shared across the agents who manage and receive health care.

Many of the concerns about communication and information politics voiced in contexts other than medicine have great relevance to this discussion. Information technology critics such as Herbert I. Schiller warn us that local, democratic uses of the Internet are precluded by the political and corporate dynamics of communications systems globally. Citing Vice President Al Gore's infamous claims of the early 1990s about the Information Superhighway, Schiller argues that deregulation and privatization of the telecommunications industry renders null arguments about the free flow of information and that, moreover, the global movement toward privatization in the industry parallels a shift toward wider income gaps among sectors of the population in developing countries (Colombia is Schiller's example). A corollary of this trend toward growth with emiseration, Schiller notes, will be multinationals' deployment of the electronic circuitry for transnational marketing, internal business operations, and other corporate agendas within countries whose communications networks are tapped into the Internet. He lists corporate data flows, home shopping venues, online gambling casinos, and video games as some of the likely consumer uses of the network that stand to benefit multinational interests (17–34). We might add to Schiller's list the virtual clinic, the delivery of health services to residents of the United States and members of a virtual global health community constituted by the telemedicine programs I have described.

I prefer to think that this view of a joint medical-telecommunications venture is a paranoid exaggeration. It is difficult to maintain this assessment, though, when I read that this is the stated objective of the telemedicine experts who are building professional alliances such as the one planned between Bell Atlantic and the 335,000-member Kaiser-Permanente Mid-Atlantic region HMO (Taylor 45). It is not unlikely that this type of telemedicine HMO service will be an option for many U.S. residents—communities whose population, geographical positioning, and economic structure together guarantee that an adequate local health care facility will not materialize any time soon. Moreover, the examples I have noted suggest that U.S. telemedicine services may be a source of health care for significant numbers of people living in developing countries where the lines for the electronic importation of health care at a distance are already in place. High-tech health care involving digital diagnostic imaging and specialized medical consultation may become a media export item alongside American network television shows. This situation indicates that we need to evaluate the place of communications concepts as they appear in telemedicine practices. As the medical consumer also becomes a media consumer, his or her agency in the medical process is also determined by his or her position within relations of spectatorship, communications, reception, and representation as they take shape in media discourses.

Still, it is not at all clear whether the constituents of this virtual global health system will wholly be overtaken by corporate agendas. The U.S. public, judging from past opinion polls rating the 1995–1996 Republican-backed health initiatives, was astute about the implications of federal and corporate health care agendas and the role of communications systems in corporate interests. Presumably, there is equally strong awareness of the politics of global health and communications systems among consumers in developing countries where telemedicine systems are being instituted. Evidence suggests that a Schillerian health communications system may be in the offing. Nonetheless, it is essential that we also look for evidence of patient agency. As telemedicine programs are only now being constructed for widespread use, it is crucial that we see the possibilities for intervention in this system and that we keep open the potential for technologically advanced models of community medicine in which the lines for patient agency remain open for people across a range of economic and cultural contexts.

NOTES

1. Other related sites include the Federal Telemedicine Gateway <http://www.tmgateway.org>; Cyberspace Telemedical Office <http://www.telemedical.com>;

Telemedicine Information Exchange <http://tie.telemed.org>; RuralNet <http://ruralnet.marshall.edu/informat/telemed.html>; Department of Defense Telemedicine Test Bed <http://www.matmo.org>; and HospitalWeb <http://www.iinchina.net/yyws/hosda/hosda_net.html>.

2. The U.S. Census Bureau defines as rural all areas that do not fit the designation of urban (a central city or cities and adjacent areas with a combined population of at least 50,000 or any area with 2,500 or more residents). By this definition, 27 percent of the U.S. population is rural. The Office of Management and Budget offers a different definition: rural areas are counties that do not have a central city of 50,000 or a countywide population of at least 100,000. Following this definition, currently approximately 25 percent of the U.S. population (60 million people) live in rural areas. In areas of the country designated as remote, even basic health care services are difficult to obtain. Use of the term "remote" sometimes overlaps with the use of "frontier," a term that describes a county with a population density of less than six people per square mile. On average, residents of frontier areas must travel more than one hour to receive any health care (Simpson and Simpson 503). In the telemedicine literature, the term "remote" encompasses a more diverse set of communities and circumstances. It indicates something about the way that geographical distance becomes both a key factor in and a compelling metaphor of cultural difference. The term refers not only to the geographical distance between telemedicine site and center but to the economic or developmental relation of these two locales. Between the two poles of a telemedicine system, the site with less technology and the lower economic status is the "remote" site. Similarly, ideological, cultural, and economic differences are as much a factor as geographic distance in the use of the term.

3. But the NII vision is not as far out of synch with reality as these statistics might indicate. One 1998 study in the journal *Business Week* predicts that by 2002, sixteen million households will subscribe to a fast network service such as digital subscriber lines (provided by telephone companies) or Internet provider (IP) services (provided by cable companies), making them eligible for high-tech services at home. In 1999, Congress allocated $400 million to upgrade the communication infrastructure for telemedicine in rural areas, guaranteeing that the statistics from mid-decade will change significantly, and quickly. The wiring of poor rural populations thus becomes a direct solution to the problem of bringing medically underserved communities into the health care loop and getting these communities up to speed with the newest in medical technology.

4. *Telemedicine* has become the industry standard.

5. Brauer describes the use of the jointly operated U.S.-Canadian Hermes satellite in 1976 to deliver diagnosis and treatment at a distance to northern Canada's medically underserved and poor rural communities, and the more recent initiatives of SatelLife, a nonprofit international organization members of the International Physicians for the Prevention of Nuclear War founded to address health communication and information needs in developing countries. Whereas the northern Canadian project exemplifies the nascent field's ties to the goal of democratizing domestic health care, SatelLife's HealthNet, a telecommunications network linking health care workers in Africa to medical literature, databases, consultants, and colleagues in Europe and North America, is typical of telemedicine's links to a more global agenda of Western technology and information transfer.

6. On federal cutbacks in health care funding and the devolving of authority over health issues to state and local governments, see Inglehart; and the summer 1995 volume of *Health Affairs,* which is devoted to the impact of managed care. On the shrinkage and closing of public hospitals and clinics, see Kassirer 1348–49; and the correspondence regarding this essay (Correspondence).

7. Arguments linking telemedicine to expansion of the managed-care market are frequent in the literature. For example, an item from the News Desk in a 1995 issue of *Telemedicine* states that telemedicine is one way managed-care companies can expand market share into rural areas, but contracting issues create barriers to market penetration. In the same issue, John Patterson of New England Medical Center in Boston is quoted as stating that telemedicine is critical to facilities that want to be the best high-quality, low-cost provider in an integrated delivery system ("Medical" 1, 7).

8. For a survey of private funding sources for telemedicine projects, see "Private Sources" 8.

9. Woolhandler and Himmelstein's editorial responds to an essay that appeared on pages 1678–87 of the same volume.

10. From personal correspondence with Marvin P. Mitchell, administrator of the Mayo TeleHealthcare Center, 24 March 1999. See also "New Global"; Goldberg et al.; "Mayo"; Rao; Ferguson, Doarn, and Scott.

11. This experimental system, designed to demonstrate the use of voice-grade phone lines to transmit high-resolution digital images in near real time, was conducted in Saudi Arabia and the United Arab Emirates. See "New Global"; Goldberg et al.

12. The concept had been developed over a number of years as a part of NASA's Integrated Medical and Behavioral Laboratory Measurement System, a space technologies applications research division. My information here and below is taken from Justice and Decker; and Fuchs. On Western medicine in Navajo culture generally, see Kunitz.

13. See *Telemedicine* for a discussion of who constitutes telemedicine's rural and remote clients.

14. Rafuse. See also Roberge et al. 707–9; Dunn et al. 484–85.

15. On the Georgia program, see Sanders and Tedesco.

16. See Taylor 47. See also Smits and Baum 139–42.

WORKS CITED

Allen, Ace, and Jeannie Hayes. "Patient Satisfaction with Telemedicine in a Rural Clinic." *American Journal of Public Health* 84.10 (1994): 1693.

Brauer, Gerhard. "Telehealth: The Delayed Revolution in Health Care." *Medical Progress through Technology* 18 (1992): 151–63.

Carrell, Steven. "Overview and Summary of the First National Conference on Consumer Health Informatics." Consumer Health Informatics: Bringing the Patient into the Loop. Proceedings of the First National Conference on Consumer Health Informatics. University of Wisconsin, July 1993.

Correspondence. *New England Journal of Medicine* 334.9 (1996): 597.

Delaplain, Calvin B., et al. "Tripler Pioneers Telemedicine across the Pacific." *Hawaii Medical Journal* 52.12 (1993): 338–39.

Dunn, Earl, et al. "Telemedicine Links Patients in Sioux Lookout with Doctors in Toronto." *Canadian Medical Association Journal* 122.4 (1980): 484–87.

Ferguson, Earl W., Charles A. Doarn, and John C. Scott. "Survey of Global Telemedicine." *Journal of Medical Systems* 19.1 (1995): 35–46.

Fuchs, Michael. "Provider Attitudes toward STARPAHC." *Medical Care* 17.1 (1979): 59–69.

Goldberg, Mark A., et al. "Making Global Telemedicine Practical and Affordable: Demonstrations from the Middle East." *American Journal of Radiology* 163 (Dec. 1994): 1495–1500.

Health Affairs. Summer 1995.

"Health Care and the NII." *Putting the Information Infrastructure to Work: A Report of the Information Infrastructure Task Force Committee on Applications and Technology*. Washington, D.C.: Information Infrastructure Task Force, 1994.

Hopper, Kenneth D., Thomas R. TenHave, and Jonathan Hartzel. "Informed Consent Forms for Clinical and Research Imaging Procedures: How Much Do Patients Understand?" *American Journal of Radiology* 164.2 (1995): 493–99.

House, Maxwell, et al. "Into Africa: The Telemedicine Links between Canada, Kenya, and Uganda." *Canadian Medical Association Journal* 136.4 (1987): 398–400.

Inglehart, John K. "Health Policy Report: Politics and Public Health." *New England Journal of Medicine* 334.3 (1996): 203–8.

Justice, James W., and Peter G. Decker. *Telemedicine in a Rural Delivery System*. New York: Academic, 1979.

Kassirer, Jerome P. "Our Ailing Public Hospitals: Cure Them or Close Them?" *New England Journal of Medicine* 333.20 (1995): 1348–49.

Keepnews, David. "The Role of Nurses in the New Health Care Marketplace." Letter. *Health Affairs* (Fall 1995): 280–81.

Kunitz, Stephen J. *Disease Change and the Role of Medicine: The Navajo Experience*. Berkeley: University of California Press, 1983.

Latour, Bruno. *Science in Action: How to Follow Scientists and Engineers through Society*. Cambridge: Harvard University Press, 1987.

"Mayo Linking to Jordan via Intelsat." *Global Telemedicine Report* Apr. 1994: 14–15.

McCarthy, Daniel. "The Virtual Health Economy: Telemedicine and the Supply of Primary Care Physicians in Rural America." *American Journal of Law and Medicine* 21.2 (1995): 111–30.

"Medical Center Forges New Markets in Competitive Health-Care Climate." *Telemedicine* 3.5 (1995): 1, 7.

"New Global Telemedicine Network Scores First Middle East Contract." *Telemedicine* 2.3 (1994): 2–7.

News Desk. *Telemedicine* 3.5 (1995): 2.

Norton, Scott A., C. Eric Lindborg, and Calvin B. Delaplain. "Consent and Privacy in Telemedicine." *Hawaii Medical Journal* 52.12 (1993): 340–41.

Perednia, Douglas A., and Ace Allen. "Telemedicine Technology and Clinical Applications." *Journal of the American Medical Association* 273.6 (1995): 483–88.

"Private Sources Offer Wealth of Grant Funds for Health-Care Projects." *Telemedicine* 3.4 (1995): 8.

Rafuse, Jill. "University of Alberta Program Links Northern Hepatitis Patients with Southern Consultants." *Canadian Medical Association Journal* 151.5 (1992): 654–55.

Rao, U. R. "Global Connectivity through Telemedicine." *Journal of Medical Systems* 19.3 (1995): 295–304.

Roberge, Fernand A., et al. "Telemedicine in Northern Quebec." *Canadian Medical Association Journal* 127.8 (1982): 707–9.

Sanders, Jay, and Francis J. Tedesco. "Telemedicine: Bringing Medical Care to Isolated Communities." *Journal of the Medical Association of Georgia* 82.5 (1993): 237–41.

Schiller, Herbert I. "The Global Information Highway: Project for an Ungovernable World." *Resisting the Virtual Life: The Culture and Politics of Information*. Ed. James Brook and Iain A. Boal. San Francisco: City Lights, 1995. 17–33.

Simpson, Christopher, and Martha A. Simpson. "Complexity of the Healthcare Crisis in Rural America." *Journal of the American Osteopathic Society* 94.6 (June 1994): 502–8.

Smits, H. L., and A. Baum. "The Health Care Finance Administration and Reimbursement in Telemedicine." *Journal of Medical Systems* 19.2 (1995): 139–42.

Taylor, Kathryn S. "We're (Almost) All Connected." *Hospitals and Health Networks* 68.18 (1994): 43–47.

"Telemedicine: Fad or Future?" *Lancet* 345.8942 (1995): 73–74.

"Telemedicine Gets a Piece of Medicare Pie." *Telehealth Magazine.* 10 Apr. 1999 <http://www.telemedmag.com>.

U.S. Department of Commerce. "Falling through the Net: A Survey of the 'Have-Nots' in Rural and Urban America." July 1995 <http://www.ntia.doc.gov/ntiahome/fallingthru.html>.

Woolhandler, Steffie, and David Himmelstein. "Extreme Risk—The New Corporate Proposition for Physicians." *New England Journal of Medicine* 333.25 (1995): 1706–1708.

Chapter 12

Biotechnology on the Margins

A Haitian Discourse on French Medicine

PAUL E. BRODWIN

In this essay, I focus not on a specific apparatus or procedure but rather on the panoply of biomedical technologies and the meanings it acquires as an ensemble for a given community. These meanings are not determined by the technologies or their clinical applications. They emerge instead from the historical trajectory of the community in question, the social contradictions it faces, and the strategies to resolve them. I examine this process through an ethnography of the Haitian migrant enclave in Guadeloupe, French West Indies. Haitians arrived on the island no more than twenty-five years ago, and, as a group, they constitute the poorest sector of the population and are subject to occasional public harassment as well as immigration sweeps and deportation. At the same time, they have surprisingly good access to state-of-the-art French biomedical services at the island's largest hospital. The way Haitian migrants talk about French medicine sheds light on more general questions: What are the discourses of biotechnology in an era of globalization? What cultural and political work do these discourses accomplish?

BIOTECHNOLOGY AND GLOBALIZATION: TWO CONTRASTIVE LOGICS

In the world at present, biotechnologies continually move away from their point of origin in First World laboratories and hospitals and enter communities with far less political and economic power. This movement demands enormous investments of money and energy, and it proceeds under several banners: liberal schemes to improve the health of developing countries or underserved populations, the search for new markets, and the international expansion of biomedicine's authority. However, these paternalistic and professional rhetorics about biotechnologies are less convincing for people living at a great distance—economic, political, or geographic—from their point of origin.

The talk about biotechnologies in marginal communities indexes the embattled relations between them and the dominant social order, between their particular interests and self-image and the overarching signs which conceptualize and inscribe them, usually to the communities' disadvantage. This discursive framework weaves together several sorts of commentaries: about illness and healing, the limits and powers of biomedicine, and the shifting social relations and cultural identity of a given community. As a discourse—an organized set of statements, terms, categories, and beliefs—it is one of the main sources for the cultural and political meanings of biotechnology as it crosses the boundaries of nation, region, and class. Of course, there does not exist a singular, univocal discourse used by all "marginal" or "peripheral" groups (even leaving aside how to define such categories).[1] The way people talk about biotechnology is calibrated to their specific historical and political position and the details about who introduced which forms of medical technology and at what period, who has access to them, what other forms of healing they displaced, et cetera. Nonetheless, discussions about these issues arise from the same general conditions: living at a distance from biotechnologies' origins and receiving them through complex relations of dependency. There are also systematic differences between the discourse of medical technology in marginal or peripheral groups compared to industrialized nations.

Despite the pervasive trope "technology transfer," biotechnologies are never merely transferred from one setting to another. Their cross-cultural movement shifts the context of the object or instrument, and this changes its value, meaning, and ultimate usefulness. Philosophers have approached this problem through a cultural hermeneutics of technology, which traces how people make use of a newly introduced artifact by inserting it into an already existing praxis (see Ihde). Consider the Hindu prayer wheel—a wind-powered device that automatically sends prayers to the gods as it spins (White, cited in Ihde 70, 127). When it was introduced to Europe, it was refigured into the windmill and drove industrial operations in the early modern period. This illustrates how the meaning and broad effects of technological artifacts change according to the new cultural field where they are implanted. The global diffusion of pharmaceuticals offers countless similar vignettes: for instance, in Sri Lanka and south India, people judge contraceptive pills to cause leanness, heating, and drying, according to local anatomical beliefs (Nichter and Nichter); in Sierra Leone, certain Mende take red-coated One-A-Day vitamins to replace blood lost during illness (Bledsoe and Goubard). These examples privilege the local matrix in which new or imported technologies become embedded. Through this process, people creatively re-author the technology in question and deploy it for the cosmological and historical purposes which are most compelling to them (the drive for secular power and practical applications, in the case of the windmill; the humoral regulation of the disordered body, in the case of current pharmaceuticals).

It is dangerous, however, to insist too strongly on the cultural distinctiveness of the groups receiving particular technologies. After all, the same history of contact, exchange, and political rule through which the artifact arrived in the first place reforms people's lives in other ways. The local discourse about technology, therefore, typically does more than re-interpret it according to familiar structures of knowledge and practice. Especially in the case of new medical technologies, people often frame them as genuinely, even shockingly, different from familiar techniques of healing and bodily intervention. Their discourse then centers on the contrast between the familiar and the unknown, the stable and the potentially subversive. If this refers on one level to the restricted domain of medicine and healing, at another level it addresses broader questions of historical change and the disjunction among a particular community, the surrounding society, and the global forces which shape them.

In their discourse about new biotechnologies, people do more than embed them in local and long-standing cultural frames. They also articulate the technology's ramifying and novel effects in social life. They accomplish this by relying on a logic of contrasts, which portray contemporary biotechnologies as one item in a contrasting pair, often linked to several other contrasting pairs as a homology. This rhetorical device marks the discussions of experts and non-experts: policy makers, scholars, critics, and people who pragmatically engage with biotechnology during routine or emergency medical care. However, the discussions proceed along different lines in urban industrialized societies compared to the developing world and marginalized communities in the West. In the first case, the contrastive logic is based on time. Biotechnologies seem notable largely because they are new. They signify or even advance other profound historical transformations, and commentaries focus on their future benefits or horrors. In the developing world, the relevant contrast concerns space. Biotechnologies are significant because they come from somewhere else, most often wealthier societies which extend their political and ideological influence, via biotechnologies, into the intimate bodily experience of sickness and healing.

Biotechnologies thus always enjoy the aura of the modern, but in one case people contrast this to the past and in the other case to the local. In a contrastive logic based on time (evident throughout this book), biotechnologies threaten to outstrip legal precedent and prior moral constraints. They unsettle notions of the body and selfhood which, for better or worse, had been accepted for years. In a contrastive logic based on space, biotechnologies join the other signs and practices that infiltrate a given society from outside, and they retain some of the authority and cultural cachet of their origin. Obviously, "time" and "space" here are not abstract or neutral qualities. They are gradients of both difference and power which people use to explain how biotechnologies unsettle conventional knowledge and magnify social contradictions. Moreover, people invoke these contrasts not just to describe the impact of biotechnologies but also to advocate strategies of resistance, acceptance, or accommodation (rhetoric is, after all, speech designed to persuade). Again, this process unfolds differently in the West and in developing societies.

Most European and North American rhetorics about biotechnologies invoke the shock of the new. To begin with the academic literature, certain technologies are held to produce cyborgs—hybrids of the organism and machine—as both a material reality (for example, technology-dependent individuals) and an innovative cultural icon which displaces previous notions of the body as an organic whole (Haraway). Simultaneously, bio-technologies inject uncertainty about the "facts of biology" previously considered self-evident. This uncertainty, in turn, can subvert other forms of cultural knowledge once justified as natural and hence indisputable (Strathern; Franklin, "Postmodern"). These two basic arguments, of course, entail each other in several ways, and they underlie many of the scholarly debates.

The most appropriate chronological contrast will depend on the specific technology in question. For instance, the routine use of pacemakers, titanium hips, and other prostheses suggests the contrast between the older organic and current bionic body (Williams). Ethical debates about organ transplantation turn on the shift from earlier models of the body as intrinsic to one's identity to modern notions of the body as objectified commodity (see Joralemon, this volume).[2] Genetic screening and diagnosis set in motion the contrast between two responses to suffering: traditional religious discourses of fate and present schemes for techno-rational control (see Cranor; Wexler, this volume). Laparoscopy and fetal imaging take once invisible bodily processes and render them visible (for example, Balsamo 80–116; Taylor, this volume), which helps transform pregnancy from a private to a public experience. Taken as a whole, contemporary biotechnologies mark a historical shift in our control over the body. They offer a privileged site to witness the passage from one form of bodily knowledge to another, and diverse scholars read this as symptomatic of larger transitions from modern to postmodern hierarchies and schemes of social control (for example, Clarke; Hogle).

The scholarly conversations about biotechnologies intersect with popular discourse at several points, particularly through the twin rhetorics of hope and fear. These rhetorics are essentially narratives about alternative futures, and they pervade legislative, journalistic, and literary discourses, often bridging these genres in surprising ways. Parliamentary debates about embryo research in England, for instance, have typically invoked optimistic claims about imminent scientific achievements as well as dystopian scenarios of future moral decline and social disorder (Mulkay). Even non-invasive, rather minor technologies, such as a portable microchip patient card, come surrounded by a marketing rhetoric of promise and expectation (which nonetheless does not preempt popular resistance; see Godin). The mass media, of course, emphasize the novelty of biotechnologies and portray them as newfound ways to satisfy long-standing needs and desires (for example, Franklin, *Embodied*). Mass media accounts also draw freely upon the rhetorical resources of science fiction and horror movies to suggest that the future has suddenly irrupted into the present (Mulkay 736ff). Conversely, science fiction writers such as Margaret Atwood and Octavia Butler supply rich scenarios of the future for us to debate the moral and political risks of contemporary technologies. The rhetorics about biotechnologies thus blur the lines among journalistic, literary, and scientific discourses. Positions mapped out in one discourse are borrowed by another, and scenarios of a hopeful or nightmarish future seem compelling to all (see Van Dyck).

Consider now the rhetorical frameworks which organize debates in the developing world and marginalized communities in the West. In these settings, new medical technology is just one of the many globally circulating practices which reshape people's lives (Lock and Kaufert). The contrasts people draw, therefore, concern not the new and the old but rather the global and the local, along with its homologies: core and periphery; "advanced" and "backward" societies; and metropolis and dependent colony, ex-colony, or client state. Biotechnologies emerge from arenas of knowledge and power far beyond the face-to-face communities of everyday life, and this is why they generate debate (compare Ginsburg and Rapp 8–9). People may either embrace or oppose specific technologies, but their rhetoric always addresses the distant but dominant political forces that lie behind them.

In some cases, this rhetoric galvanizes community opposition through appeals to nationalism or local sovereignty. In Egypt, for instance, women activists led the popular rejection of internationally financed clinical trials of the contraceptive techniques of Norplant and Depo-Provera. They made explicit connections among the drug trials, women's subordination, and the policies of the International Monetary Fund, as well as the long history of development schemes which have threatened public health (Morsy). Their rhetoric thus embedded specific technologies within the panoply of other top-down projects of "modernization" and "Westernization," in addition to the alliance of Egyptian elites with First World development agencies. In this discourse, spatial metaphors about medical imperialism and the colonization of the body became startlingly literal. At the same time, Egyptian government ministers promoted these contraceptive drug trials as an honor bestowed on their country and a chance to introduce or upgrade medical services (Morsy 92). This official talk of benign technology transfers and international scientific cooperation was obviously rejected by the activists. The same technology thus became wrapped in rhetorics of celebration and resistance, but both are essentially commentaries on global dependency relations.

This suggests why biotechnologies in the developing world often become the site of intense political struggle. Voices of support and opposition emerge from groups with unequal power and different conceptions of the national interest. Often the groups rely on disparate conceptions of the body and claim distinct types of cultural authority. In northern Canada, for instance, debates over obstetric technology pit the Canadian government health service against Inuit communities. Building their case on epidemiological data and medical discourse on risk, health professionals support the policy of "evacuating" expectant Inuit women hundreds of miles to give birth in well-equipped urban hospitals. (Kaufert and O'Neil). Inuits counter this policy by claiming the right to give birth in their community, and thereby both restore traditional knowledge of midwifery and resist their growing dependency on government assistance (O'Neil and Kaufert). Not surprising, debates between these two positions are riddled with miscomprehension and distrust. Physicians' talk about the risks of home birth are rooted in the techno-rational authority of epidemiology and personal fears of legal liability. Inuits frame risk in terms of the local history of political disenfranchisement and the personal experience of natural disasters (compare Balshem). Rhetorical disputes thus repeat the more general contradictions of internal colonialism.

However, rhetorics about new medical technology in the developing world do not always split along the lines of outright resistance or paternalistic approval.[3] After all, most communities in the world today are already open to cosmopolitan flows of knowledge and practice (compare Gupta and Ferguson 8). Negotiations between global and local forces, therefore, proceed in more complex and ambivalent ways than in the above examples. What types of rhetoric about biotechnology emerge in such communities, inescapably if ambivalently caught up in the global exchange of persons, commodities, ideas, and techniques for ordering life? The arrival of biotechnologies on the local scene sets off a series of commentaries about the difference between local ideologies and ways of life and the global forces which reformulate them. People use this rhetoric to mark and sustain cultural distinctiveness; nonetheless, they do not do this through a knee-jerk rejection of the foreign and nostalgia for the local forms of traditional healing. Their talk about biotechnology still comments critically on global/local relations, but these are the more subtle politics of negotiation and appropriation.

Rhetorics of health development in Nepal illustrate this strategy. Both villagers and government officials regard biotechnologies as synonymous with modernization. Many Nepalis embrace these icons of modernity along with corollary notions of objective scientific truth and individual responsibility for health. However, this is not mere mimicry of the modern scientific West or a calculated bid for legitimacy (compare Nandy). For middle-class professionals, it is part of an explicit attempt to forge a uniquely Nepali modernity (Adams). For instance, many Nepali doctors frame current public health practice in both scientific, rational terms and as part of a broad movement against corruption and favoritism. They enthusiastically support technologies such as contraception, immunization, and pharmaceuticals as the tools of a new, pro-democratic political regime. This reformist and nationalist rhetoric foregrounds the clinical effectiveness and aura of impartial truth connected to medical technology. Through this rhetorical stance, medical professionals articulate their identity as modern democratic citizens in a Nepal committed to both sacred cosmologies and specific Western-style political reforms.

Ordinary Nepali villagers frame biomedical technology in a different way, one calibrated to the specific social contradictions they face (see Pigg). Villagers know that, according to foreign experts and the Nepali middle class, they are immersed in traditional, premodern, and supposedly incorrect beliefs. In this context, some people's vocal support of biotechnologies is a form of symbolic capital, a shorthand means to identify with the modernizing vanguard. More often, however, villagers debate the relative merits of medical technologies and local healing practices with a complex mixture of skepticism and belief. Their rhetoric juxtaposes the stethoscopes and thermometers of dispensary workers with the spirits of shamans without dismissing either one. The result is not a counter-ideology that resists modernity, but rather it is an interwoven identity of local and cosmopolitan meanings (Pigg 192–93). From the villagers' perspective, shamanic cures along with the technological interventions of biomedicine can be either authentic or inauthentic, a healing art or a sham. This rhetoric articulates villagers' uneasy position as the "least developed" sector of a developing country. By negotiating between local and cosmopolitan forms of healing, villagers defend against denigrating stereotypes and respond to their specific marginalized situation. This rhetoric about medical technology sets the stage for a similar discourse among the Haitian diaspora community of Guadeloupe.

HAITIAN DISCOURSES ABOUT BIOMEDICINE

About twenty-four thousand Haitians currently live in Guadeloupe, French West Indies, an "overseas department" of France.[4] This migrant community is much smaller than major nodes of the Haitian diaspora in New York City, Miami, and Montreal. In Guadeloupe, there is virtually no Haitian middle class, no Haitian newspaper or television station or radio station, and most migrants lack the proper visas. These migrants are thus especially vulnerable to the negative stereotypes which seem to accompany Haitians wherever they travel (see Lawless). Forced to live in the worst housing and to work in the lowest paid jobs, Haitians face suspicion and at times outright hostility from the majority Guadeloupean population. At first glance, this seems surprising. The majority population in this case is French West Indian—descendants of the Africans brought first as slaves and later as indentured laborers to work in the sugar fields of what was then a French colony. Long-time residents of Guadcloupe thus resemble the recently arrived Haitian minority in terms of historical origins, race, color, and language (both French and Creole). Nonetheless, for reasons I will examine below, Haitians in Guadeloupe are widely portrayed as uncivilized, chaotic, and a threat to the island's self-image as a well-functioning, modern outpost of European society. This negative image, and the way that Haitians respond to it, is the broadest context for their rhetoric about medical technology and Western biomedicine.

When these Haitian migrants in Guadeloupe discuss sophisticated biomedical services, they typically contrast them to traditional forms of Haitian religious healing. These discussions, moreover, address the Haitian community's marginal position and the denigrating clichés about it in the dominant society. This rhetoric resembles the talk in Nepali villages about scientific medicine and its "others" (particularly shamanism). However, unlike the Nepali and Inuit cases, Haitians are emphatically not considered part of the national society. They are not "country cousins" in need of development from the paternalistic state. They are instead treated as foreigners, at times even as unwanted intruders, and they are marginalized both in tangible economic and political terms and in the Guadeloupean collective imagination. For this reason, the rhetorical contrast between sophisticated biomedical technologies and the practices of traditional Haitian healers becomes even more politically loaded.

The way Haitians in Guadeloupe contrast medical technologies and religious healing also has a deeper history which pre-dates the diaspora in Haitian society (see Brodwin). Therefore, the ethnographic story must start in the Haitian homeland, the source of the rhetoric which migrants creatively adapt to the diaspora situation. Throughout urban and rural Haiti, people are familiar with a range of biomedical technologies. Most open-air markets feature itinerant vendors of pharmaceuticals: ampicillin, tetracycline, and other antibiotics manufactured by multinational companies as well as caffeine and analgesic-based preparations produced in Haiti or elsewhere in the Caribbean. Since World War II, a network of health posts and dispensaries has been constructed which now serves residents throughout the countryside. Funded largely by missionary groups and international agencies, these clinics provide primary care, immunization, and maternal and child health services. Most Haitian villagers talk about their encounters with these clinics not as "biomedical care" in general but rather in terms of specific technical services: X-rays and laboratory analyses of blood, stool, or urine. Moreover, many healers with no medical training (such as midwives or herbalists) will use simple technologies such as injections and pharmaceuticals, and these too enter people's stories of specific illness episodes.

Biomedical technologies are thus interwoven into the everyday practical experience of illness and healing. People freely use them along with other discrete healing systems that co-exist in Haitian society (see Brodwin; Farmer; Hess). People typically have access to several forms of therapy in the course of a single illness: commercial pharmaceuticals, primary care, more sophisticated hospital-based services, as well as herbal treatments, the services of indigenous specialists (bonesetters, midwives, et cetera), and religious healing rituals from Vodoun (the popular neo-African religion practiced to a greater or lesser extent by most Haitians). Residents of Haiti pragmatically mix and match therapies in different ways, and even healers are aware of other therapeutic practices and their respective costs and benefits.

Nonetheless, people do consider biomedicine as distinctive in a symbolic and rhetorical sense. This rhetoric does not emphasize the technical superiority of biomedicine. To the contrary, it casts biomedicine as relatively weaker than Vodoun treatments for certain kinds of sickness. Briefly, each type of therapy is considered uniquely suited to a particular class of disorder, as defined by etiology: an "illness of God" or an "illness of man." Examples of illnesses of God include dying of old age or of some visible trauma as well as self-limiting diseases. In contrast, the prototypical illnesses of man include a sudden death with no apparent cause, the death of someone young and otherwise healthy, or a previously unknown illness. This sort of illness raises the suspicion that someone is trying to kill the victim, perhaps out of his/her own jealousy and hatred, or, more disturbing, as revenge for a pathogenic attack that the victim had previously launched on someone else.

These two etiologic categories form a contrasting pair, and they structure the way people engage and speak about biomedicine. At the beginning of any illness, its cause is unclear. It only emerges gradually in the course of biomedical treatment. The key to interpretation is simple: if biomedical treatment succeeds, the problem was an illness of God. If people seek out biomedical treatments but they fail, the same illness is reclassified as caused by the victim's human enemy. The victim must then turn to the more potent forms of healing only available from a Vodoun practitioner. Taking this step, however, raises certain moral risks. If biomedicine succeeds, people are regarded as innocent and their suffering the result of natural and impersonal forces. If biomedicine fails, then the sufferer enters a morally murky domain. The sufferer must begin by engaging the services of Vodoun healing specialists who are malign and dangerous characters, according to Catholic dogma and the elite canons of respectability which still hold sway in Haitian society. Moreover, a strong suspicion of guilt attaches to the sufferer. Perhaps she merited the attack in the first place by sending a prior sickness upon someone else, and she is now the victim of the intended victim's revenge. Only someone absolutely convinced of his moral rectitude (and wanting to proclaim it publicly) would forswear the use of traditional Vodoun religious healing once biomedicine fails.

This entire discourse about illnesses of God and of man turns on the basic contrast between biomedicine and Vodoun healing. Which type of therapy one uses (or claim to use) speaks volumes about one's moral status: people often vehemently denied to me that they ever needed Vodoun healing because, as they put it, they were innocent of any wrongdoing (hatred, jealousy, and the like). Moreover, the type of therapy one prefers also reveals one's position in the cultural landscape of Haiti. Using Vodoun healing means participating in the more African-inflected domains of Haitian society. Some people do this openly, some in secret, and others publicly abjure this realm entirely; all these are meaningful options given the colonialist origins of Haitian society and its history since independence in 1804. The country has long been wracked by divisions between the urban, Francophilic elite and the rural peasantry, more attuned to African-derived practices and beliefs. In this context, people draw contrasts between the effectiveness of biomedicine and religious healing in order to take up a definitive position within these conflicting arenas of cultural prestige. They may publicly reject Vodoun healing in order to attach themselves more securely to the centralized Catholic Church and its formal canons of respect. They may seek out Vodoun specialists because they acknowledge their informal, more immediately tangible power and also because they share the implicit repudiation of the privilege and prestige of Catholicism.

This rhetoric of medical power thus frames a long-standing opposition in Haitian society. The contrast between biomedicine and religious healing replicates older contrasts between the urban elite and the peasantry; between the formal, centralized Catholic Church and the local power of Vodoun healers; and between the French and African components of Haitian society in general. The rhetoric is pre-adapted to situations of hegemony and subordination, and for this reason it takes a recognizable form in Guadeloupe. Nonetheless, people must reshape it to articulate the particular form of marginality they face in this diaspora setting. I will first describe the specific contradictions which affect this community, and then I will outline how people's rhetoric about biomedical services intervenes in and re-imagines their position in Guadeloupean society.

THE HAITIAN DIASPORA IN GUADELOUPE: MARGINALITY AND MEDICAL DISCOURSE

Haitian migrants have been marginalized since their arrival in Guadeloupe in the mid 1970s. At that time, the Guadeloupean sugar industry was involved in a protracted labor dispute, a symptom of the long, slow death of the island's colonial economy. Haitians were brought to Guadeloupe by the growers, essentially as scab labor to cut sugar cane which would otherwise rot in the field. Tragically, these Haitian men were ignorant of the political situation they were entering and of their role as strikebreakers. The situation unleashed a bitter opposition to Haitians' presence in Guadeloupe (referred to now as a *chasse aux Haitians,* or hunt for Haitians), culminating in public violence (including lynch mobs). Although the violence was fairly quickly quelled by local intellectuals and activist Catholic priests, it left an enduring image of Haitians as foreigners who are opposed to the interests of Guadeloupean workers.

The next stage in the formation of a hegemonic image of Haitian migrants in the Guadeloupean imaginary occurred in the early 1980s. At that time, Haitian immigration increased from the controlled and documented deployment of cheap agricultural labor to a wave of immigrants who came without documentation or who stayed on past the date of their visas (from the early 1980s to the present). Haitian migrants found work in the building boom in Guadeloupe and on the dependent island of St. Martin, which by the beginning of the 1980s had emerged as a destination for North American tourists. This wave is the chief source for today's transnational Haitian community in Pointe-à-Pitre (Guadeloupe's commercial center), where most men work in the construction industry and most women become independent vendors who sell clothing, housewares, and other commodities on the streets in Guadeloupe's major towns.

This particular economic adaptation gives rise to three widespread images of Haitians. The first portrays the Haitian migrant as an economic drain on society: someone who takes in money (through his daily wages on construction jobs or her sales on the street) and then sends it all back to Haiti. "They work, they take our money, but then they never spend it here" is the gist of this reading of the Haitian as greedy foreigner who is not willing to settle down in Guadeloupe, who maintains a continued allegiance to his or her country of origin (and whose allegiance harms Guadeloupean society in some unspecified way). The second dominant image emphasizes not the greed of Haitians but their sheer numbers. "They are crowding us out" is the popular expression for this cliché, and the metaphor is rooted in the tangible experience of street life in Pointe-à-Pitre. Most Haitian market women do not have the capital to open their own stores. They display their wares on the sidewalks, and older Guadeloupean residents consistently complain about being pushed out of their own city and of the chaos of street vendors who subvert their desires for an orderly and clean urban space. (The old downtown was laid out in the colonial era and contains numerous well-preserved and restored examples of Creole architecture.) Finally, there is a third hegemonic image of Haitians which dovetails with their reputation for chaos. This is the notion, created by news reports of crime and political instability from Haiti, that Haitians are essentially a disorganized people who cannot rule themselves effectively. The fear just below the surface of this particular image is that Haitians will bring this disorder with them to Guadeloupe.

The contrasts which migrants draw between state-sponsored biomedicine and Haitian religious healing are calibrated to these specific images of denigration and creatively intervene in them. To begin with, the balance between these two forms of therapy changes from Haiti to Guadeloupe. Hospital-based biomedical treatment becomes relatively more important in people's overall health care, compared to the norm in Haiti. Most women in Guadeloupe give birth in hospitals; hence, the practice of lay midwives has largely died out in this setting. The importance of domestic herbal medicine is also vastly reduced. People simply do not have the time to prepare and administer these infusions, compresses, et cetera, and they do not have access to the right plants. Although the island has the proper climate and soil conditions, Guadeloupeans have largely neglected the Creole herbal tradition. The island is fully part of consumerist French society, and commercial pharmaceuticals have replaced the once-thriving practice of herbal healing. Moreover, the majority of Haitians lack the legal documents to travel securely back and forth to Haiti, and few people have the capital to open stores that directly serve the diaspora community. Thus, the circulation of herbs (leaves, buds, roots, et cetera) between Haiti and certain shops in New York City, Miami, or elsewhere does not extend to Guadeloupe.

Migrants therefore take most cases of serious illness to the state-run Centre Hospitalier Universitaire (University Hospital Center). Like all cosmopolitan hospitals, it is a centralized and bureaucratically controlled institution which demands that patients present their documents before admission. Formal state institutions such as this are potential threats to the Haitian community, since about 75 percent of migrants do not have the proper papers. Under French law, they are subject to immediate deportation without legal counsel or the right of appeal. I thus expected that Haitians' rhetoric about biomedical services would seize on the Centre Hospitalier Universitaire as a potential site of formal legal control—an institution on which they depend (because of the relative absence of other forms of healing) but which symbolizes the political weakness of their community and their personal vulnerability.

People's talk about the hospital does index their marginalization, but not in the way I expected. I heard of no cases where people were refused health care because of their undocumented status or where their identity papers were passed to the immigration police. Simply having a Haitian passport, regardless of other documents, is enough to receive hospital-based care. However, people did complain that they received a lower level of care or were forced to leave before they were cured, and these comments testify to several forms of legal and economic discrimination. French socialized medicine provides a highly subsidized level of care to all legal residents, but Haitians whose papers are not in order must pay the entire cost in cash. These high out-of-pocket costs severely limit the kinds of hospital-based care undocumented migrants can receive. Moreover, even those with valid work permits encounter the same problem, and the reason is again rooted in French socialized medicine. This medical system operates somewhat like American social security; employers contribute a certain amount to a national fund for each employee who works a minimum period of time. Construction bosses routinely fire Haitian workers a few weeks before this minimum, and this locks them out of health benefits.

These complaints about the shortcomings of sophisticated hospital services thus articulate some of the immediate and tangible insecurities which migrants face. However, their marginalization is not only a matter of poverty and uncertain legal status. As we have seen, it is also produced through discrete images of denigration in the collective imaginary of Guadeloupe. Haitians are cast as an essentially chaotic people who threaten to subvert the orderly conduct of life on the island. They are portrayed as embodying the polar opposite of French norms of civility and rationality which many Guadeloupeans self-consciously uphold. These images as well come under scrutiny in Haitians' rhetoric about biomedicine. Migrants rhetorically launch a counter-critique of the dominant Guadeloupean self-image as possessors of French culture. They do not enunciate this counter-critique directly; it emerges, rather, through the widespread contrasts between French biomedicine and Vodoun healing.

Recall that in the Haitian scheme, biomedicine is considered weaker than Vodoun healing for a certain category of illness: illnesses of man. People who are judged to suffer from this category must turn to African-derived healers, even though they may lose cultural prestige in the process. Haitian migrants assume that Guadeloupeans follow the same logic of medical decision making and suffer the same risks. The following exchange with a Haitian friend shows how this rhetoric operates. I asked what Guadeloupeans would do if they go to the hospital but are not cured. "They go to an *houngan* [the Haitian Creole term for a Vodoun healing specialist]. They find one here or they go to Haiti." Surprised, I asked whether Guadeloupeans believe in this sort of healing power. "They believe in it more than we do! But they won't tell you. You can ask them, but they keep it hidden."

A similar theme appears in discussions about the range of non-biomedical healers available in Guadeloupe. Haitian migrants are aware of a class of local healers (called *quimboiseurs* and *gadezafès* in Creole) whose practices overlap those of Haitian Vodoun specialists. These healers enjoy certain spiritual gifts; they perform exorcisms and lead prayer groups; and they specialize in illnesses caused by social conflict (see Bougerol; Ducosson).

They emerge essentially from the same historical matrix as Haitian Vodoun: the legacy of plantation slavery, the intermixing of West African and French Catholic religious practices, and the history of suppression by the Catholic Church. These traditional religious healers in Guadeloupe are, in a historical sense, the equivalents of those in Haiti. However, Haitian migrants judge them as far weaker than the Haitian counterparts. The following conversation gives the flavor of this dismissal. "Here in Guadeloupe," a thirty-year-old Haitian migrant told me, "it is easier to treat an illness of God, but if it is an illness of man sent upon you, it is more difficult to find treatment." The first half of this statement means that biomedical care is simply much more available. The difficulty of finding cures for "an illness of man" means that it is harder to find effective or powerful religious healers. When I asked why, the answer was self-evident: Haitian religious healers are the sole source of successful treatment—they are stronger than biomedicine and more efficacious than the local *quimboiseur*. The latter, I was also told, learns his trade through an apprenticeship with a Haitian.

Comments such as these often come up when Haitian migrants discuss their Guadeloupean employers and neighbors. Certain wealthy Guadeloupeans, it is said, owe their wealth to a Faustian bargain with a Haitian Vodoun practitioner. Migrants claim that the same Guadeloupean who cheats them on the job or disrespects them in the street will nevertheless run to a Haitian healer when their own biomedical treatments fail. Set against the negative clichés attached to Haitians, this is a resistant and cynical counter-image which migrants hold of the dominant society. It asserts that Guadeloupeans also are afflicted by illnesses of man, and French biomedicine is worthless in this case. Despite their French legal and cultural citizenship, Guadeloupeans then have recourse to healing with distinctly African roots, whether they admit it or not. Moreover, even the religious healers from Guadeloupe are less powerful than their Haitian counterparts.

What do Haitian migrants accomplish by drawing these contrasts? They have ferreted out the Achilles heel in the Guadeloupean self-image. The rhetorical contrast between sophisticated biomedical services, imported from the French metropole, and Haitian religious healing plays up the uneasy and ambivalent relationship toward European modernity on the part of Guadeloupeans. This is a powerful critique of the Guadeloupean self-image and, indirectly, of the images of denigration which flow from it.

Perhaps the major cultural contradiction among residents of Guadeloupe is the widespread ambivalence about assimilating into the French state and accepting its claims to a universal culture. This ambivalence was first enunciated by the founders of Negritude, the intellectual and literary movement begun in the 1940s by, among others, Antillean and African thinkers trained in European philosophy. Speaking from the position of African-descended French colonial subjects, the founders of Negritude affirmed an absolute difference between their society's values and the European norm, a racial and ontological distinction (for this summary I am indebted to Burton 141ff). In this model, the surface layer of Frenchness concealed and held captive an African or black "substance," and this made assimilation a contradictory and ultimately wrong-headed process. The relation between the French cosmopolitan surface and an Antillean or African core became a central topic for Frantz Fanon, the most well-known writer about the psychic alienation wrought by three hundred years of French colonialism. In the most recent attempts to resolve this contradiction of culture and identity, writers such as Edouard Glissant and Patrick Chamoiseau portray Antillean identity in less monolithic terms. "Creoleness"—the privileged term for them—is not a single substance but a "principle of mixture" (Burton). It is an unlimited combination of diverse cultural materials, and its spirit comes across in the words of a recent manifesto: "Neither Europeans, nor Africans, nor Asians, we proclaim ourselves Creoles" (Bernabé, Chamoiseau, and Confiant).

However, based on my ethnographic research in Guadeloupe, today's literary discourse on Creoleness is restricted to an elite intellectual class. The unresolved inner tension and ambivalence Fanon exposed provide a better guide to most people's tacit images of racial and national belonging, and also to their explicit discourse about Haitians. Guadeloupeans remain enormously ambivalent about all things coded "African," and this emerges in several arenas of everyday life. The use of Creole language is punished in school and at home, for learning Creole raises fears that one will not learn how to express oneself in French. There is a complicated preference for Caucasian physical features (not only and not necessarily lighter skin color but also certain types of hair and facial structures). Indigenous forms of arts and artisanship—drumming, Creole cooking, the production and sale of madras cloth dresses—are weak and constantly struggling for popular and governmental support.

This ambivalence also manifests in Guadeloupeans' fear of Haitians. As several Guadeloupeans, mostly anthropologists and sociologists, commented to me, "we are afraid of Haitians because they are closer to Africans. They are the mirror that we don't want." I interpret this to mean that Haitians take the place of the rejected but still desired African "other" in Guadeloupe. Haitians, who are coded as darker, as African, as more savage and less civilized (and these are the words used in French), reflect the attributes that Guadeloupeans have been trained to suppress. "We are afraid of them," one of my Guadeloupean colleagues continued, "precisely because they have their own language, their own music, their own history, whereas we have nothing that is our own." In this context, she was referring to the absence of any artistic or cultural form that does not emanate from France. "The Haitian already knows his value, he has pride. We have no foundations, no *base d'identité* [basis for an identity]."

The tenuous, conflicted self-image that many Guadeloupeans hold thus produces both a grudging respect for Haitians and, more insidious, the denigrating stereotypes of Haitians as a chaotic, disorderly threat. Popular discourse usually casts Haitians as the polar opposite of Guadeloupeans: secure and self-controlled modern French citizens. In this scheme, Haitians represent the unruly, uncivilized past that Guadeloupeans judge themselves to have left behind. But Haitians also represent an alternate and feared future.[5] There is a long-standing public debate in Guadeloupe about the island's possible independence from France. Although quiescent now, in the 1960s and 1970s there was an active and occasionally violent independence movement. The opponents of independence still invoke Haiti as the best reason to remain a French department. "What will we become as a sovereign nation?" goes the rhetorical question. The usual response is "Another Haiti: resource poor and disorganized and politically corrupt, independent but at an unacceptable price." At the same time, Haitians retain the sorts of pride and cultural self-confidence that Guadeloupeans believe they have lost by their acceptance of French cultural citizenship. Haitians hence become objects of both desire and fear, who play two roles at once in the Guadeloupean collective imagination. Haitians symbolically contain what Guadeloupeans believe they lack, and they symbolically threaten what Guadeloupeans believe they have achieved.

Haitians know about the ambivalence over French identity which bedevils the residents of Guadeloupe, and it is the target of their rhetoric about French biomedicine and Haitian healing. Haitians possess this knowledge on a practical not a theoretical level, simply from observing with an outsider's eye how Guadeloupean society works: where Creole is spoken and where it is suppressed, the endless debates over independence, the consumerist lifestyle, and the dependency on the French welfare state coupled with a studied indifference to Bastille Day and other icons of French civic life. Nonetheless, Haitians are fixed in the dense constellation of images this ambivalence generates. They cannot easily escape the designation as primitive, disorderly, and too African. Haitians are a small minority community with few legal rights, no neighborhood of their own, and no newspaper or radio station or television station to project counter-images.

There are only a few rhetorical ways to win some room to maneuver. Haitians can ferret out the precise point of ambivalence, amplify it, and turn it back on itself. This is what their rhetoric about medical power accomplishes. By talking about the weakness of French biomedicine, Haitians chip away at the prestige of one of the major French institutions in their midst. By joking about Guadeloupeans who, like it or not, are forced to consult Haitian Vodoun healers, Haitians undercut the usual claims to cultural superiority and portray as hypocrites those who make these claims. In this way, the contrasts migrants draw between French biomedicine and Haitian religious healing accomplish important rhetorical work in this besieged diaspora community. Admittedly, this resistant rhetoric does not alter the practical constraints under which people live. It provides nothing more than a different way for this diaspora group to imagine its place in the host society. However, imagination is a cultural resource (Appadurai). It contains a storehouse of alternative and more favorable identities than the ones imposed upon a particular group. Perhaps this work of the imagination will someday enter social life in more tangible ways in the shifting and fragmented situation outlined here.

CONCLUSION

The discourse about French biomedicine that Haitian migrants craft stands at the intersection of two important and much-debated features of the contemporary world: (1) the flow of cosmopolitan forms of knowledge and practice across national, regional, and class boundaries, and (2) the transnational flow of people—migrants, laborers, refugees, displaced persons of all types—and the diapsoras and enclaved communities they construct. The anthropologist Arjun Appadurai discusses the first as "technoscapes," the global fluid configurations of technology which "now move at high speeds across various kinds of previously impermeable boundaries" (34). "Ethnoscapes," in turn, comprise groups of people constantly moving between stable polities and, on the level of collective representations, the "imagined worlds" which link together spatially dispersed populations. Appadurai's category of ethnoscape and the more expansive notion of technoscape (to include medical as well as industrial and communication technologies) highlight the core dynamics of my argument: the process by which medical technologies diffuse away from their origin in industrialized urban centers and, at the same time, migrant laborers come to inhabit the lower-class sectors of much wealthier societies. In the case of Haitians in Guadeloupe, these processes co-occur and feed off each other, and this makes their situation emblematic of some of the pervasive features of globalization.

The discursive strategy Haitians use to criticize French biomedicine and thereby re-imagine their place in Guadeloupean society also suggests the specific role of biotechnology within these larger and shifting currents. Globalization does not imply the emergence of a homogenized global society or a worldwide assimilation to a single cultural template (typically cast as North American or European). Indeed, the ongoing and intertwined movements of technology and people (along with media, money, and ideologies) have a surprising outcome. Global facts take local forms; cosmopolitan practices become indigenized; and communities of people united by sentiment, real or imagined histories, and political interests emerge in the midst of (and at times are reliant upon) de-localizing trends such as migration, instant communication, rapid shifts in the organization of labor and capital, et cetera.

Biomedicine participates in this general transformation. Although a product of the Enlightenment West, biomedicine has by now spread throughout the world, but not as a monolith or juggernaut, erasing previous forms of therapy and preempting new ones. We find instead innumerable tensions and exchanges between local and global forms of medical knowledge (Good 461). Despite the shorthand that I and others have used, cosmopolitan medicine does not exist as a singular, homogenized institution. There are instead local worlds of biomedical practice and many local economies of meaning which endow specific medical technologies with a new political and moral valence. Medical anthropologists have long discussed how specific groups outside Euro-America appropriate Western medicine and its technological apparatus according to local logics of sickness and healing. We should now start to link this process to the other ideologies and social forms that move across borders and the way that dispersed, dislocated communities improvise with all of them at once. In such settings, the discourse about biotechnology is not a one-dimensional refusal or celebration, and it concerns more than clinical effectiveness. It is rather one of the many resources which people use to address local contradictions and blockages, and this rhetoric is one of the numerous optics utilized to re-imagine their identity and debate alternative futures in inherently unstable circumstances. Haitian migrants freely make use of whatever modern biomedical services they can, while resisting the stereotypes forced upon them from the dominant Guadeloupean narrative of modernity. Haitians' discourse about biomedicine subverts that narrative, and in their seriously comic picture of French Guadeloupeans rushing to consult Haitian healers, we see a resistant commentary perfectly calibrated to a globalized world.

NOTES

1. Tsing reviews some current anthropological engagements with "marginality" as an analytic term. "Literary critics crafted the marginal as an intervention into Western humanism; margins are the sites of exclusion from this tradition from which its categories and assumptions can be seen more clearly. . . . [A]ttention to the marginal has opened discussion of linked cultural constructions of domination and difference" (14). As Tsing notes, people join these discussions from several positions and in various registers: critical, scholarly, and practical. The discourses of biotechnology examined here are one such discussion. They derive from and participate in specific regional processes of social exclusion, enclaving, and troubled incorporation.

2. As the example suggests, this rhetorical device often vastly simplifies a com-

plex historical situation. Slavery and prostitution bear ample witness to the commodification of the body throughout history, and the sale of blood plasma or hair occupies a mid-zone between the two poles of this contrast. Nonetheless, both scholarly and popular polemics about organ transplantation rely on the stark contrast between the older organic and the newer partible and commodified body. In general, a similar caveat is needed for the provocative argument that contemporary technologies (among other fundamental changes in our society) are producing "post-human" bodies (for example, Halberstam and Livingston). As Henrietta Moore warns, "the idea that the modern world is producing individuals who are no longer fully human . . . is misleading if we are trying to suggest that people in other times and places have been simply fully human" (8).

3. In many instances, people use newly introduced biotechnologies simply because they are effective, and they pragmatically combine them with local non-biomedical therapies (herbal treatments, religious healing, et cetera). Their critical commentary, praise, and warnings about sophisticated medical interventions are part of a larger utilitarian calculus to choose among medical options. This sort of rhetoric, however, is not unique to developing societies. It does not specifically index the complicated and often subordinate relations between developing societies and distant centers of knowledge and power.

4. Guadeloupe was colonized by France in the 1630s. It became a department of the French Republic in 1946, at which time all residents received full French citizenship and metropolitan administrative law was extended to the island. It is one of four overseas departments, along with Martinique, French Guiana, and Réunion.

5. Although this formulation relies on a contrast between past and present, it has nothing to do with the contrastive logic discussed earlier in this essay. Guadeloupeans attempt to deny their coevality with Haitians in several ways, and this underlies their notion that Haitians represent an earlier stage of Antillean society. Guadeloupeans distance themselves from the time period which Haitians supposedly inhabit, one that resembles, in some murky fashion, Guadeloupeans' imagined past. In general, this way of conceptualizing time helps to enact relations of power and domination between members of two societies (see Fabian). Guadeloupeans, nonetheless, cannot fully banish the notion that they are coeval with the Haitian migrants in their midst. This creates the fear that a future fate of an independent Guadeloupe is somehow contained in the "present" of Haitians.

WORKS CITED

Adams, Vincanne. *Doctors for Democracy: Health Professionals in the Nepal Revolution.* Cambridge: Cambridge University Press, 1998.

Appardurai, Arjun. *Modernity at Large: Cultural Dimensions of Globalization.* Minneapolis: University of Minnesota Press, 1996.

Atwood, Margaret. *The Handmaid's Tale.* New York: Ballantine, 1985.

Balsamo, Anne. *Technologies of the Gendered Body: Reading Cyborg Women.* Durham: Duke University Press, 1996.

Balshem, Martha. *Cancer in the Community: Class and Medical Authority.* Washington, D.C.: Smithsonian Institution, 1993.

Bernabé, Jean, Patrick Chamoiseau, and Raphael Confiant. *Eloge de la Créolité* [*In Praise of Creoleness*]. Trans. M. B. Taleb-Khyar. Paris: Gallimard, 1993.

Bledsoe, Caroline H., and Monica F. Goubard. "The Reinterpretation and Distribution of Western Pharmaceuticals: An Example from the Mende of Sierra Leone." *The Context of Medicines in Developing Countries: Studies in Pharmaceutical Anthropology.* Ed. Sjaak van der Geest and Susan Reynolds Whyte. Dordrecht: Kluwer, 1988. 253–76.

Bougerol, Christine. *La Médecine Populaire à la Guadeloupe.* Paris: Karthala, 1983.

Brodwin, Paul E. *Medicine and Morality in Haiti: The Contest for Healing Power.* Cambridge: Cambridge University Press, 1996.

Burton, Richard D. E. "The Idea of Difference in Contemporary French West Indian Thought: Negritude, Antillanité, Créolité." *French and West Indian: Martinique, Guadeloupe, and French Guiana Today.* Ed. Richard D. E. Burton and Fred Reno. Warwick University Caribbean Studies. London: Macmillan Caribbean, 1995. 137–66.

Butler, Octavia. *Dawn.* New York: Warner, 1987.

Clarke, Adele. "Modernity, Post-Modernity, and Reproductive Processes, ca. 1890–1990, or 'Mommy, Where Do Cyborgs Come from Anyway?'" Gray 139–55.

Cranor, Carl F., ed. *Are Genes Us? The Social Consequences of the New Genetics.* New Brunswick: Rutgers University Press, 1994.

Ducosson, Dany. "La Folie, les Esprits, et Dieu." *Le Phénomène Religieux dans la Caraïbe.* Ed. Laennec Hurbon. Montreal: CIDIHCA, 1989. 241–62.

Fabian, Johannes. *Time and the Other: How Anthropology Makes Its Object.* New York: Columbia University Press, 1983.

Farmer, Paul. *AIDS and Accusation: Haiti and the Geography of Blame.* Berkeley: University of California Press, 1992.

Franklin, Sarah. *Embodied Progress: A Cultural Account of Assisted Reproduction.* London: Routledge, 1997.

———. "Postmodern Procreation: A Cultural Account of Assisted Reproduction." Ginsburg and Rapp 323–45.

Ginsburg, Faye D., and Rayna Rapp, eds. *Conceiving the New World Order: The Global Politics of Reproduction.* Berkeley: University of California Press, 1995.

Godin, Benoît. "The Rhetoric of a Health Technology: The Microprocessor Patient Card." *Social Studies of Science* 27 (1997): 865–902.

Good, Mary-Jo DelVecchio. "Cultural Studies of Biomedicine: An Agenda for Research." *Social Science and Medicine* 41.4 (1995): 461–73.

Gray, Chris Hables, ed. *The Cyborg Handbook.* New York: Routledge, 1995.

Gupta, Akhil, and James Ferguson. "Beyond 'Culture': Space, Identity, and the Politics of Difference." *Cultural Anthropology* 7.1 (1992): 6–23.

Halberstam, Judith, and Ira Livingston, eds. *Posthuman Bodies.* Bloomington: Indiana University Press, 1995.

Haraway, Donna J. "A Cyborg Manifesto: Science, Technology, and Socialist Feminism in the Late Twentieth Century." *Simians, Cyborgs, and Women: The Reinvention of Nature.* New York: Routledge, 1991. 149–81.

Hess, Selinda. "Domestic Medicine and Indigenous Medical Systems in Haiti: Culture and Political Economy of Health in a Disemic Society." Diss. McGill University, 1983.

Hogle, Linda F. "Tales from the Cryptic: Technology Meets Organism in the Living Cadaver." Gray 203–16.

Ihde, Don. *Technology and the Lifeworld: From Garden to Earth.* Bloomington: Indiana University Press, 1990.

Kaufert, Patricia A., and John D. O'Neil. "Analysis of a Dialogue on Risks in Childbirth: Clinicians, Epidemiologists, and Inuit Women." *Knowledge, Power, and Practice: The Anthropology of Medicine and Everyday Life.* Ed. Shirley Lindenbaum and Margaret Lock. Berkeley: University of California Press, 1993. 32–54.

Lawless, Robert. *Haiti's Bad Press.* Rochester: Schenkman, 1992.

Lock, Margaret, and Patricia A. Kaufert, eds. *Pragmatic Women and Body Politics.* Cambridge: Cambridge University Press, 1998.

Moore, Henrietta, ed. *The Future of Anthropological Knowledge.* London: Routledge, 1996.

Morsy, Soheir A. "Not Only Women: Science as Resistance in Open Door Egypt." Lock and Kaufert 77–97.

Mulkay, Michael. "Rhetorics of Hope and Fear in the Great Embryo Debate." *Social Studies of Science* 23 (1993): 721–42.

Nandy, Ashis. "Introduction: Science as a Reason of State." *Science, Hegemon,y and Violence: A Requiem for Modernity.* Ed. Ashis Nandy. Delhi: Oxford University Press, 1989. 1–23.

Nichter, Mark, and Mimi Nichter. "Cultural Notions of Fertility in South Asia and Their Impact on Sri Lankan Family Planning Practices." *Human Organization* 46.1 (1987): 18–28.

O'Neil, John D., and Patricia A. Kaufert. "Irniktakpunga! Sex Determination and the Inuit Struggle for Birthing Rights in Northern Canada." Ginsburg and Rapp 59–73.

Pigg, Stacy Leigh. "The Credible and the Credulous: The Question of 'Villagers' Beliefs' in Nepal." *Cultural Anthropology* 11.2 (1996): 160–201.

Strathern, Marilyn. *Reproducing the Future: Anthropology, Kinship, and the New Reproductive Technologies.* New York: Routledge, 1992.

Tsing, Anna Lowenhaupt. *In the Realm of the Diamond Queen: Marginality in an Out-of-the-Way Place.* Princeton: Princeton University Press, 1993.

Van Dyck, José. *Manufacturing Babies and Public Consent: Debating the New Reproductive Technologies.* New York: New York University Press, 1995.

White, Lynn, Jr. *Medieval Technology and Social Change.* New York: Oxford University Press, 1962.

Williams, Simon J. "Modern Medicine and the 'Uncertain Body': From Corporeality to Hyperreality?" *Social Science and Medicine* 45.7 (1997): 1041–49.

CONTRIBUTORS

PAUL E. BRODWIN is associate professor of anthropology at the University of Wisconsin–Milwaukee, and adjunct assistant professor of bioethics at the Medical College of Wisconsin. He is the author of *Medicine and Morality in Haiti: The Contest for Healing Power* (1996) and the coeditor (with Mary-Jo DelVecchio Good, Byron J. Good, and Arthur Kleinman) of *Pain as Human Experience: An Anthropological Perspective* (1992).

LISA CARTWRIGHT is associate professor of visual and cultural studies and English at the University of Rochester. She is the author of *Screening the Body: Tracing Medicine's Visual Culture* (1995), the coauthor (with Marita Sturken) of *Practices of Looking* (2000), and the coeditor (with Paula A. Treichler and Constance Penley) of *The Visible Woman: Imaging Technologies, Gender, and Science* (1998).

THOMAS J. CSORDAS, professor of anthropology at Case Western Reserve University, is the author of *Language, Charisma, and Creativity: The Ritual Life of a Religious Movement* (1997) and *The Sacred Self: A Cultural Phenomenology of Charismatic Healing* (1994), as well as the editor of *Embodiment and Experience: The Existential Ground of Culture and Self* (1994). He is preparing a book comparing styles of religious healing in contemporary Navajo society and pursuing work on the relation between religion and globalization.

GILLIAN M. GOSLINGA-ROY is a doctoral candidate in the History of Consciousness Program at the University of California, Santa Cruz. The producer and director of several ethnographic documentaries, she currently is preparing for dissertation field research in South India on local practices of assisted reproduction and pre-producing her next ethnographic film, "The Wives of Pandi," about a deity reputed for his powers to grant infertile women children.

DEBORAH GRAYSON is assistant professor of English at the Georgia Institute of Technology. She has contributed chapters to *Having Our Way: Women Rewriting Tradition in Twentieth-Century America* (1995) and *Black Women, Spectatorship, and Visual Culture* (1996).

DONALD JORALEMON is professor of anthropology at Smith College. He is the author of *Exploring Medical Anthropology* (1999) and the coauthor (with Douglas Sharon) of *Sorcery and Shamanism: Curanderos and Clients in Northern Peru* (1993). He has contributed articles to *Medical Anthropology Quarterly, Journal of Anthropological Research, Social Science and Medicine,* and *American Anthropologist.* His present work focuses on the anthropology of organ transplantation.

HANNAH LANDECKER is a postdoctoral fellow at the Max Planck Institute for the History of Science in Berlin. Her contribution to this volume comes from her dissertation, "Technologies of Living Substance: Tissue Culture and Cellular Life in Twentieth-Century Biomedicine" (MIT, 1999). She currently is working on an analysis of the early years of microcinematography in Europe and the United States.

THOMAS W. LAQUEUR, professor of history at the University of California, Berkeley, is the author of *Making Sex: Body and Gender from the Greeks to Freud* (1990) and *Religion and Respectability: Sunday Schools and Working-Class Culture, 1780–1850* (1987), and the coeditor (with Catherine Gallagher) of *The Making of the Modern Body* (1987). He is working on the history of sexuality and on a book about death and modernity.

ROBERT M. NELSON is Associate Professor of Anesthesia at the Children's Hospital of Philadelphia and the University of Pennsylvania School of Medicine. He has contributed articles to the *Journal of Medicine and Philosophy* and the *Journal of Law, Medicine, and Ethics.* Currently he is working on a project exploring the issues of parental permission, child assent, and risk assessment about research participation from the perspective of children and parents.

SUSAN M. SQUIER is Julia Gregg Brill Professor of Women's Studies and English at The Pennsylvania State University. Her most recent book is *Playing Dolly: Technocultural Formations, Fantasies, and Fictions of Assisted Reproduction,* edited with E. Ann Kaplan (1999). She is the author of *Babies in Bottles: Twentieth-Century Visions of Reproductive Technology* (1994) and *Virginia Woolf and London: The Sexual Politics of the City* (1985). She currently is completing the book-length manuscript "Liminal Lives: Replotting the Human" as well as an edited collection of essays entitled "Radio Century, Radio Culture."

JANELLE S. TAYLOR received her doctorate in anthropology at the University of Chicago in 1999 and is currently assistant professor in the Department of Anthropology at the University of Washington. She is the author of a chapter in *Reproducing Reproduction* (1998) and of articles in *American Journal of Sociology, Technique et culture,* and *Science as Culture.*

ALICE RUTH WEXLER is the author of *Mapping Fate: A Memoir of Family, Risk, and Genetic Research* (1995), *Emma Goldman in Exile* (1989), and *Emma Goldman in America* (1984). Currently a research scholar at the UCLA Center for the Study of Women, she has been awarded fellowships from the American Council of Learned Societies and from the John Simon Guggenheim Memorial Foundation to work on a history of Huntington's disease.

INDEX